王绍增文集 1

王绍增 著

中国建筑工业出版社

图书在版编目（CIP）数据

王绍增文集1/王绍增著．—北京：中国建筑工业出版社，2018.1
ISBN 978-7-112-21675-8

Ⅰ.①王… Ⅱ.①王… Ⅲ.①园林设计－中国－文集 Ⅳ.①TU986.2-53

中国版本图书馆CIP数据核字（2017）第310792号

责任编辑：管　粟　杜　洁　李玲洁
版式设计：锋尚设计
责任校对：王　瑞

王绍增文集1
王绍增　著
*
中国建筑工业出版社出版、发行（北京海淀三里河路9号）
各地新华书店、建筑书店经销
北京锋尚制版有限公司制版
大厂回族自治县正兴印务有限公司印刷
*
开本：787×1092毫米　1/16　印张：19½　字数：425千字
2018年4月第一版　　2018年4月第一次印刷
定价：68.00元
ISBN 978－7－112－21675－8
（31511）

版权所有　翻印必究
如有印装质量问题，可寄本社退换
（邮政编码100037）

王绍增简介

　　王绍增（1942年10月8日～2017年1月27日），生于北京，祖籍河北吴桥，汉族，无党派人士。1960～1964年就读于北京林学院（现北京林业大学）园林专业；1964～1979年先后在成都市园林局、成都市青白江区工作；1979～1982年在北京林学院攻读园林专业硕士学位，师从郦芷若教授、程世抚教授，是中国在"文化大革命"后培养的首位园林规划设计硕士；1983～1989年先后在四川省城乡规划设计研究院、四川省建委工作；1989～2002年任华南农业大学林学院（现林学与风景园林学院）高级工程师、教授、风景园林系主任及首席教授；1998～2016年先后任《中国园林》杂志编委、副主编、常务编委、主编；2006年始任华南农业大学风景园林设计研究院院长；兼任北京林业大学客座教授、东南大学客座教授、华南理工大学亚热带建筑科学国家重点实验室客座教授等职务，是首届广东园林学会"终身成就奖"获得者。

　　王绍增先生是我国著名的风景园林思想者与践行者，一生致力于风景园林的学科建设、行业发展、教学科研及设计实践，出版专著或教材3本，发表论文数十篇，主持过多项环保监测、城市绿地系统规划、风景区规划设计项目，主要有成都青白江区环境本底调查、四川九寨沟风景区规划、长江三峡四川段旅游及文物保护规划、广东汕尾遮浪海滨总体规划、广州市生态绿地系统规划研究、江门市绿地系统规划、开平市绿地系统规划、台山市绿地系统规划、广东省和平县核心区城市设计、广州市龙头山森林公园规划、新丰云髻山风景旅游规划、成都紫竹院居住区环境设计、肇庆市鼎湖山广场规划设计、海口市中华恐龙园总体规划设计等。

序1

2017年春节前夕，王绍增教授永远地离开了我们。王绍增教授对中国风景园林学会以及中国风景园林事业做出了卓越的贡献，他的逝世是中国风景园林学会的巨大损失，更是我国风景园林事业的重大损失。

我与王教授相识、相交已十年有余。我来学会工作时，王绍增教授是学会的常务理事、《中国园林》杂志主编，我们在工作中有很多交流，我对他的学术水平、工作作风、道德情操有深入的了解。王绍增教授是风景园林行业著名的学者，是"文化大革命"后培养的第一位风景园林规划设计研究生，师从郦芷若、程世抚两位大家，他非常睿智，思想深刻，思维活跃，在理论和实践方面都有很高的造诣，他完成了数十篇论文，在业界具有广泛的影响，启发着大量行业工作者；他主持过成都青白江区环境本底调查、广州市生态绿地系统规划研究等几十个规划和设计项目，都很有前瞻性，为项目所在地的城市与环境建设提供了标杆。

王绍增教授是杰出的教育家，他在华南农业大学从教多年，并担任系主任和首席教授，为华南农业大学的风景园林教育做出了巨大贡献，桃李满天下。他参与起草的《增设风景园林学为一级学科论证报告》最终使风景园林学从三级学科成为与建筑学、城乡规划学并列的一级学科，提升了风景园林行业在国家建设大局中的地位，提升了风景园林从业人员的自豪感。

王教授在担任《中国园林》编委、副主编、主编期间，兢兢业业，成绩有目共睹。他提出的"关注国际动态，立足本土特色"的思路不但是《中国园林》杂志的发展方向，也是中国风景园林事业的发展方向。最近几年他还在酝酿"营境学"，要建立中国自己的风景园林理论体系。他对稿件都认真给予意见和建议，无论与他的观点相同与否，作者们都获得了启迪，对他充满敬意。他对工作极为负责，在医院的重症监护室里，他还在考虑着工作，思考着风景园林的重大事项。王教授是一个有着崇高品格的人，他正直，敢于直言，无论是广东省的绿道建设还是海绵城市建设，他都积极向有关部门建言，力求使这些政策更好地被执行。明代大学者杨继盛所写的名联"铁肩担道义，辣手著文章"，是对王教授的最好写照。

王绍增教授为中国风景园林学会做了大量工作，除了负责《中国园林》杂志的主编工作外，他还积极促成了学会理论与历史专委会的成立，参与学会年会优秀论文奖的设立与评选。他也是广东园林学会的顾问，对广东园林学会给予了大量的指导，广东省是全国风景园林建设最好的省份之一，有很多重要的园林上市公司，像岭南园林、棕榈园林、普邦园林，都曾得到他的建议和指导，他对广东风景园林事业的贡献也是卓越的。他非常重视人才梯队建设，积极向学会推荐青年英才，是名副其实的优秀导师。

出版王绍增教授纪念文集，具有非常重要的意义，有助于学界同仁、年轻学子深入了解王绍增教授的学术思想，共同缅怀这位为我国风景园林事业做出卓越贡献的老风景园林工作者。文集包括了他的论著、感言和同事、亲友、学生对他的回忆文章，能够使读者从多方面感受到王绍增教授的学术与人格魅力。纪念文集的编排非常巧妙、合理。第一册收录了他在《中国园林》杂志社工作期间所撰写的《主编心语》。这些文章篇幅不长，但是非常凝练，其内容也不仅仅限于每期主题的介绍，而是体现了王绍增教授晚年对于风景园林甚至社会发展的诸多深层次思考，非常精彩，我是每期必读，很受启发。这些文章无论对于风景园林的资深专家，还是初学者，亦或"门外汉"都值得阅读，行业外的读者可以借此对风景园林行业有一个初步全面的了解，初学者可以从中窥得学习风景园林的途径，资深学者则可以细细体会王绍增教授的深意，思考行业发展的未来，这是一个很好的开篇。之后的第二册主要是王绍增教授的论著，王绍增教授学识渊博，对于风景园林行业的方方面面都有涉猎。这些论著时间跨度有几十年，读者们可以从中看到王绍增教授学术思想体系的形成过程，对于深入研究王绍增教授的学术思想非常重要。第三册主要是对王绍增教授的回忆文章，对于研究王绍增教授的学术思想和生平具有补充价值，读者可以从多个角度看到作者眼中在不同时间段内作为同事、朋友、师长的王绍增教授。我想读者们必然能有所收获。

  古语说"高山仰止，景行行止"，这用在王绍增教授身上极为恰当。缅怀王绍增教授，往事历历在目，借为《王绍增文集》作序的机会，再次向王绍增教授致敬。

住房和城乡建设部原总规划师，
中国风景园林学会原理事长

序2

纸上园林
——送王绍增先生

若非王者
何以山河深雪
梅花飘一程
雪花送一程

推开窗,是一面湖水
水岸线曲折,水中
是飞鸟、游鱼
殒落的星辰

在林荫道散步
谈起新开的那个小园
园中之园,园外之园
变幻的沙画在一个时辰

不能演尽你的平生之园
噢,春天的桃园
蹊径上尽是青青学子
你端坐亭中如一幅画……

再唤你一声老师天色已晚
再铺开一页新纸
无法照见你亲切的容颜
霞光中的园林多美

用汉语说出的园林多美
在一本书里,我们折叠了日月
在芬芳的书页里
你留下一个个纸上园林

金荷仙

《中国园林》杂志社社长、常务副主编,
浙江农林大学教授

前 言

王绍增先生的学术研究始于20世纪70年代。当时他在成都市青白江区环境保护办公室工作，其间代表农民和那些污染大厂谈判赔偿，负责筹建环保监测站，并组织大兵团式的包括空气、土壤、水质、农产品的环境本底监测。1979~1982年在北京林学院园林系（现北京林业大学园林学院）攻读规划设计专业外国园林史方向研究生。1984年担任长江三峡（四川段）旅游和文物保护总体规划纲要负责人。1986年发表的《系统工程与园林效益》和次年发表的《园林与哲理》两篇论文初步勾勒了他对风景园林学科的学术思考，也清晰地表述了他的风景园林生态观和价值观。

20世纪末，王绍增先生又从生态科学的角度开创了城市绿地系统规划的"生态机理"理论，提出了城市空气污染降解机理，区分不同气候条件下的空气污染机制（微风：2~6m/s天气的空气污染机制；晴天静风天气的城市空气污染机制；阴、雨、雾天气的城市空气污染机制），在此基础上提出城市绿地系统布局的生态机理，包括氧源及氧平衡机理、降温及制造城市环流、城市组团的生态化规划布局模式、城市空气通道与绿地集中布局原则等。

为应对全球化浪潮对我国风景园林的冲击，王绍增先生提出要打破对西方Landscape Architecture的迷信，建立属于中国的风景园林学科体系。2015年与学界同仁一起倡议创立"风景园林中国学派"的构想，认为中国风景园林是以辩证的综合性思维和实事求是的方法为传统特长，在摆脱封闭守旧文化环境的羁绊和掌握现代科技发展的成果后，中国风景园林必将超越以个人表现和分裂式思维为特征的所谓西方工业时代景观设计，相信中国风景园林完全有可能开辟一条不同于西方、不同于传统的新路，并且从方法上提出了风景园林规划设计的营境理论体系，同时又从价值观上提出了风景园林行业的善境伦理导向。

《王绍增文集1》的内容是从2007年开始在《中国园林》杂志撰写的署名文章"主编心语"，是他对当期杂志的重点文章进行点评并提出了个人思考，将近十年的"主编心语"集结出版，可以勾勒出王绍增教授在应对风景园林全球化浪潮当中的思考起点和立场转向，辩证地批判西方文化的体系，为了人类长远利益而更倾心于弘扬中国文化，主张"为了人类长远利益，从全球西化转向平等对待中西文化"。

《王绍增文集2》（上册）的内容是王绍增教授的生态思考与实践。1986年发表的《系统工程与园林效益》和次年发表的《园林与哲理》两篇论文初步勾勒了先生对风景园林学科的学术思考，提出了"人与天调"的生态观，并在2001年发表《城市开敞空间规划的生态机理研究》，从生态科学的角度开创了绿地规划的"生态机理"理论系统。最后一篇论文《风景园林学科承担着生态文明建设的历史使命》认为中国人应该提出一套比当下流行的各种生态理论和生态工程方法更具有生态智慧、后患最小、投入产出比最高的真正可持续发展的生态理论体系和生态建设模式。

《王绍增文集2》（下册）的内容涵盖了《园林、景观与中国风景园林的未来》、《论风景园林的学科体系》、《风景园林学的领域与特性——兼论Cultural Landscapes的困境》、《建立完整的中国风景园林理论体系》、《论"境学"与"营境学"》、《论不过分张扬的风景园林师——尊重科学，理解人性》等重要文章，体现了他"中体西用"的价值导向和思维体系，指出了向中国传统回归的路径和方法，坚信在摆脱封闭守旧文化环境的羁绊和掌握现代科技发展的成果后，中国风景园林必将超越以个人表现和分裂式思维为特征的所谓西方工业时代的景观设计，开辟一条不同于西方、不同于传统的新路。

《王绍增文集3》的内容是对王绍增先生的回忆文章，对于研究王绍增教授的学术思想和生平具有补充价值。

目 录

2007 年

2007 年 1 期	003	2007 年 5 期	011	2007 年 9 期	019
2007 年 2 期	005	2007 年 6 期	013	2007 年 10 期	021
2007 年 3 期	007	2007 年 7 期	015	2007 年 11 期	023
2007 年 4 期	009	2007 年 8 期	017	2007 年 12 期	025

2008 年

2008 年 1 期	029	2008 年 5 期	037	2008 年 9 期	045
2008 年 2 期	031	2008 年 6 期	039	2008 年 10 期	047
2008 年 3 期	033	2008 年 7 期	041	2008 年 11 期	049
2008 年 4 期	035	2008 年 8 期	043	2008 年 12 期	051

2009 年

2009 年 1 期	055	2009 年 5 期	063	2009 年 9 期	072
2009 年 2 期	057	2009 年 6 期	066	2009 年 10 期	074
2009 年 3 期	059	2009 年 7 期	068	2009 年 11 期	076
2009 年 4 期	061	2009 年 8 期	070	2009 年 12 期	078

2010 年

2010 年 1 期	083	2010 年 5 期	091	2010 年 9 期	100
2010 年 2 期	085	2010 年 6 期	093	2010 年 10 期	102
2010 年 3 期	087	2010 年 7 期	096	2010 年 11 期	104
2010 年 4 期	089	2010 年 8 期	098	2010 年 12 期	106

2011 年

2011 年 1 期	111	2011 年 5 期	119	2011 年 9 期	129
2011 年 2 期	113	2011 年 6 期	121	2011 年 10 期	131
2011 年 3 期	115	2011 年 7 期	123	2011 年 11 期	133
2011 年 4 期	117	2011 年 8 期	127		

2012 年

2012 年 1 期　137	2012 年 5 期　145	2012 年 9 期　154
2012 年 2 期　139	2012 年 6 期　147	2012 年 10 期　156
2012 年 3 期　141	2012 年 7 期　150	2012 年 11 期　158
2012 年 4 期　143	2012 年 8 期　152	2012 年 12 期　160

2013 年

2013 年 1 期　165	2013 年 5 期　174	2013 年 9 期　183
2013 年 2 期　167	2013 年 6 期　177	2013 年 10 期　185
2013 年 3 期　169	2013 年 7 期　179	2013 年 11 期　187
2013 年 4 期　171	2013 年 8 期　181	2013 年 12 期　189

2014 年

2014 年 1 期　195	2014 年 5 期　207	2014 年 9 期　220
2014 年 2 期　199	2014 年 6 期　211	2014 年 10 期　224
2014 年 3 期　201	2014 年 7 期　215	2014 年 11 期　226
2014 年 4 期　205	2014 年 8 期　218	2014 年 12 期　230

2015 年

2015 年 1 期　235	2015 年 5 期　247	2015 年 9 期　258
2015 年 2 期　237	2015 年 6 期　249	2015 年 10 期　260
2015 年 3 期　241	2015 年 7 期　253	2015 年 11 期　263
2015 年 4 期　245	2015 年 8 期　255	2015 年 12 期　267

2016 年

2016 年 1 期　271	2016 年 5 期　280	2016 年 9 期　289
2016 年 2 期　274	2016 年 6 期　282	2016 年 10 期　291
2016 年 3 期　276	2016 年 7 期　284	2016 年 11 期　294
2016 年 4 期　278	2016 年 8 期　286	2016 年 12 期　297

2007年

2008年　2009年　2010年　2011年　2012年　2013年　2014年　2015年　2016年

2007年1期

本期主题：城市空间规划设计

2006年，经国务院批准中国风景园林学会正式加入国际风景园林师联合会（IFLA），本刊也荣幸地成为IFLA在出版领域的合作伙伴，负责IFLA通讯中文版的发布工作。由此，本刊将更自觉地担负起中外学术交流，特别是中国风景园林复兴优秀传统、推动现实进步和创建和谐环境的任务。

2007年来了，新的一年自然有新的期望，我们用全新的版面预祝2007年成为《中国园林》学刊又一个新的起点。

我们的工作领域囊括从城市核心区到自然保护区之间的极为广大的领域。在扫描这个广阔的领域时，本期选择了"城市空间规划设计"作为专题。

我们周围的空间从纯人造到纯天然之间，存在着无数可能性。《自然的含义》是一篇内涵十分丰富的文章，提供了很多值得深入思考的材料。

近年来的规划设计介绍文章，大体成为创意和美图的大比拼。《采石场上的记忆——日照市银河公园改建设计》是一篇很平实的文章，文中谈的都是实际的思考过程和解决方法。

城市内部的滨河地区不仅应有大部分文章所关注的生态、景观、休闲之类的功能，还应根据城市总体规划的分配具有文化教育、商业购物、交通办公、生活居住和产业开发等功能。《滨水城市概念性规划及设计方法研究——常州新运河地区概念规划及中心区城市设计》为我们提供了一个从总体规划入手在空间上处理城市复杂功能体系的范例。

《融外滩建筑精髓 立海派文化特色——上海市十六铺地块规划及城市设计国际征集德国SBA公司方案研究》从地块规划和城市设计角度介绍了上海的一项重要工作，其思路和方法有供其他城市的参考之处。

此外，《风景名胜区总体规划环境影响评价的程序和指标体系》、《从和谐社会·和谐环境所想到的——是创造一个环境，还是培育一个环境》、《北京私家园林的建筑艺术》等都是作者的鼎力之作，值得向读者推荐。

Phase 1, 2007

Topic of This Issue: Urban Space Planning and Design

In the year of 2006, the Chinese Society of Landscape Architecture (CHSLA) formally joined the International Federation of Landscape Architects (IFLA) with the approval from the State Council, and our journal was also honored to be one of the partners of IFLA in the area of publishing, responsible for the releasing of the Chinese version of IFLA newsletters. And our journal takes the task more consciously of the foreign academic exchanges, especially chinese landscape architecture in reviving fine traditions, promoting actual development and creating a harmonious environment.

The new year of 2007 comes and brings new expectations as well. With the new layout and design, we wish this year would be the new milestone in the history of the journal.

Our working area covers the wide range from city center to natural reserve. And the urban space planning and design are chosen as the topic of this issue when inspecting this broad domain.

Our surrounding space is between the pure man-made one and the pure natural one, and countless possibilities exist in it. *The Meaning of Nature* is of extremely rich connotations and provides a lot of materials worth of in-depth thinking.

The articles introducing planning and designs in recent years has largely become the mixture of ideas and fine photos. *The Reconstruction Design of Yinhe Park, Rizhao City* is a very down-to-earth article, it talks about the real thinking process and solutions.

The river-side areas should not only have the functions of ecology, landscape and recreation which have been the concerns of most articles, but also be functional in culture and education, business and shopping, communication and office, residence and living, and industrial development according to the master plan of the city. *The Concept Planning for New Canal District & Urban Design for the Center Area of Changzhou* provides us an example of dealing with complicated spatial urban function systems from the master plan.

The Case Study on the Urban Design & Planning of the Shiliupu Parcel Land in Shanghai by the Germany SBA Co. introduces the important work of Shanghai from the perspective of parcel land planning and urban design, and the ideas and methods can be reference for other cities.

Besides, the articles like *The Procedure and Indicator System of the Environmental Impact Assessment (EIA) of the General Management Plan (GMP) for Chinese National Parks, Thinking from the Harmonious Society and Harmonious Environment — To Create an Environment, or Nurture an Environment, and Buildings in the Private Gardens in Beijing* are of the authors' great efforts and worth to be recommended to our readers.

2007年2期

本期主题：城市公园规划设计

人们建造园林的一个基本原理是：园林服务于当代生活。因此，很多我国20世纪五六十年代建造的城市公园，由于社会的巨大变迁已经基本不适应当代的需要了，我国存在着大量的公园改造任务。北京为了配合2008年奥运，近年来做了许多公园改造，本期就此组织了一组稿件，包括《北京百旺公园设计》、《古都风貌保护与大型集中绿地建设——以北京北二环城市公园建设为例》、《北京德胜公园项目建设》、《北京月坛公园总体改造设计》等，以供各地参考。需要注意的是，在我国的"老公园"中有一些老一代城市园林工作者呕心沥血的精品，也是一个历史时代的文化记忆，首先需要挑选出来作为保护对象，不要被某些创造"政绩"的愿望匆匆毁掉。曾经被认定为中国第一个城市公园的上海黄浦公园的消失就是一个惨痛的教训。

中西和古今风景园林文化的结合，是21世纪必须基本解决的课题，否则意味着中国古老传统的消亡，或者中国园林重新彻底孤立。对此，从已有的理论研究看存在两种基本出发点：或者将中国文化看作基于西方文化的当今强势"世界文化"的一个部分，或者将西方文化当作东方古老"宇宙观"的一个特例。至于真正跨出这两个体系的思考，非得有大慈悲大智慧而不能得及，似乎迄今未见有此功力者。依靠当前流行的基于个人自我实现的各种理论体系是无望解决这个问题的。《古代文人宅园诗意景观对当今住居审美价值取向的启示》虽然比较难读，实际上已经碰触到这一方面，有志于园林历史和理论的学子或可从中得到启发。

中国风景园林的振兴，不但需要功力深厚的理论，更需要自己的实践大师。要成为这样的大师，第一要务是踏踏实实地创造精品，为此，必须谦虚地学习人类的文化成果，体验人类文化积累的艰辛，而不是自我狂妄地否定一切。《永恒·轴线——清华大学核能与新技术研究院中心区环境改造》一文为我们提供了这样一个样板。作为一个IFLA设计大赛的金奖获得者尚且如此虚心，我国上万学子当可从中获益匪浅。《上海滨江森林公园规划设计研究》则是一篇不玩弄概念、充满着实事求是的精神的好文。《北京私家园林的掇山艺术》、《苏州古典园林铺地纹样实例分析》等也在这方面做着切实工作。

园林节水、中国风景园林教育评价体系、绿道建设等都是当前大家的关注点。可以说，本期发表的很多文章都是令人深思爱不释手的高作，或是应时需要的佳品，有望对于推进中国风景园林事业有所助益。

Phase 2, 2007

Topic of This Issue: City Park Planning and Design

One of people's basic principle of landscape creation is that the landscape serves the modern life. Thus the many Chinese city parks built in 1950s and 1960s can not meet the contemporary needs because of the great social changes, which require a lot reconstruction. To be prepared for the 2008 Olympic Games, many parks have been rebuilt in Beijing. This issue groups a serious of papers of this topic, including The Design of Beijing Baiwang Park, Ancient Landscape Preservation and Large Concentrated Green Space Construction, The Construction of Beijing Desheng Park, The Overall Planning and Design of Beijing Yuetan Park, for readers' reference. We should also notice that some of the old parks are masterpieces and efforts of landscape people of the older generation. These parks bear the cultural memories of certain times, and they should be protected first, not destroyed quickly out of the desire of bureaucratic achievements. The disappearance of Shanghai Huangpu Park, regarded as the first city park in China, is a bitter lessen.

The blending of the Chinese and western, and the traditional and modern landscape cultures should be seriously dealt with in this new century, otherwise the Chinese tradition will extinct and the Chinese landscape will be isolated again. There are two kinds of considerations on this topic: regarding the Chinese culture as part of the contemporary world culture based on the western culture, or regarding the western culture as a special case of the ancient eastern world outlook. Stepping over these two considerations requires great benevolence and wisdom and we have seen no one achieve this. The popular theories based on the personal achievements have no hope to solve this problem. Inspiration from the Poetic Landscape in the Ancient Literati's Private Gardens for the Aesthetic Values of Modern Residence seems hard to read, but it has actually touched this aspect and provides inspirations for those of landscape history and theory studies.

The revitalization of Chinese landscape requires practitioners as well as solid theories. The first step of trying to be a master is to create masterpieces in a down-to-earth way. Eternity and Axis — The Environment Transformation of the Central Area of Tsinghua INET sets an example. The author is so modest even as a gold-medal winner of the IFLA Design Contest, which should be beneficial to our students. Research on the Planning and Design of Shanghai Riverside Forest Park does not juggle with concepts and is a good paper of practical and realistic spirits. Rockeries in the Private Gardens in Beijing, and An Analysis of Ground Pavement Designs of the Classical Gardens in Suzhou are also practical in this aspect.

Topics like landscape water-saving, the accreditation system of Chinese landscape education, and greenway construction are current focus. Many papers in this issue are interesting and trigger deep thinking, and others are timely and excellent, which are expected to be helpful for the development of Chinese landscape undertakings.

2007年3期

本期主题：城市滨水景观规划设计

　　本期围绕城市滨水地区的建设，从多角度选择了一组案例。《水城明尼阿波利斯的公园体系》讲的是美国的一个样板，《厦门筼筜湖及水岸公园系统整治总体规划》显然是受国际潮流影响较大的佳作，《京城滨水新景观——北京通州运河城市景观设计》是在流行与传统之间进行的一种艰苦探索，《传承古越文脉　展示水乡风情——绍兴运河园建设的理念和实践》在传承基础上对适应时代的变化做出了重大的努力，《礼乐相迎山水相映——铁岭新城核心区规划设计》则体现了一种以我为主的大气，而这种大气是振兴中国风景园林并进而达到领先世界所必需的一种精神。我们这样做的目的，是向世人表明排外绝不是我们的选择，但是自信是我们更重要的品质。兼收并蓄，以我为主，改革创新，领袖群伦，是中国风景园林的应有之义。

　　需要我国在现行方针、政策、法规、流程、规范等方面不断进行完善，是实践工作中经常遇到的问题。本期若干文章对此从不同方面和不同深度上进行了一些研讨，包括涉及土地分类的《寻求生态环境改善与土地资源利用高度平衡的风景规划方法——北京市海淀北部地区四个风景地规划实例研究》，探讨在我们行业科学地贯彻节约方针的《节约型社会城市园林绿化可持续发展对策研究》，全面介绍上海科研工作战略战术，对其他城市颇有参考价值的《加强上海绿化体系研究升华园林城市科技内涵》，进一步深入探讨城市规划"生态优先"问题的《关于城市开放空间优先的思考》等。虽然国外在这些方面有许多好经验，需要不断介绍过来，但是我国的国情、国体的确有很强的自身特点，完全照搬几乎是不可能的。上述这些文章都是密切结合我国国情和实际需要的，希望从事行政管理和政策研究的同志就此继续深入研究和发表意见。

　　2006年我有幸参加了中国建筑学会历史与理论分会召开的一次中国古典园林国际研讨会，内心感到很紧张，因为比起建筑界对中国传统园林的重视，风景园林界似乎在反其道而行之，界中不少人能读英文而看不懂古文，却以彻底反传统而自豪。中国风景园林学会开始筹建历史与理论专业委员会是对应此种倾向的重要举措，为了与之配合，我刊将适当强化"园林历史"与"风景园林论坛"栏目。

Phase 3, 2007

Topic of This Issue: City Waterfront Landscape Planning and Design

The issue focuses on the construction of urban waterfront area and a series of cases are selected from various perspectives. *Minneapolis Park System and Its Water Space* elaborates an American prototype. *The Overall Planning of Xiamen Yundang Lake and Waterfront Park System Rejuvenation* is a masterwork obviously influenced by international trends. *New Waterfront Landscape in Beijing – Tongzhou Canal City Landscape Design* conducts a tough exploration between the trend and the tradition. *To Continue the Cultural Tradition and Reveal the Waterfront Sceneries – On the Concept and Practice of Shaoxing Canal Park Construction* tries its best to make changes to get with the time while continuing the heritage. *Welcoming Music with Mountain and Water Landscape – The Planning and Design of the Core Area of Tieling New Downtown* embodies a self-confidence spirit, which is a must to re-energize Chinese landscape architecture and put it into a leading role in the world. Our purpose is to show that the exclusive policy is not our choice while the self-confidence is our more important quality. Incorporating things of diverse natures, self-confidence, reform and innovation, and the leadership spirit should be the inherent meaning of Chinese landscape architecture.

China needs to improve its current guidelines, policies, regulations, procedures, standards and other aspects, which is frequently encountered in our practical work. In this issue some papers discuss on this from different aspects and depths, including *Explore the Landscape Planning to Balance the Ecological Environment Improving and Land Resource Use – Case Study on the Four Tourism Planning Projects for the North Part of Haidian District, Beijing* on land classification, *Study of the Urban Landscape Greening Sustainable Development Measures in the Conservation-oriented Society* on the scientific implementation of the conservation policies in the landscape industry, *Reinforce the Research of Greening System in Shanghai to Enrich the Scientific Connotation of Landscape City* introducing the strategies and tactics of landscape research in Shanghai and of valuable reference for other cities, *On the Priority of Urban Open Spaces* on the "priority of ecology" in urban planning, and etc.. There are many good foreign experiences need to be introduced, while China has its own national conditions and these experiences cannot be simply copied. The above-mentioned papers have close links with Chinese conditions and actual needs, and we welcome colleagues in administration and policy research to continue their research and comments on this.

Last year I had the honor to attend the International Symposium on Chinese Classical Gardens held by the History and Theory Branch of the Architectural Society of China and I was very nervous, because the architecture circle pays much attention to Chinese classical gardens while the landscape architecture circle is doing exactly the opposite. Many people in the landscape architecture circle can read English documents other than ancient Chinese ones, but they are proud of the so-called thorough anti-tradition. Chinese Society of Landscape Architecture prepares to establish the History and Theory Specialized Committee, which is an important measure to deal with such tendency. In coordination, our journal will reinforce the columns of Landscape Architecture History and Landscape Architecture Forum accordingly.

2007年4期

本期主题：风景名胜区规划研究

国际风景园林师联合会（IFLA）将4月定为风景园林月，以向全球推广风景园林事业，这是个很好的创意，必将有利于地球环境和人类生活的改善。

在我看来，风景园林首先是一个空间（或地域）的概念，它涵盖了除纯粹人工和纯粹自然以外（即出了建筑大门到自然保护区之间）的整个空间（当然在大门里面和自然保护区里面还有少量的灰色地带，如室内园林和自然保护区的可进入区。同样，在风景园林空间中也有纯人工或纯自然科学可以渗入的灰色地带，如建筑包围的广场、公园里的自然群落）；其次，风景园林是一个关系概念，它是对应着人在生活上与自然发生的关系，也包括了生产和生活交织在一起时的关系，后者如观光农业、工业遗址、大地景物规划等。

风景区和园林，作为代表和典型占据着风景园林领域的两端。

国外的国家公园，是这一大片空间中最靠近纯自然的一块。我国的风景名胜区可以类比为国家公园，但实质上有着比国家公园长得多的历史和深刻的人文渗入。这句话有两重意思：一方面显示了与国家公园相比，我国风景名胜区向着人文这一边在空间上伸延了很长，在文化上伸延了很厚，人与自然的交融关系更为鲜明；一方面在对待与人类不那么友好的荒野一类的纯自然的态度上，风景名胜区显得有些漠然。风景名胜区的第3个表象特征是时代性的，由于我国经济条件、社会管理和科学技术还不如西方，在对待风景名胜资源的态度、风景名胜区的科学规划和管理上还存在着大量问题有待解决。

当前我国风景名胜区存在的错综复杂的矛盾，令很多有识之士十分担忧，纷纷出谋献策。本期为此组织了一批文章，各自闪烁着独特的思想火花：有的是多年工作的结晶，有的从现实存在的问题出发进行研究，有的着重与国外对比，也有纯理论的探讨。本刊本着一向贯彻百花齐放的原则，一视同仁，欢迎讨论，以期引起社会的注意和促进学术的发展。

李嘉乐先生是我国风景园林事业的开创者之一，他的为人和业绩令我们怀念，他的离去令我们哀伤，也警示我们必须加快向世人介绍中国风景园林大师的工作。毛泽东主席曾经批评过"言必称希腊"的作风，我们现在也存在着言必称国外某某大师的现象，好像大师都是外国的。如果我们照此认真研究和总结一下中国风景园林界的前辈和当代活跃的后来者，一定可以推出一批自己的大师！

Phase 4, 2007

Topic of This Issue: Study on Famous Scenic Sites Planning

The International Federation of Landscape Architects (IFLA) sets the April as landscape architecture month in order to promote the landscape architecture cause globally. This is a good idea and will help improve the world environment and human lives.

In my view, landscape architecture is a spatial (or geographical) concept at first, which contains the whole space except the pure artificial or pure natural parts (namely the area between the buildings and natural reserves) (Of course, there are small grey spaces inside the building and natural reserves, such as the indoor garden and the accessible space in the natural reserves. Similarly, there are also grey areas with pure artificial or natural features, such as the buildings-surrounded square and the natural community in the park). Secondly, landscape architecture is a concept of relationship, which reflects the relations between human lives and the nature and also includes the ones with living and production interweaved, such as sight-seeing agriculture, industrial ruins and ground landscape planning.

Scenic sites and landscape architecture, as the representatives and models, occupy both ends of the field of landscape architecture.

Foreign national park is the part of a large piece of space closest to the pure nature. China's scenic sites may be assimilated to the national park, but it actually has a much longer history and profounder cultural contents. This contains the dual meaning: on the one hand China's scenic sites is rich in culture and humanism which shows the more interactive relations between human and the nature, while on the other hand China's scenic sites treats the pure natural but no so friendly wilderness with unconcern. The third character of scenic sites is to be epochal, and there are a lot of problems in the attitude toward scenic sites resources, and the planning and management of scenic sites need to be dealt with due to the less advanced economy, social management and technologies of China.

Some complicated problems and contradictions exist in the scenic sites of China, and far-sighted people are quite worried about this and give a lot of proposals and suggestions. A serious of papers is collected in this issue on this topic. Insights spark in these papers, which include the experiences of many years work, the study of the existing problems, the comparison with foreign counterparts, and the pure theoretical exploration. Our journal welcomes and gives equal treatment to various viewpoints so as to attract the social attention and promote the academic development.

Mr. LI Jiale is one of the pioneers of the Chinese landscape architecture cause. We cherish his integrity and achievements and we feel sorry about his death. And we should quicken our step in introducing Chinese landscape architecture masters to our people. Chairman Mao once criticized the always-talk-about-Greece style of work, and nowadays the phenomenon also exist that foreign masters are always talked about, which seems that masters are all foreigners. If we do seriously study and summarize the pioneers and active contemporary architects of Chinese landscape architecture, a number of masters will certainly be recognized.

2007年5期

本期主题：风景园林教育

本期中心是教育，主要有3篇，放在"风景园林教育"栏目里，其他有些文章的深处也会涉及教育领域的问题。特别感谢教育部学位管理与研究生教育司欧百钢等同志撰写了《风景园林研究生教育改革与发展对策研究——以北京林业大学园林学院为例》一文，相信关心此事的读者在细读以后会浮想联翩，引发许多建议和感想。

我作为一个曾经的教育工作者（1989~2005年），回过头来反思亲历过程，亲切、得意、懊悔、遗憾等思绪齐涌上来。亲切是因为亲历，得意是因为培养出若干不错的学子，懊悔是因为在中国高等教育最不景气的近十年间（那时我所在的学校每年经费仅仅够发工资）几乎一无作为，遗憾的是在中国高等教育和风景园林事业发展最为迅猛的又一个近十年间风景园林专业偏偏被取消了。当前风景园林教育的一些问题是历史造成的，不全怪学术，这是我想叙说的第一个心绪情结。

俯瞰风景园林教育，首先可分为素质教育和专业教育；专业教育又可分为职业型教育和学术研究型教育；职业教育又可分为规划设计型和生产实践型等等。有了这种胸襟，就可以正确评议各国各校做法的得失，就可能把教育里面的各种问题放到恰当的位置去思考，这也许是避免看问题时得出偏激、片面性结论的基本手段。拿风景园林素质教育来说，我觉得它首先是指所有从事这个行业的人的基本素养，无此大家就缺乏共同语言，就无法充分交流。其次，在上述的各亚类中（如规划设计型）又有各自亚类的共同素质要求。把这些理清，就比较容易获得共同的认识。

至于精英教育，那是另一回事，可能需要专题讨论。

我始终相信，"先进"文化中不一定一切都先进，"落后"民族不一定一切都落后。曾经有幸参与意大利某大学的交流，令我惊讶的是他们让大体相当于景观三年级的学生去作一个陡峭山区的规划。我想，这些年轻人对地震、滑坡、崩岩、山洪、泥石流以及相关的地质学、土壤学、气象学、生态学、流体力学等毫无认识，仅凭一点景观学基础训练能承担这种任务吗？想当初，1993年做九寨沟第一次规划时，是四川省政府规划工作组集中了相关各个厅局的多方面专家，许多人对树正沟的一大片滩地很感兴趣，觉得可以用来建设旅游基地，我表示反对。问我为什么，我说很简单，你看旁边的藏寨盖在山脚的山坡上，而不利用这块平地，必有原因。后经了解，果然是一片泥石流冲积滩。所以，风景园林是一门正在发展的庞大的学问体系，虽然入门看起来不难，精深却不易，于兹有志的年轻人是可以大有作为的。

Phase 5, 2007

Topic of This Issue: LA Education

The very topic of this issue is education. There are mainly three articles centering on it under the column of "Topic on Landscape Architecture Education". We owe boundless gratitude to OU Baigang, Minister of Academic Degree Administration & Postgraduate Education Department of Educational Ministry and other comrades who collaborated Research on Reform and Development Strategies of *Study of Revolution and Development Countermeasures on Postgraduate Education of LA – With Beijing Forestry University Landscape Architecture School as the Example*. Obviously, it willgreatly arouse a great deal of imagination in those readers who are concerned with it, and stimulate much thinking and offer of suggestion after a close reading of it.

Having been a teacher before (1989-2005), I feel an upsurge of geniality, pride, regret and pity and so on while reviewing what I have gone through personally. I feel genial for it is all my own experience; I feel proud to have brought up some excellent students; while I also feel regretful for having accomplished almost nothing valuable during the most difficult decade of Chinese higher education (at that time, the annual budget of my university could only afford to pay salaries of teachers), and pitiful for the annulment of LA as a major during the next decade when both Chinese higher education and landscape architecture thrived vigorously. Some recent problems of landscape architecture education have lasted for quite a long period of time, so they cannot be wholly traced back to academic causes. This is my first complex.

LA education can first be divided into all-round education and professional education; then professional education can be further divided into vocational education and academic education; and vocational education can even further divided into planning & design mode and practice mode and so on. Only with this overall idea, can we make a fair judgment on the gain and loss of different countries and schools, can we have a proper attitude towards various questions in the educational field. This is probably a basic means to avoid drawing some radical or one-sided conclusion while dealing with problems. Taking all-round education of landscape architecture for example, I think it should first be referred to the basic qualities of all the practitioners in this field. Without them, we will be lack of the common ground to communicate with each other. Second, there are also requirements of common qualities in the subdivided fields (like planning & design mode).

I always hold the belief that "advanced" culture is not necessarily advanced in every field, while "undeveloped" nation is not necessarily fell behind in every aspect. I have ever been very fortunate to attend an exchange with an Italian university. To my surprise, they asked those students having the qualifications of about a junior to do a design of a steep area. The youngsters had less idea of the earthquake, landslide, rockslide, mountain torrents, mud-rock flow or related knowledge of geology, pedology, meteorology, ecology, and fluid dynamics and so on.

I wondered how they were capable of shouldering the responsibility of this kind only with limited basic training of landscape. I still remember that the year 1993 when we were making the program of Jiuzhai Valley for the first time. Sichuan provincial governmental programming group gathered many experts of various fields in related departments, many of whom showed great interest in the vast beach in Shuzheng Valley and thought it could be constructed into a tourist resort. I opposed, because it was quite simple. It must be some reasons for local people to build the Tibetan hamlets on the slope at the foot of mountains, rather than making use of this land for other purpose. Then after some research, it was finally tested to be an alluvial beach formed by mud-rock flow just as expected. Therefore, landscape architecture is a developing and huge system of knowledge. Though it seems easy to start to learn it, but rather difficult to be proficient in it, so only those with great ambition can have great achievements.

2007年6期

本期主题：中国传统园林研究

　　作为中国传统园林代表的苏州园林成为世界文化遗产，是中国风景园林史上的一件大事，充分说明了中国传统园林在世界人们心目中的高尚地位。曾经有一个青年作者在其文章中写道：中国园林产生于封建社会、小农经济、专治体制下。我给作者的建议是：不要过于武断，如果说现代园林是产生于商人社会、流行文化、资本控制下，你喜欢吗？所以，对于事物不可做简单判断，对于潮流要有辨别能力。自信而谦虚，包容而持中，这正是中国传统文化教给我们中国君子做人的应有态度。在全球化冲击下的世界文化，即使像中国园林这样曾经的强势文化也遭遇到从未有过的挑战，包括评价、批评、批判甚至攻击。为此本期集中发表了包括《试论我国风景园林建设的继承和发展——为〈世界文化遗产——苏州古典园林〉序》、《奠基人之奠基作——赞汪菊渊院士遗著〈中国古代园林史〉》、《小巧简约之典型　南北过渡园林之范例——十笏园园林艺术风格研究》等几篇讨论中国传统园林的文章。深入探讨这个问题，应该是《中国园林》这本杂志义不容辞的义务。中国风景园林学的创始者们在这个阵地上打下了很好的基础，当前需要的是除了在原有基础上继续深入研究，还要开拓新的视野，包括研究地方性和民族性的园林，利用哲学、历史和美学研究的新成果研究园林等，特别应强化探讨中国传统与现代社会融合的途径与方法，以及中国园林对更远的人类社会适应性的展望。我欣喜地看到，当前许多学者，包括一批在读的博士生和他们的导师，正围绕着传统园林开展他们的研究，意味着这方面的研究将很快出现一批高水平的成果，人们对中国园林的理解将出现一个飞跃。本期的《徽州古村落与水口园林的文化景观成因探颐》、《关于北宋皇家苑囿艮岳研究中若干问题的探讨》二文可以属于此类文章。

　　继北京奥林匹克森林公园之后，上海世博园的规划建设成为界内的注视焦点。上海的同仁们为本刊撰写了一批文章，以飨读者。其中《上海世博公园规划设计国际征集方案评析》、《城市尺度空间的塑造——上海世博公园的空间研究》、《打造和谐城市客厅——上海世博公园建设构想》等文可以满足大家的好奇心，而《上海世博公园设计任务书中技术要点解析》和《上海世博公园深化设计方法与内容》更具有供各地参考的实用价值。

　　现任IFLA主席戴安妮·孟斯女士应邀为本刊撰写了《新西兰乡村风景的管理》一文，很多内容对我国的风景园林发展具有深刻的启示意义，我以为这是本期的另一个亮点，值得推荐。

　　应该承认，我国的城市绿地系统规划还有许多问题存在。生态优先，绿地规划优先或与总规同步，此项工作做好了，意义十分重大。北京在这方面迈出了重要的一步，本刊就此发表了两篇文章，并将继续跟踪和推介。

Phase 6, 2007

Topic of This Issue: Study on Chinese Traditional Gardens

Suzhou gardens, as the representative of traditional Chinese garden, being recognized as the world cultural heritage is a big event in Chinese landscape history. It fully demonstrates the lofty status of Chinese traditional gardens in the eyes of the world people. Once there was a young scholar writing in his article that "Chinese gardens came under the feudal society, small peasant economy and autocracy." My advice to the writer is that "Don't be too assertive. Do you like that the modern landscape not so is that I give the author's recommendations are produced from the business society, popular culture and capital controls?" So we should not put simple judgments on things and should have ability to see the trends in their true light. Being confident, modest, tolerant and persistent, this is exactly what the Chinese traditional culture has put in our integrity. With the world culture under the impact of globalization, the influential cultures like Chinese landscape architecture is also facing never-encountered challenges, including evaluation, criticism, and even attack. Regarding this, a series of papers, like *On the Succession and Development of China Landscape Architecture Construction – Preface for The World Cultural Heritage – Suzhou Classical Gardens, Cornerstone by the Founder – Commending Academician WANG Ju-yuan for His Posthumous Work The History of Ancient Chinese Gardens,* and *A Model of Small and Concise Style, and An Example of the Transitional Garden between the South and the North – Study of the Landscape Style of Shihu Garden,* are collected in this issue for the in-depth discussion of traditional Chinese gardens, which is the bounden duty of our journal. The pioneers of the Chinese landscape architecture studies have laid a good foundation in this field. Currently in-depth study should be continued on the original basis, while the new vision should be extended, including the study of local and national landscape, and the study of landscape with the research achievements of philosophy, history and aesthetics. The discussion on the ways of blending the Chinese tradition with the modern society, as well as the prospect of the adaptability of Chinese landscape to the human society, should be strengthened. I am pleased to see that many scholars, including a number of doctoral students and their supervisors, are doing their research on the classical gardens, which means some high-level results will come out soon and people's understanding of Chinese gardens will have a leap. The two papers, *The Causes of Origin of Cultural Landscape of the Huizhou Ancient Village and Shuikou Garden* and *Discussion on Some Issues in the Study of Genyue Royal Garden of the Northern Song Dynasty,* are in this category.

Following the Beijing Olympic Park, the planning and construction of World EXPO Park Shanghai has become the focus of attention in the industry. My colleagues from Shanghai have contributed several papers on this. Papers like *Evaluation of the Internationally-collected Designs for the World EXPO Park Shanghai, Spatial Creation Based on Urban Model: Study on Landscape Space of World EXPO Park Shanghai*, and *The Urban Hall of World EXPO 2010 Shanghai: An Image of World EXPO Park Shanghai*, can satisfy everyone's curiosity, while papers like *Requirements and Solutions – Special Research on Project Description of Design of World EXPO Park Shanghai* and *The Method and Content of Design for World EXPO Park Shanghai, The Planting Landscape Design Idea of World EXPO Park Shanghai* are of useful reference and practical value.

Dr. Diane Menzies, IFLA President, contributed the article *New Zealand Rural Landscape Management* at the invitation of our journal, which is of deep inspiration reference for the landscape development in China. I think this is another bright spot of this issue and worth recommending.

We should admit that there are many problems in urban green space system planning in China. It is very important to have the ecology as the top priority, and the green space planning prior to or together with the master plan. Beijing has step a significant step forward in this aspect. Two papers on this topic are published in our journal, and we will continue the follow-up and introduction.

2007 年 7 期

本期主题：防灾避险绿地

有一次我突发奇想：如果我们耗费大量资源造出一个又大又美但是毫无用处的东西，人类该怎样评价它？这个看来简单的问题，却让我想了很久，因为有多种答案。譬如说，以现在的观点看建造金字塔（包括长城）有什么用？但它们在旅游方面却有了大用；又譬如对庄子"吾有大树，人谓之樗"的寓言有各种阐释，以我看来，它的意义在于庄子洞穿了在"有用和无用"之辩背后是人的功利心，当人抛弃了功利，那就会是另一个境界。问题是我们不是神人、圣人，甚至连至人都达不到，能够真心为人民服务，那就很了不起了，所以还是要回到有用无用的问题上来。

人类花费巨大的土地、资本、生态、材料等资源建设室外环境，首要不是好看，是有用。风景园林和建筑这类艺术与纯艺术（如绘画）的巨大区别是它要耗费巨大的资源。纯艺术可以（或应该）让艺术家们充分自我发挥，但如果耗费了巨大资源的项目不被用来服务于人类，而被用来服务于某些"大师"的个人主观意愿（包括艺术创造），这是不是社会的悲哀？如果把这个问题想清楚了，人们会对当前很多工程决策的程序和原则做出新的审视。

城市园林绿地的一个大用途，是避灾。不知是否受到单纯景观（顶多加上自然生态）观念的影响，对此已经多年不怎么提到了，甚至我们的城市绿地系统规划编制办法中都没有将其列入必须做的工作内容。其实所有城市都有避灾问题，即使真的没有自然灾害的城市，人为灾害（战争、恐怖主义行动等）也是城市规划和建设者应该考虑的。乘奥运之风，北京又将此事提上了议事日程，我们觉得这是个大事，需要为其宣传一番，以期引起各界的注意，为此本期推出了"防灾避险绿地"这个专题。专题的文章以介绍外国经验为主，特别是日本在这个方面做得很好，人多地少的国情也有与我国相近的地方，故作了较多的介绍。北京在研究避灾绿地分布的方法上有着新的创造，新东西不一定完美，但这种创新精神值得推崇，有了这种精神，我国的风景园林事业达到与现代科学接轨的目标并不遥远。

韩国首尔的清溪川工程，是近年来世界著名的城市重建工程，其中有着大量的城市生态和生活内涵，正是我们这一行工作的核心。据我所知，我国不少城市刚刚把城市的小河覆盖掉，有更多的城市还在准备这样做，主要的借口是节约土地。首尔的做法为我国提供了借鉴。

我愿在此重申，"百花齐放，百家争鸣"是我刊的宗旨，除了必要坚持的科学和政治原则，在我刊发表的文章主要看是否有利于中国风景园林学科的发展，是否有新意。对于学术发展中出现一些偏差、漏洞，本刊并不求全责备。读者对刊出的文章如果有不同意见，欢迎提出商榷和讨论。

Phase 7, 2007

Topic of This Issue: Emergency and Disaster-prevention Green Space

Once I suddenly had a strange thought that: how people would evaluate what is created to be useless while big and beautiful with plenty of resources? It seems to be a simple question, but I thought for a long time since there are many kinds of answers. For example, what is the use of building pyramids (or the Great Wall), in current point of view? However they mean a lot in tourism. There are a lot of explanations for Zhuangzi's saying"I have a great tree but people say that's a tree of heaven", but in my view its significance lies in that Zhuangzi pointed out the utilitarian mind behind the debate of "useful or useless". It would be a realm of loftiness if the utilitarian mind is thrown way. The problem is that we are not gods, saints, or even a perfect man. It is already amazing to truly serve the people, so we still have to come back to the question of "useful or useless".

Enormous land, capital, ecology, building materials and other resources have been put in the building of outdoor environment in order to be useful, and it is not the priority to be pretty. The tremendous difference between the pure art, such as painting, and the art of landscape and architecture is that the latter costs plenty of resources. Pure art can, or should, give the artists a full self-realization. But if the project costing enormous amounts of resources is used to serve the individual subjective desires, or artistic creation, of some so-called masters rather than in the service of human, is this not a social tragedy? If this issue is clear, people will have a new look at the decision-making procedures and principles of many present projects.

One big use of urban green space is to prevent disaster. I do not know whether it is influenced by the concept of pure landscape, or plus natural ecological, and this has not been mentioned for many years. Even our urban green space system planning rules have not included this in the necessary working contents. In fact, all cities face the problem of disaster prevention. Even in the cities without natural disasters, the man-made disasters (wars, terrorist actions, etc.) should also be considered by the planners and builders. Taking advantage of the coming Olympic Games, Beijing puts this issue on the agenda, and we feel this is a big event which needs to be clamored and publicized so as to arouse the attention of all sectors. The special topic of "disaster-prevention green space" is put in this issue of the journal. The papers of this topics mainly focus on the introduction of overseas experiences. Japan has done very well in this aspect and it is similar with China in the national conditions of big population on little land, so there are more introductions of Japanese experiences. The study of the distribution of disaster-prevention green space in Beijing has its innovative method. Even though the new is not always perfect, but the innovative spirit should be praised. The convergence of Chinese landscape architecture course with modern technology will not be far away with this spirit.

The Cheonggyeoicheon project of Seoul, South Korea, is a world-renowned urban renewal project in recent years, and there are a lot of contents of urban ecology and life, which is the core of our profession. As far as I know, many cities in China just covered the urban streams and more cities are preparing to do so, with the main excuse of saving land. The Seoul practice provides reference for China.

I would like to reiterate here that the free development of different varieties and styles and encouraging different thoughts contending are the aim of our journal. In addition to adhering to the scientific and political principles, the papers published in this journal should be conducive to the development of landscape discipline in China and original. The journal is not nitpicking on some deviations or mistakes in the academic development. Readers are welcome to join the discussion if they have different opinions on the papers published in the journal.

2007年8期

本期主题：乡土地被植物研究与应用

　　45年前，陈俊愉先生在给我们年级讲课时就谈到了地被植物在园林里的重要性，但此后我在长达30多年的时间里，几乎见不到这个理念的实践。自从景观设计的概念特别是平面构形的手法流行以来，地被植物得到前所未有的普遍应用。当初，随着这种"全球化"风潮一起进入开放的中国的，还有大量的国外地被植物材料。从情理来讲，这是必然的，因为我们没有为这股潮流预作准备，所以也就没有自己足够的地被材料供使用，照搬国外设计，包括其使用的材料，就成了捷径。随即，舶来品类的地被植物材料也就占据了我国的市场和苗圃的大部分空间。

　　其实，我们有充足的理由宣布应该大力发展自己的乡土地被材料产业：科学发展观、节约型社会、绿色GDP、物种多样性、民族产业、民族文化……不一而足。在此我想说的是，将植物材料委托给国外设计师不但常常行不通，有时还存在危险。咱们有个大城市建设了一个湿地公园，方案是国外中标的，外国设计师给了一个植物名录，经中科院水生植物研究所的专家审核，不但有大量不适应当地条件的物种，还有好几种已被列入我国《外来入侵物种名单》。本期组织了一批有关乡土地被植物的研究文章，除了可供各地参考使用，更重要的是期望能够引起行业内同仁的重视。对地被的研究，除了用观察的方法，我国比较发达的地区已经具有了作进一步科研的条件，这也是今后的发展方向，所以也发表了两篇这方面的研究文章。

　　对于"80后"的一代来说，他们中的大多数开始接触园林时就已经满眼是"现代化"的景观了，当然对风景、园林、景观的概念的理解和老一辈的人有所不同。但对"80年代的新一辈"和更老的人来说，应该知道20世纪80年代以前我国基本没有这种大面积平面图案式的设计，能够种点草，不放任地面自由生长就已经很不错了，当时深圳出现的所谓"抽象式园林"还被全国看作新鲜事物。道理不复杂，那时国家很穷，即使有人想搞这种耗能费钱的地面也没有条件，何况中国传统也没有这种做法。所以我不明白为什么有人把后来出现的浮华性风气归罪于中国园林。本期有的文章提得好，我们对地面的观念从材料到做法是否也需要变一变？其实国外也有很多明智人士和聪明作法，我在悉尼领地公园就看到人家用灭火用的风筒将道路上的落叶吹到草地里面去。值得我们思考的是，为什么这些好东西介绍不进来，反而那些浪费的、高消费的东西很快就在我国普及开来了呢？观念的背后是什么东西在操纵呢？是不是资本的力量在后面作祟呢？

Phase 8, 2007

Topic of This Issue: The Research and Application of Native Groundcover Plants

45 years ago, Mr. CHEN Junyu talked about the importance of groundcover plants in landscape architecture when giving lectures to our grade, but I have seen almost no practice of this concept through the 30 years since then. Since the concept of landscape design, especially the plane configuration skill gets popular, groundcover plants have been in unprecedented universal application. Originally, together with the entering of the globalization trend in China, plenty of foreign groundcover plants also came. Reasonably speaking, it is inevitable. We do not have enough groundcover materials of our own to use since we have not get prepared for this trend. Simply coping foreign designs, including their use of materials, become a shortcut. Immediately, foreign groundcover plant materials occupy a large share of the market and nursery of China.

Actually we have ample reason that we should vigorously develop our own industries of native groundcover materials, such as the scientific concept of development, conservation-oriented society, green GDP, biological diversity, national industries, and national cultures. I would like to remind everyone that entrusting plant materials to foreign designers often does not work, and sometimes even risky. Once there was a wetland park built in one of our city, and the design plan is from overseas bidding. The foreign designer proposed a plant directory, which included plenty of species unsuitable for local conditions as well as some species listed in the *List of Invasive Alien Species* of China, as examined by the experts of the Institute of Aquatic Plants of the Chinese Academy of Sciences. Some papers on native groundcover plant research are organized in this issue, for our readers' reference all over the country and to draw attention of our colleagues of the industry. Some developed areas of China have the abilities to do further research on the groundcover plant beside the observation approach, which is the future direction of development and there are two papers on this aspect.

For the generation of the 1980s, the modernization landscape has in their eyes when most of them began to meet with the landscape architecture. Their concept of scenery, garden, and landscape are of course different from the older generation. But the new generation of the 1980s and the older ones should know that there was basically no such large-area plane design in China before the 1980s. It had been quite good that grasses were planted, under control, on the ground then. The so-called abstract-style garden in Shenzhen at that time was regarded as a new thing in China. The reason is not complicated that the country was poor then and there were no conditions even if someone wanted to establish such energy-consuming and costly grounds, and there were no such practices before. So I do not understand why some people put the blame of the subsequent ostentatious trend on Chinese landscape architecture. Some of the articles in this issue put forward well that our concept of the materials and practices of ground might need a change. In fact, there are many foreign smart approaches, and I saw in The Domain in Sydney that the hairdryer of the fire-extinguisher was used to blow the fallen leaves into the grassland. We should think about why these good things do not come in while those wasting and high-consuming things quickly spread in China? What is controlling behind the concept? Is it the mischief of the capital?

2007 年 9 期

本期主题：第六届园博会（厦门）风景园林师园

我国的展览花园最早出现于20世纪80年代后期的香港花展。1993年广州市春节花展出现了20来个被称为"小园圃"的展品，以后成为每年的定例，基本上是由各园林单位送展。今年厦门国际园林花卉博览会上的风景园林师园是一个非常好的开端，是我国首次让"风景园林设计师们在没有任何外界人为干预限制性下的自主设计，设计师可以凭借这样的一种契机完整地表达个人的设计理念，这是一种前所未有的独立设计行为"（朱建宁语）。我们相信，未来"会有越来越多的设计师参与到展园和小花园的设计中，在此过程中，中国的设计师会成长、园林产业会发展，行业会更加成熟"（王向荣语）。

回顾历史可以看到，19世纪以后没有再出现过以民族国家命名的风格流派，新流派大体以"主义"分野，而"主义"的创始人则成为这种流派的代表。在当今这个世界，我国的风景园林事业要想在世界上再领风骚，必须培养出一批著名的中国设计师，我刊将此当作主要使命之一。本期借陈俊愉院士90寿辰之机，介绍了陈先生对中国和世界风景园林事业的贡献，并且以介绍风景园林师园作为主题。今后，我们还将着力推介一批成就卓著的中国设计师。

我觉得中国当代的风景园林设计师是幸运的，因为他们有祖先创造的中国在世界上最优秀的艺术门类——园林——作为基底，并完全可以以此为自豪，但我们同样需要谦虚地学习。在建设和谐世界的过程中，学习尊重别人（包括祖先，也包括外国，甚至自然）的经验，是一个重要的社会建设过程。现在的地球，人类与自然的关系已经相当紧张、没有了让人横冲直撞的余地，《对自然的倾听：解读南非园林》一文介绍的南非风景园林设计师的态度可供我们参考。

本期的《西湖风景名胜区新农村建设的实践与思考》、《我国观赏植物种质资源流失的原因及对策》、《城市绿地系统规划塑造城市特色》等篇都抓住了我国当前城乡建设工作的一些难点，提出了一些可供参考的解决方案；《面向LA专业的景观生态教学体系改革》和《从硕士学位论文写作看中国主要高校风景园林学科的研究发展——中国10余年来风景园林学科硕士学位论文回顾与展望》则抓住了专业的高等教育中的一些关键；而种植设计施工图近来带来的问题越来越严重，《园林种植设计施工图中的几个问题》对此做了有益的探讨。上述文章值得向关心这些问题的读者推荐。

Phase 9, 2007

Topic of this Issue: The Landscape Architects' Gardens of the 6th China International Garden & Flower EXPO (Xiamen)

The earliest exhibition gardens of China appeared in Hong Kong Flower Show in late 1980s. There are around 20 exhibits called small nursery garden, mainly presented by landscape organizations and units, in the 1993 Guangzhou Spring Festival Flower Show, which became a usual practice in the following years. The landscape architects' gardens in this year's China International Garden & Flowers Exposition, Xiamen, is a very good start, and it is the first time in China for the landscape architects to design without any human intervention and restrictions and to fully express their own design concepts with this opportunity, which is an unprecedented independent design action. (Remarks of ZHU Jianning) We believe that in the future more and more designers will join in the exhibition garden and small garden design, and in this process Chinese designers grow, the landscape industry develops, and the trade becomes maturer(Remarks of WANG Xiangrong).

It is shown by reviewing the history that there was no longer a nation-state named style or school after the 19th century. New schools were generally divided by doctrines, while the founders of the doctrines became the representatives of the schools. A number of Chinese designers should be trained into well-known ones if the Chinese landscape course is hoped to take a leading role in the present world. It is also one of the major missions of our journal. With the opportunity of the 90th birthday of Academician CHEN Junyu, his contribution to Chinese and world landscape course are introduced, and the introduction of the designers' garden is taken as the theme. We will also put forth efforts hereafter to introduce a number of well-accomplished Chinese designers strive to promote the achievements of a group of outstanding Chinese designer.

I think the contemporary Chinese landscape designers are lucky because they have the most outstanding art category created by their ancestors, Chinese traditional garden, as the basement and they can take pride in this, but we should also need to learn modestly. In building the harmonious world, learning to respect the experiences of others (including ancestors, foreign countries, and even the nature) is an important social construction process. On the present earth, the relationship between human and the nature is quite strained and there is no room for people to dash around madly. The South African landscape architect's attitude introduced in *Listen to Nature: Decoding the Landscape Architecture of South Africa* can be the reference for us.

The papers like *The Practice and Thinking of New Countryside Construction in West Lake Scenic Area, Hangzhou, The Causes and Countermeasures on the Running Off of Germ Plasm Resources of the Ornamental Plant in China,* and *Urban Green Space System Plan Molds City Characteristics* all grasp some difficult points of the urban and rural construction work of China and put forward some reference solutions. The papers *The Discussions on Landscape Ecology Principles and Its Application Educational System and Innovation with the Orientation of LA* and *The Research Development of Landscape Architecture in Main Colleges in China Based on Theses Statistics — A Review of Master Degree Theses on Landscape Architecture in China in Recent 10 Years and Prospect* seize some key problems in the professional higher education. The problems brought by the construction blueprint of landscape planting design are getting more and more serious, and the paper *Study on Some Problems in the Construction Blueprint of Landscape Planting Design* makes useful study on this. These articles are worth recommending for readers being concerned about these issues.

2007年10期

本期主题：儿童公园

本期的主题是儿童公园，我们一共组织了6篇文章。"合理地设计儿童游戏环境，让他们在游戏中学习，在游戏中培养他们的探索精神、敢于克服困难的勇气和通过交流创造一种团队合作的能力"（李建伟语），几乎是每一篇文章共有的精神。

澳大利亚墨尔本皇家植物园的儿童园充满着对儿童（甚至大人）极具吸引力的多样空间和新颖设计，并比较深入地介绍了儿童园的植物设计。美国苏珊·G·所罗门博士《1966～2006年美国儿童游戏场的变化》着重批判了"麦当劳模式"的游戏场地，让我们认识到仅仅依靠资本和法律并不可能给这个世界以十全十美，提醒我们还有许多事情要去探索。美国布鲁克林植物园的儿童园以让儿童主持自己的园子，主动培育和了解植物，培养一种主人翁意识为主旨。《与绿色一起成长的童年——江苏启东市头兴港畔儿童公园设计》一文介绍了作者从城市儿童的成长需要出发设计的江苏启东市的儿童公园，我特别为他的简洁、清新的图面效果所吸引。《儿童游乐所的设计目标与创意》一文批评了大人们总是以成人的眼光来看待孩子的世界，提倡用儿童的眼光看世界，并提供了一个五光十色充满魅力的范例。（日）中村攻的文章谈到了游戏对于儿童获得处理人际关系的自制能力的作用，这对我有着强烈的启示，我们是到了在培养下一代时考虑让他们能够成长为适应社会主义民主体制建设需要的时候了。文章还指出随着生活方式、政治以及经济急速西化，特别是越来越美国化，犯罪就越频繁发生，这是自然资源丰富且让人满意的公园规划设计的一块绊脚石。对此，广州一些设计已经开始注意，比如，公共场所注意必要的视觉通透性，不再盲目提倡纯植物生态主义的乔灌木比例。

令我十分尊敬的周维权先生不幸离我们而去了，这是我国风景园林界的一大损失，也是学术良心的一大损失，相信读者看到孟兆祯院士、金柏苓先生、杨锐教授分别写的纪念文章后会感同身受。8年前我曾经写过《十年再磨剑——读周维权先生〈中国古典园林史（第二版）〉有感》一文发表在本刊2000年第4期，对周先生严谨的科学精神十分钦佩。后来听说周先生以"磨剑"的精神一直在为第三版继续修改，而今竟未能毕其功而撒手人寰，哀哉！在浮躁的当今，严肃的中国园林的历史和理论建设不知还能有人承继否？

在城市居住区的面貌已经基本被操纵于房地产商之手的条件下，政府如何控制城市面貌成为新课题，对此《珠海市淇澳岛风貌规划》一文做了有益的探索。中国园林的神韵是单纯的景观二字远远不能囊括的，《留园空间的音乐美感》一文对此做出了新的贡献。

Phase 10, 2007

Topic of This Issue: Children's Park

The theme of this issue is the Children's Park, and we have six articles on this."Proper design of children's game environment will enable them to learn from games, foster their exploring spirit and courage of overcoming hardships, and build up the teamwork ability through exchanges with others", as Li Jianwei said, is almost the common spirit of every article.

The Children's Garden of the Royal Botanic Garden of Melbourne, Australia, is full of diversified space and new designs which are attractive for children (and even adults), and the plant design of the garden is introduced in depth. Dr. Susan G. Solomon's Changes in American Playgrounds 1966—2006 mainly criticized the McDonald's-model playgrounds, which makes us realize that only capital and laws can never make a perfect world and a lot of things still need to be explored. The Children's Garden of Brooklyn Botanic Garden allows the children to host the garden by themselves and nurture and understand plants positively, with the purpose of fostering their sense of ownership. The article of A Childhood in Growth Together with the Green—The Design of Touxing Port Shore Children Park, Qidong City, Jiangsu Province introduced the author's design of the Children's Park of Qidong City, Jiangsu Province, with considerations of the needs in children's growth. I am particularly impressed by the simple, clean effect of his plan. The article of The Design Goals and Creativity of Children's Playground criticized that people always treat the children's world from the adult perspective. It proposed to look at the world with children's eyes and also provided a colorful and attractive model. Japanese Osamu Nakamura's paper discussed the role of games in fostering children's self-governing abilities of interpersonal relationships. It is strongly enlightening for me, and it is the time for us to consider enabling the children to grow up as the ones suitable to the socialist democratic system construction. The paper also pointed out that crimes become frequent along with the westernization, especially Americanization, of lifestyle, politics and economy, which is a stumbling block toward a rich natural resources and satisfying park planning and design. Some designs of Guangzhou have paid attention to this aspect, for example, emphasizing the visual permeability in public places, and not blindly proposing for the pure plant ecological ratio of trees and shrubs.

My respectable Mr. Zhou Weiquan has unfortunately left us, and this is a big loss of our landscape architecture circle and the academic conscience, which I believe our readers will have the same feeling after reading the commemorative articles by Academician Meng Zhaozhen, Mr. Jin Boling and Professor Yang Rui. Eight years ago I wrote an article titled Regrinding Sword in a Decade—Some Inspirations after Reading Mr. Zhou Weiquan's History of Chinese Ancient Gardens (Second Eidtion), published in the 4th issue of our magazine in 2000, in which I expressed my admiration for his rigorous scientific spirit. Later I heard that Mr. Zhou has been revising for the third edition with the sword-grinding spirit, and it is a pity that he died before the completion of his work. In today's agitated world, I do not know if the serious Chinese garden history and theory construction could be succeeded.

In the situation that the appearance of urban residential areas has been basically manipulated by the real estate, how the government gets the city appearance under control becomes a new research topic. The article of The Landscape Planning of Qi'ao Island in Zhuhai makes a useful exploration in this respect. The spirit of Chinese garden is never what the simple word landscape can describe, and the article of The Musical Beauty of the Space of Lingering Garden (Liu Yuan) makes new contributions to this.

2007年11期

本期主题：文化景观

人文景观（Cultural landscapes）问题，近来成为国际上文物保护界和风景园林界瞩目的焦点，估计国内对此的关注会逐渐增长，有关的讨论也会逐渐兴旺。读者从本期发表的有关文章可以看出，本来中国对此应该是最有发言权的，在人与自然关系的问题上，中国人把它看成是顺理成章的事物，不仅有深厚的理论，而且有丰富的实践，但在西方语境中却偏偏成为难题。

对Cultural landscapes的翻译，本期原稿直译为"文化景观"。我觉得，用"文化景观"去与"自然景观"对应，就不如用我们过去熟悉并进入中国法规文件的"人文景观"来与"自然景观"对应来得贴切。《辞海》对文化的解释是"人文与教化"，其源出于《易经》："观乎天文，以察时变；观乎人文，以化成天下。"文者纹也，是指人类刻在事物上的痕迹，显然，其含义比之西方源于"栽培，耕耘"的culture的含义要丰富得多。

此外，由相关的文章可见，不同的国家和地区对landscape这个词的理解是有很大差别的。照习惯翻译成景观，则将出现"景观研究也从将景观纯粹视作'土地'的理念……"（见《〈欧洲风景公约〉：关于"文化景观"的一场思想革命》一文）这样混乱的概念。又如，"现在的欧洲风景园林师们可能是在所有与景观有某些相关的专业人员当中，具有最小可能性认为景观是仅仅与视觉相联系的一个概念的职业人员"（参见英文原文）一句，说明欧洲的风景园林师经过长期训练可以不认为landscape仅仅与视觉相关，但其他人即使从事相关事业却偏偏这样认为。这种混乱在汉字文化圈内可能更加严重，因为汉字"景观"比起landscape更加强调视觉。

如果循吴良镛先生的看法将landscape翻译成"地景"，则更接近这个词的本义，可以有助于校正这个偏误，但不知为何"人文地景"的说法总觉得有些别扭。我有个想法，既然地理界伸手把landscape这个词拿去使用，我们也可以把汉语"地貌"这个词拿来使用，显然"貌"比起"观"更加客观，容易进入科学和法制领域，而且"人文地貌"听起来也比较顺耳。推而广之，"地貌生态学"比起"景观生态学"也要准确得多。或者，我们干脆另造一个中国词，然后把它翻译成英文，如何？

如何处理这个问题，欢迎读者来信发表意见。这个问题与过去的"景观"与"园林"之争毫无关系。当然，其他行业如何理解landscape应由他们自己去决定。

中国风景园林的发展到了关键时期，除了专业地位，教育改革等问题外，一个重要问题是与建筑、城规等行业相比，我们的法规制度极不完善，而完善法规的重要前提是明确概念，更重要的是，对于过于主观的事物是很难立法的。由此看来，名词已经不仅仅是心态包容性的问题，需要业界更认真地对待。

限于篇幅，这次我就不在这里一一介绍其他文章了。但是我还是要推荐《"都市伊甸"——北京商务中心区（CBD）现代艺术中心公园规划与设计》一文，因为它体现了一种极为认真的设计精神，而这是当前浮躁的社会氛围中最为珍贵的精神。

Phase 11, 2007

Topic of This Issue: Cultural Landscapes

Cultural landscapes has recently become a major focus of attention of the international heritage conservation community and landscape architecture industry. It is estimated that the concern of this topic in China will gradually increase and the discussion will gradually flourish. Readers can see from the papers of this issue that China should have the most say on this since Chinese people regard the relationship between human and nature as a matter of course and there are profound theories as well as rich practices, which is a problem in the context of the Western world.

As to the term of cultural landscapes, it is literally translated in the papers of this issue. I think it is more appropriate to use " 人文景观（Renwen Jingguan)", which is more familiar to us and has been put in related regulations and documents, to correspond with "natural landscape" than use "cultural landscape". The definition of culture is "humanity and education" in the dictionary Ci Hai and it originated from Yi Jing (Book of Changes)—"judging from the astronomical to detect time-varying; judging from the humanities to turn into the world." Humanity also means marks and refers to human traces on things, and obviously its meaning is much richer than the word culture which originated from "cultivation" in the Western world.

In addition, it can be seen clearly in relevant papers that different countries and regions have different understanding of the word landscape. If it is translated according to the usual practice, there will be such idea as "landscape research also changes from the concept of landscape simply as land" (As in the paper The European Landscape Convention: A Revolution in Thinking about 'Cultural Landscapes'). Another example is this sentence "it is interesting that landscape architects in Europe are now perhaps the least likely of all those professionals involved in landscape in one way or another to consider landscape as a concept relating solely to the visual", which shows that European landscape architects can understand after long-term training that landscape is not only related with the visual landscape while others, even though they are in the related professions, just choose to think so. Such confusion is more serious in the cultural circles using Chinese, since the word landscape in Chinese more emphasizes the visual.

If the word landscape is translated as "earthscape" in Chinese according to Mr. Wu Liangyong's view, I think it is even much closer to the original intent of the word and helps correct the bias. But saying "humanity earthscape" is somehow awkward. I have an idea that since the geography circle take the word "physiognomy" in their use, we can use the word "physiognomy". "Physiognomy" seems more objective than "landscape" and can have easier access to the scientific and legal fields, and "humanity physiognomy" sounds relatively smooth. By an extension of this logic, "physiognomy ecology" is more accurate than "landscape ecology". Or, we simply create a separate Chinese word and then translate it into English, how about this?

Readers are welcome to write letters to us and express views on how to deal with this problem.

The development of Chinese landscape architecture has entered a crucial period. In addition to the issues like professional status and education reform, an important issue is that the rules and regulations on landscape architecture still need to be improved compared with the architecture and urban planning trades. The prerequisite of improving laws and regulations is to have a clear concept, but more importantly it is very difficult to conduct legislation on things that are too subjective. From this perspective, the term has not been a simple question of inclusive mentality and our professional circle should treat it more seriously.

I will not introduce other papers one by one there due to limited space. However, I still want to recommend the paper Urban Eden— The Planning and Design of the Modern Art Park in Beijing CBD Area, for the serious design spirit it shows, which is most precious in the current impetuous social atmosphere.

2007年12期

本期主题：风景名胜区研究

本期的主题是风景名胜区研究。读者可以从文章中感觉到，风景名胜和城市园林是既有强烈相关与相似（特别是对以山水为主题的中国园林来说）又有巨大区别的两块领域。对于城市园林来说，其价值主要靠造，评（相地）是其前期工作；对于风景名胜区来说，其价值主要靠评，而不是造。《作为整体的"中国五岳"之世界遗产价值》一文从世界遗产申报标准出发，对我国"五岳"的价值进行探讨，对其他关心申报自然或文化遗产的地方着手这方面的工作具有示范作用。园林与风景更大的差别在于管理，管理成为风景名胜区的难题，而恰如其分的管理基础在于规划，因而规划必须立足于管理，这是由我国建立风景名胜区制度的目的决定的。风景名胜区的管理不仅指本身的管理，更是一个综合性的社会工程，本期主题栏目的其他4篇文章，都是在这方面进行着基础理论研究或结合实践对规划理论进行了探讨。

我一直有一个强烈的想法，为什么"大师"都是外国的？如果说中国传统建筑或城市不大适应这个时代，园林也是这样吗？有人说是这样的，我却不相信，除了对中国传统园林的信心以外，还由于园林是室外生活方式的反映，我不相信当前发达国家的生活方式可以统一全球，因为地球根本就没有这样的资源让这种生活方式"全球化"。比较一下历史上的各种园林流派，中国园林的管理是最节省的。我觉得，未来的世界级大师应该主要出于中国，作为传承的关节，现在就应该有一些这样的大师。本刊除了为纪念去世的汪菊渊、周维权、李嘉乐等先生制作了专栏外，还曾经推出过介绍朱有玠、程世抚、陈俊愉先生的专栏。本期我们向广大读者全面地介绍了孙筱祥先生，文章虽然有点长，却是孙先生用一生写成的，相信读者读后会对人生、对事业、对我们这个行业都能从中获得某种启迪。

当前我国园林建设实践中，种植施工是个大问题，而施工中的许多问题的根源在于施工图，《议园林种植设计施工图中的几个问题》就此做了初步探讨。依我看，有些问题更深的根源在于方案图，比如方案图上所有树木种植点都与路缘距离三四米以外，以免树冠破坏了道路线形的美观，于是施工图照抄，照此施工后公园道路便成为阳光大道，没有遮阴，多少年后也难以改变。

《中、日、欧的被试者对河川景观注视过程的比较研究》一文使用的一种景观研究方法，值得向研究型读者推荐。应该承认，在对心理现象的定量化研究方面，我们还得虚心向国外学习。当然，我要提醒一下，这种机械式的研究探讨的只是社会鲜活事情的一个方面，并不等于事情的全部。

Phase 12, 2007

Topic of This Issue: Study on Famous Scenic Sites

The topic of this issue is the study of famous scenic sites. Readers can feel from the articles that the scenic sites and urban landscape are two areas of close relationship and similarity (especially for those Chinese landscape garden) as well as great differences. For urban landscape, its value depends mainly on creation and the site evaluation is just the front-end work, while for the scenic sites, its value depends mainly on evaluation instead of creation. The article The Outstanding Universal Value of "The Sacred Five Mountains of China" starts from the application criteria of World Heritage and discusses the value of the sacred five mountains, which sets an example for the efforts of those concerned with the application of World Natural or Cultural Heritage in this aspect. The greater difference between landscape and scenery is management, which has become a problem for scenic sites. The basis of proper management comes from planning, and the planning must take management into consideration, which is determined by the aim of setting up Chinese scenic sites system. The management of scenic sites is also a comprehensive social engineering as well as the management itself. The other four articles in this column conduct basic theoretical research or discuss the planning theory with practical cases in this area.

I always have a strong idea: why "masters" are all foreigners? If it is said that traditional Chinese architecture or cities do not fit in with the times, so is landscape? Some people say yes but I don't think so. In addition to the confidence for the traditional Chinese garden, it is also because garden is reflection of outdoor life, and I don't think that the lifestyle of developed countries can be the global one, as there is no such resource for this "globalized" way of life. Compared with various schools in the garden history, the management of Chinese gardens is the most cost-saving one. I think that the future world-class masters should mainly come from China, and as a successive joint, there should be some of these masters now. Our journal has had special columns for the introduction of Mr. Zhu Youjie, Mr. Cheng Shifu and Mr. Chen Junyu, as well as the columns in memory of Mr. Wang Juyuan, Mr. Zhou Weiquan and Mr. Li Jiale. In this issue we have full introduction of Mr. Sun Xiaoxiang. Although the article is somewhat long, it is of his whole life, and I believe our readers will acquire some inspirations for life, for work, and for our landscape industry.

In present landscape construction of China, planting is a big problem and the root of many problems in construction lies in the construction plan. The article On Several Problems of Landscape Planting Construction Plan makes preliminary discussion of this. As I think, the deeper root of some problems may come from the program plan. For example, on the plan all trees are planted 3 or 4 meters away from the road to avoid the crowns disrupting the road line, and this is then copied into the construction plan and the road will have no tree shades, difficult to change for many years, if planting like this.

The landscape research method used in the Study on Fixation Behavior for River-landscape by the Subjects of Chinese, Japanese and European is worth recommendation for research-oriented readers. It should be recognized that we need to learn from foreign countries in the area of the quantitative research of psychological phenomenon. Of course, I would like to remind that this mechanical study is only one aspect of things and it does not mean all.

2008年

2007年　2009年　2010年　2011年　2012年　2013年　2014年　2015年　2016年

2008年1期

本期主题：风景园林教育与实践

新年伊始，万象更新！祝本刊广大读者身心健康，学业有成，工作愉快。

改革开放已经30年，子曰：三十而立。可以预见，我们国家今后的发展必将更加稳健，坚实，无可阻挡。

年前我们几个主编盘点了一年的工作，觉得去年的努力基本适应了行业发展的形势，听到的称赞多于批评。其实我们希望多听到点批评，听到的批评正是改进我们今后工作的良药。

本期内容有两大热点：主题与论坛。

算起来"文化大革命"后招收的第一届风景园林大学生入学迄今也已整整30年。中国风景园林学会教育分会（筹）2007年11月在南京林业大学召开了年会，集中研讨了改进我国风景园林高等教育的问题，本期即以此为主题。会后我国学科的几大主力高校（北林、清华、同济、南林）都在本刊阐述了自己的理念或方法，华中农业大学副校长高翅教授也发表了他的意见，并介绍了国际风景园林师联合会和联合国科教文组织（IFLA–UNESCO）共同拟定的《风景园林教育宪章》，相信这些对其他院校的教学改革会有很好的借鉴作用，三五年后必见成效。此外，我们也请中国农业大学的李树华教授介绍了同属东亚的日本的情况作为参考。

本期论坛的主要发言者是IFLA的现任主席孟赛斯女士，题目是Landscape stewardship，意思是风景大管家的工作（包括内容与职责）。我体会孟赛斯主席觉得用stewardship来表达我们这个行业比用architecture更加准确，我也有同感。我们这些人到底是以设计和建造landscape为主，还是以管护这个地球表面为主，实际上是存在不同见解的。在当前全球都关注于气候变暖之时，孟赛斯的提醒非常重要，我想这才是我们这个行业的国际主流和今后主要发展方向。朱建宁教授的《做一个神圣的风景园林师》也包含着相似的精神。如果大家都能够在这个见解上达到统一，许多过去的争议都会化解为无形。

2008年1月1日起，我国的《城乡规划法》开始实施，城乡统筹规划是该法的主要特征之一，认真学习和贯彻《城乡规划法》不但是风景园林行业的任务，也肯定有利于推进风景园林行业的发展，比如谁来做乡村规划已成为当前困惑之一，希望大家关注这方面的发展，积极来稿。本期关于城市风景区的一些问题的谈论，实际上已经在某种程度上牵涉到相关问题。

Phase 1, 2008

Topic of This Issue: LA Education and Practice

At the beginning of the New Year, everything takes on a completely new look. We wish our readers good physical and mental health, great achievements, and happy work.

It has been thirty years since the reform and opening up of China. Confucius said that man should be well established at the age of thirty. It can be foreseen that the future development of our country will certainly be more stable, solid, and invincible.

We chief editors reviewed the one-year work right before the new year, and we feel that our efforts in the last year basically fitted in with the development trend of the industry and we received more commendations than criticism. Actually we hope to have more critical opinions and advice, which would be the panacea to improve our work from now on.

There are two hot spots in this issue: the topic and the forum.

It has also been thirty years since the students of landscape architecture were enrolled for the first time after the Cultural Revolution. The Education Branch (preparatory) of Chinese Society of Landscape Architecture held the annual meeting in Nanjing Forestry University in November 2007 and discussed in focus of how to improve the higher education of landscape architecture in China, which is the topic of this issue. Several major universities (Beijing Forestry University, Tsinghua University, Tongji University, and Nanjing Forestry University) in this discipline expound their own ideas or methods in our journal. Professor GAO Chi, Vice President of Huanzhong Agricultural University, also expresses his opinion and introduces the IFLA-UNESCO Charter for Landscape Architectural Education. We believe these will be good consultation for the educational reform of other institutions and there should be good effects in near future. Besides, we also invite Professor LI Shuhua from China Agricultural University to introduce Japanese cases as reference.

The key speaker of LA Forum is Dr. Diane MENZIES, IFLA President, and the title is Landscape Stewardship, including the contents and duties. I think Dr. Menzies must feel it is more accurate to use stewardship than use architecture to describe our industry, and I feel the same way. There are different viewpoints on whether we should focus on designing and building landscape or focus on managing and protecting the surface of the earth. Dr. Menzies' reminding is very important during the current global concern of climate warming, which I think is the global main stream and future development direction of this industry. Professor ZHU Jianning's To Be A Sacred Landscape Architect also contains the similar spirit. If we can achieve unity on this view, many disputes will be resolved.

The Town and Country Planning Act is enforced form January 1 2008, and the overall town and country planning is one of its main features. To earnestly study and implement the law not only is the task of this industry, but also will benefit the development of this industry. For example, one of the current confusions is who will conduct the rural planning. We hope our readers pay attention to this aspect and papers are welcome. The discussions on some issues of urban landscape area actually have involved some related issues.

2008 年 2 期

本期主题：废弃地改造与复建

"废弃地"一词是对brownfield或brown land的意译，也有直译为"棕地"或"褐地"的。废弃地是工业化的产物，废弃地问题是工业后时代的问题，因为其新，所以国内外的名称还没有统一。我在Google上检索了一下，国内"棕地"的使用率稍微大于"褐地"，国外brownfield的使用率则明显优于brown land。对于这种与中国传统无关的东西，我主张以直译或音译为好，以简洁者为佳。鉴于土壤界将棕壤翻译为brown earth，我的感觉是使用棕地的前景可能较大。

我国总体上虽然还没有进入这个时代，但是现代工业也有100多年了，近年来在改革开放的先行地带开始出现产业转型现象，所以废弃地也开始出现。利用老企业用地建设绿地，是改善城市环境，特别是旧居民区环境的重要出路，为此，我们介绍了一些国内案例。我觉得，在技术上这种事情没有多少难题，特别是如果我们善于和相关专业（如环保、土壤、市政）的专家合作，重要的是指导思想和工作程序。相比之下，欧美在这方面的工作更多，时间更久，法规也比较完善。本期的2篇译文很有趣，《康涅狄格州雷丁镇的原吉尔伯特和本尼特金属丝工厂的再开发——废弃地再开发中公私合作的成果》是介绍最大的工业国美国的一个工作案例，《工业化国家的城市改造政策：文化的标准和环境的关注》则将最老的工业国家英国和美国在基本思维和法律体制上作了对比，都可以供我们借鉴。不管如何，他们的共同特点是法制健全，前期论证和技术准备很严格，而我国在这方面的法制还有点欠缺，经常出现匆匆上马的政绩工程，这与我们社会主义国家的根本利益是不相符的。

从我国行业总体发展上看，我们在2个突出方面需要更紧密地跟踪世界的发展：标准化和虚拟现实技术。《ISO风景园林标准体系现状研究》介绍了行业的标准化情况，《虚拟现实技术在风景园林规划与设计中的应用研究——几种常见虚拟现实技术的应用评价分析》介绍了虚拟现实技术在我们行业应用的进展，希望引起各地行业领导和有关技术人员的关注。

《上海市滨海盐渍土改良技术的研究与实践》一文介绍了老一辈专家主导的一项踏踏实实的长期科研工作获得的成果。此外，我尊敬的前辈，我们的首任主编余树勋先生以轻松的笔调写了一篇散文式的《园林丛谈——向往的"老人公园"》，我读了之后感慨万千，自己也算是过了一个甲子，但对于已达和超过"从心所欲"层次的老人之需，并不了解。除了推广科研成果和设计思想的本身，我还期望年轻一代的科研和设计工作者从中得到某些启示。

Phase 2, 2008

Topic of This Issue: Renovation and Restoration of Abandoned Land

"Abandoned land (废弃地)" is the free translation of the term "brownfield" or"brown land", which could also be literally translated as "brown land (棕地 / 褐地)". The abandoned land is a product of industrialization and the problem of abandoned land is a post-industrial issue. The Chinese and foreign terms for this new thing have not been unified. As I searched via Google, the frequency of the term"brownfield (棕地)"being used is slightly higher than"brown land (褐地)"in China, while "brownfield" is more often used than"brown land" in foreign countries. For things of this kind that have nothing to do with Chinese traditions, I advocate the literal translation or transliteration as well, preferably the concise one. In view of "brown soil (棕壤)" being translated as "brown earth" by the soil sector, I feel that the prospect of the use of"brownfield (棕地)"may be larger.

Generally speaking, China has not entered this era yet. However, it has been over 100 years for the modern industry in China, and the recent industrial transformation in the forerunning areas of reform and opening up also brings the abandoned land. Building green land by using the sites of old enterprises is the important way to improve urban environment, particularly the environment of old neighborhoods, on which we introduce some domestic cases. I think there is not much technical problems for this, especially if we cooperate well with experts of relevant fields (such as environmental protection, soil, and municipal service), while the guiding idea and working procedures are more significant. Comparatively, European and American countries have more and longer experiences as well as more perfect regulations in this area. The two translated papers in this issue are very interesting. The Redevelopment of the Former Gilbert and Bennett Wire Mill in Redding, Connecticut (USA) — the Results of a Public/Private Partnership in Brownfield Redevelopment introduces a case of United States, the largest industrial country, while Urban Regeneration Policy in Industrialized Nations: Cultural Norms and Environmental Concerns compares the fundamental thinking and legal systems of United States and United Kingdom, the oldest industrial country, both of which are good reference for us. Anyway, their common characters lie in the sound legal systems, and strict first-phase feasibility studies and technical preparations. Our legal system in this area still need to be improved and some achievement projects are started in hurry, which are not consistent with the fundamental interests of our socialist state.

In view of the overall development of this industry in China, we need to more closely track the development of the world in the two salient aspects: standardization and virtual reality technology. Study on the Current Situation of ISO Landscape Architecture Standard System introduces the conditions of standardization of the industry, while The Application Research of Virtual Reality Technology on Landscaping Planning and Design — The Evaluation of Virtual Reality Technology Applications introduces the progress of application of virtual reality technology in landscape planning and design, expecting to draw the attention of the leaders and technical staff of relevant industries.

The paper Research and Practice on Amelioration Technologies of Coastal Saline Soil in Shanghai introduces the achievements of the long-term down-to-earth scientific research work led by the older generation of experts. In addition, Mr. YU Shuxun, my distinguished predecessor and the first editor-in-chief of our magazine, wrote in relaxed tone the prose-style Garden Essays — Yearning for the Elderly Park. It makes me sigh with emotion after reading, for I have been over sixty years old, but I had no understanding of the needs of the elderly in even older ages. In addition to promoting the scientific research and design thinking, I also expect the younger generation of scientific researchers and designers get some inspiration from it.

2008 年 3 期

本期主题：传统商业街区规划设计

　　自从1953年梁思成的建筑美学思想被批判为"大屋顶主义"、"复古主义"以后，很长时间内"民族形式"的建筑几乎成为园林建筑的"专利"，应该说，这是园林界的一大贡献，可以设想，如果当时连园林中的传统建筑都像北京城门楼一样被毁掉，中国的古建筑文化现在会是一幅如何悲惨的景象。时至今日，园林界"传统风格"的建筑设计、施工和维护队伍仍然是业界最强大的，于是形成国内大量的传统步行商业街由园林界规划设计和施工建设的局面。我想，这种情况在世界上也是罕见的，这是中国风景园林界的一笔宝贵财富，值得好好总结。本期主题"传统商业街区规划设计"即服从于此目的，内容包括商业街本身的景观与设施的设计，也包括其周围环境的处理，特别是后者，一般的建筑院对此是比较忽视的。

　　中国园林的近代史，过去比较忽视，其实它反映的是处于"三千年未有之大变局（李鸿章语）"下的中国园林的活态。在这个历史时期，沿海和沿江一带的变局率先发生，而位于江、海交汇处的江苏省，创造了许多中国第一，《江苏近代园林几多中国之最》对此进行了探讨。一个民族的和整个人类的文化是无数先辈辛勤积累创造的，多了解点历史，有助于识破那些自称从天而降创造历史的"神仙"们编造的神话，破除利用浮躁社会弱点制造的迷信。

　　成都的新繁东湖，位置偏僻，对其起源的考证，我推荐房锐先生的《新繁东湖缘起考辨析》一文（载于《西华大学学报（哲学社会科学版）》2005年第5期）。我在成都工作时，它还保留着唐代园林的基本格局（日本称之为"寝殿造"）。这样一份珍贵遗产，始终未得到上级行政部门足够的重视，但又担忧地方行政部门对它进行建设性破坏。幸有学界还在不懈地对它关注，为此本期发表了《晚唐名园——新繁东湖》。

　　杭州近年来的变化，有目共睹，也有不同看法。过去，本刊有的文章从生态学角度提出过意见；本期的《自然与人文的交融——探索杭州西湖风景名胜区可持续发展的保护整治之路》一文，则更多地从人文角度做了肯定的阐述。其中的是非曲直，蕴含着天地人生的大道理，如果中国人真正把这个大道理厘清了，可以说是对人类功德无量的贡献。

　　西方文化延续着它的逻辑，逐步深入到对人类思维和心理本身进行科学研究的阶段，包括系统化和量化，行为科学是其中的一个分支。《行为科学对园林学发展的影响》一文对此进行了介绍。我对西方的这个发展一方面欢迎，因为它有助于人类了解自己，有助于更好地建设人类的生活环境；一方面是警惕，因为这种事过头了就有技术专制倾向，这就是海德格尔思考的问题：技术上的专制必定导致语言的技术化，从而使技术语言流行开来并被广泛接受，使语言失去了语义的丰富性和鲜活的生命力。

Phase 3, 2008

Topic of This Issue: Planning and Design of Traditional Commercial Street Area

Since 1953 when Liang Sicheng's architectural aesthetics thinking was denounced as "big roof doctrine" and "antique doctrine", the buildings with "national forms" have almost become the main content of landscape architecture, which could be regarded as a great contribution to this sector. It could be envisaged that ancient Chinese architecture culture would be tragic if the traditional buildings were overturned then like the destroyed Beijing city gatehouses. Up to now the design, construction and maintenance teams of the "traditional style" buildings are still the strongest in this sector, resulting in that numerous domestic traditional commercial streets were designed and built by landscape architecture people. I think this situation is quite rare in the world and it is also a valuable asset of China landscape architecture sector which needs to be well summarized. The topic of this issue "Planning and Design of Traditional Commercial Street Area" is right for this purpose, and the contents include the design of the landscape and facilities of commercial streets as well as handling the surrounding environment, while the latter is usually neglected by architecture institution.

The modern history of Chinese gardens, nearly neglected in the past, actually reflects the living situation of Chinese gardens in the period of "great change never met in 3 000 years (Li Hongzhang's words) ". In this historical period, the coastal area and the area along the Yangtze River encountered the change first, and Jiangsu Province, at the interchange of rivers and seas, has been a forerunner. The paper of *Some of the Best Aspects of Jiangsu Modern Landscape in China* discussed on this topic. The cultures of a nation and the whole mankind were hard accumulated by countless predecessors' creation. More understanding of the history helps see through the myth of "God makes things happen" and get rid of superstition generated from impetuous social weaknesses.

Xinfan Donghu of Chengdu is in a remote location, and for the study of its origin I recommend Mr. Fang Rui's paper of *The Discrimination Study of the Origin of Xinfan Donghu (originally published in Xihua University Journal* (Philosophy and Social Sciences Edition), 5th Issue 2005). When I worked in Chengdu, it retained the basic pattern of the Tang Dynasty garden. Such a precious heritage has never been given sufficient emphasis from higher administrative departments, and we are also worried that it will be destroyed during the construction work by local governments. Fortunately there are still unremitting academic concerns of it, and for this we publish the paper of *The Famous Park in Late Tang Dynasty – Xingfan Donghu Park*.

Changes of Hangzhou in recent years are perfectly obvious, but here are also different views about it. Some papers in our journal have expressed opinions from the ecological viewpoint, while the paper of *The Blend of the Nature and Humanity – Exploring the Protection and Renovation Approach for the Sustainable Development of West Lake Scenic Area, Hangzhou* in this issue makes favorable comments from the humanity perspective. The rights and wrongs of this case contain the great truth of the world and life. It would be a great contribution to human beneficence if the Chinese people really clarify the truth.

The western culture continues its logic and gradually goes deep into the systematic and quantitative scientific study of human thinking and psychology, while the behavioral science is one of the branches. The paper of *Influence of Behavioral Sciences on Landscape Architecture* makes introduction of this. On the one hand I welcome this western development since it helps human beings to understand themselves and contributes to the better building of human living environment, but on the other I am vigilant because of the technical dictatorship tendencies if these things are overdone, which is also what Heidegger questioned: the technical dictatorship certainly leads to the language being technical, which makes the technical language popular and widely accepted and makes language lose the richness of semantics and the fresh vitality.

2008 年 4 期

本期主题：城市公共绿地设计

　　工业革命以来人类所做的硬体工程，主要是一件事情：建设城市和延伸城市。以公园为代表的城市公共空间建设成为我们这个行业在这个时代的一个亮点，关于这方面的来稿也特别多，为此，本期的主题定为"城市公共绿地设计"。

　　城市发展的基本动力，在于资本的扩张。近现代资本的扩张大体有2个阶段，工业资本和金融资本，前者创造了贫民集中的近代工业城市，后者创造了金钱高度集中的现代大都会。以这个观点观察近二三百年的城市发展，包括公园的发展，会有些新体会。近10来年，我一直在思考一个令人困惑的难题：公园设计只要环境良好，看起来舒服，功能适合居民需要就应该算是基本合格，为什么现在"创意"忽然变得如此重要？以至一些让人在那里待着难受，无事可做，看也看不懂的设计大行其道？最近，似有所悟：如果说重视理性的现代主义是工业资本掌控条件下的产物（因为工业生产需要尊重自然规律），那么抛弃理性的后现代主义正好适应金融资本的需要，金融资本只需依靠着狡猾的策略便可大量"搂钱"，故而，对于现代这个社会的资本来说，创意比遵从规律更为有用。当否，愿与诸君共商讨。

　　其实，资本无论是以工业形态还是以金融形态出现，都离不开人，也敌不过自然规律，总之过不了人与自然的关系这一关。从人与自然的关系看，现代城市的本质并不是为人，而是为资本服务。过分歌颂现代城市（包括其公园）并以此攻击所谓小农经济的中国传统园林，反映出的是精神境界的狭隘，而这种境界的源头，就是资本世界的本身。

　　本期《诗意栖居——中国古典园林的精神内涵》是一篇不错的文章，虽然荷尔德林诗句的原意是人因为服膺了神而纯真善良，从而获得精神的绝对自由，故虽然劳累，仍然诗意地栖居在大地上，但"诗意栖居"已成为建筑界和设计界的流行语，作为一种借用似无不可，我相信以后还会有大量此类的文章出现。

　　在市域范围内构建生态廊道和康体廊道体系是一件大事，我国一些城市已经开始着手，可喜可贺。本期《城乡统筹，构建和谐生态廊道研究——以成都市绿色生态健康廊道为例》介绍了成都的工作。这种事情主要的难点是如何与国土规划以及其他规划相衔接，对此如有经验，欢迎来稿。

　　采石或挖山造成的边坡，像癞疤一样损害着我们美丽的山河，在有了一定经济基础后，各地都在想法治疗这些伤疤，近来也不时有一些这方面来稿。本期选发了《木渎景区金山采石宕口植被恢复与可持续景观设计》与《威海景石边坡绿化技术》两篇，供各地参考。

Phase 4, 2008

Topic of This Issue: Design of City Public Green Space

The concrete work that human beings have done since the industrial revolution is mainly about one thing: building and extending the city. The building of urban public space represented by parks becomes the bright spot of our trade in this era and we have received many papers about this topic, which brings the theme of our magazine for this month: Design of City Public Green Space.

The expansion of capital is the fundamental driving force of urban development. The expansion of modern capital generally has two phases, the industrial capital and the financial capital, and the former created the modern industrial city with concentration of the poor while the latter brings the modern metropolis with high concentration of money. There are some new understandings when reviewing the urban development, including park development, in recent two and three hundred years with this perspective. I have been pondering upon a puzzling problem over the past 10 years: if the park design is regarded qualified simply when it has fine environment and looks comfortable and its functions can meet the needs of residents, why "creative idea" suddenly becomes so important now? Why the designs that people feel uncomfortable to stay, have nothing to do and can not understand become so popular now? Recently I seem to understand somehow: if the rational modernism is the outcome under the controlling conditions of industrial capital (since the industrial production should follow the order of nature), the irrational postmodernism is precisely what the financial capital wants, in that the financial capital can greatly "rake up money" simply relying on a crafty strategy, and therefore, for the capital of modern society, the creative idea is more useful than following the law of nature. Whether it is right or not, I would like to discuss with you.

In fact, whether the capital comes in the industrial or financial patterns, it can not be without human beings and can not outdo the law of nature, which means it can not go beyond the relationship between human and nature. From the relationship between human and nature, the nature of modern city is not for human, but the capital. Too much eulogy of modern city (including its park) and attacking the traditional Chinese garden for its so-called small peasant economy reflect the parochialism of spiritual shackles, which is originated from the capital itself.

The paper Poetic Dwelling—The Spiritual Concept of Traditional Chinese Garden is a good article. Although the original intent of Holderlin's verse is that human beings become pure and good and acquire absolute spiritual freedom and they still dwell poetically on the earth despite of tiredness,"poetic dwelling", seemingly a fair borrowing word, has become a catchword in the architecture and design circles. I believe that there will be a large number of such articles appear in the future.

It is a great event to construct the ecological and health corridor system in the city scale, and it is gratifying that some of our cities have already started this work. The paper Urban-Rural Integration to Build a Harmonious Ecological Corridor—Case Study of Chengdu Green Ecological Health Corridor introduces the experience of Chengdu. How to integrate with land planning and other planning is the major difficulty of this kind of project, and readers who are experienced in this aspect are welcome to contribute papers.

The slopes caused by quarrying or excavating on mountains damage our beautiful mountains and rivers like scars, and the local governments with a certain economic foundation are trying to find the treatment of these scars. Recently there are frequently some papers on this topic. The papers of The Plant Restoration and Sustainable Landscape Design for Jinshan Quarry Pits in Mudu Scenic Area and The Scenic Stone Slope Green Technology of Weihai are published in this issue for the reference of local governments.

2008 年 5 期

本期主题：城市水环境

　　本期的主题"城市水环境"是配合2008年IFLA大会主题"water transformation"而设置的，得到作者们的大力支持，在此学刊对他们深表谢意。

　　当前对水环境的探讨大体分为3个方面：水体景观、水质改善、保水设计，对此本期皆有所体现。《富营养化观赏水体的生物—生态修复技术》是有关利用生物进行水质改善发展的综述，可能对各地较有参考价值。《杭州市区河道景观体系规划初探》《上海淀山湖沿岸带生态修复景观模式研究》和《上海世博园区景观水系生态要素的构建》都介绍了作者在这个领域具有独创性的努力探讨，无论探讨是否成功，首先要看到的是我们的学科正是在同仁们的共同探讨中发展前进的。我原来以为干旱地区才有推广保水设计的必要，《基地保水评价指标于风景园林中的应用——以中国台湾逢甲大学校园为例》告诉我即便是雨水丰富的台湾地区，也在提倡保水，看来这将是个大趋势。至于《城市过水区复原·湿地修复·湿地公园建设的三位一体——美国拉斯维加斯城市过水区湿地建设的启示》、《美国圣安东尼奥河改造项目回顾》两篇文章，至少在建设程序的规范和细致入微地考虑问题两方面，值得我们借鉴。

　　当我们行业从园子或garden的篱笆墙里跨出来后，我们才发现原来自己还需要做许多改变：审美习惯、思维方式、决策程序、科学素养等。Landscape architecture的本意就是对大地面貌进行塑造的学问，因此就是从事一种对自然的改造（当然，现在正在成为潮流的与其说是改造，不如说是看护）。通过长期在广阔大自然里工作，我们发现除了出自本性的热爱自然，我们更需要了解和遵从自然规律，而这对于长期习惯于形象思维的landscape architect们来说，是一项挑战。诚然，所谓景观生态学，图形学之类的新学说为图形化、模式化思维提供了一些理论依据，但以严格的科学态度来看，依靠图形来解决问题大概只能暗示一种方向，连定性的层次都没有达到，更何况定量。大家都在说地球生态出了大问题，甚至说到危机的边缘，也就是说平衡即将被打破。平衡是什么？是量的稳定，所以没有量的意识，保持平衡最多只是善良的愿望。如果我们学科真正把解决人类生态作为自己的责任，就一定要上升到定量，这就需要大量的而且结论经得起批判和检验的科研成果。

　　此外，我们特别欢迎紧密结合专业特性的园林植物方面的来稿，例如，植物观赏特性的改变、植物抗逆特性的研究、植物生态作用的量化、植物栽培、施工和养护技术等。

Phase 5, 2008

Topic of This Issue: Urban Water Environment

The theme of "urban water environment" of this month is in line with the theme of "water transformation" of IFLA conference of this year, and has received our authors' strong support. We would like to express our sincere gratitude to them.

The current discussion of water environment can be divided into three areas: water landscape, water quality improvement, water preservation design, and they are all included in the papers of this issue. An Ecological Restoration Technology to the Eutrophication of Ornamental Water Body is the summary of using bios to improve and develop water quality, which might be of reference value for many places. Preliminary Study of Hangzhou City River Landscape System Planning, Study on Landscape Patterns of Ecological Recovery in Waterside of Dianshan Lake, Shanghai, and Ecological Element Construction of the Water Landscape of Shanghai Expo introduce the authors' initiative explorations in this field, and whether they are successful or not, we should notice that it is the joint exploration which makes the discipline move ahead. I had thought that only the arid areas need to promote the water preservation design, but the paper of The Application of "Raining Storage in Site" Assessment Index in Landscape Architecture—A Case Study on Feng Chia University Campus, Taiwan, China tells me that even Taiwan region with rich rainfall is also promoting water preservation. It seems that this will be a big trend. Regarding the Trinity of Urban Wash Restoration, Wetland Renovation and Wetland Park Construction—Inspiration from the Construction of Las Vegas Wash Wetland and Review of the Rejuvenation of San Antonio River, U.S.A., at least the two aspects of standard construction procedures and carefully considerations are what we should learn.

When our industry steps out of the garden and the fence, we find that we still need a lot of changes, such as the aesthetic habits, ways of thinking, decision-making process, and scientific accomplishment. The original meaning of landscape architecture is the knowledge of shape the appearance of the earth, namely a kind of reshape (of course, the word in trend is stewardship) of the nature. Through long-term work in the nature, we find that, in addition to the natural love of the nature, we also need to understand and comply with the laws of nature, which is a challenge for landscape architects who have long been accustomed to thinking in images. Indeed, the so-called new theories like landscape ecology and graphics have provide some theoretical basis for graphical and pattern thinking, but in strict scientific approach, relying on graphics to solve the problem probably only hints a direction, even not up to the qualitative level and let alone the quantitative one. People are all saying that the earth has a big ecological problem, or even mentioning that it is on the verge of crisis, namely the balance is about to be broken soon. What is the balance? It is the stability of quantity, and without the quantity consciousness keeping balance would only the good wishes. Our discipline should go up to the quantitative level if it takes solving the ecological problem as its responsibility, which requires a lot of scientific research achievements that can withstand criticism and testing.

In addition, we welcome in particular papers on garden plants in close connection with this discipline, such as the change of ornamental features of plants, resilience characteristics of plants, quantification of plant ecological role, and plant cultivation, construction and conservation technologies.

2008 年 6 期

本期主题：园林史研究

  上个月我在初读本期稿件时，就已经很难抑制自己的激动心情，本期内容十分充实，真有花团锦簇，篇篇珠玑之感，通过两年的主编经历，我知道这种机会是很难得的。起初我想干脆哪篇也不推荐，任由读者自选，后来细想一下这样不好，于是决定还是特别推荐威比·奎台特（Wybe Kuitert）的《借景——中国园冶（1634）理论与17世纪日本造园艺术实践》、苏珊·斯科特的《"中国（艺术）风格"与中国园亭之西渐》和南希.J.沃克曼的《中国对美国造园设计的影响》3篇。文章读起来虽然不太容易，论点也可以商榷，但可以给我们很多启示。我觉得其中最主要的是只要你认真做出真正属于自己并站得住的东西，就会引起别人的注意，而不是一定要老是跟在别人的后头尽做些东施效颦的事情。我相信，凡是对理论和历史感兴趣的读者，读过这3篇文章必定仁者见仁，智者见智，引发丰富的感想。

  在这里，我得特别感谢杨鸿勋先生。本期发表的3篇外国人研究中国园林的文章，都是来自杨先生亲自操持的2006年潍坊中国古典园林国际研讨会，经过杨先生特许和作者同意，有的文章还是经杨先生亲自校对，本刊才得以发表，以飨读者。

  5月份中国的最大事件，是12日发生的汶川大地震。这次大地震对中国的社会发展和民族精神必将产生重大影响，肯定会永久载入史册。我曾在成都工作生活了近20年，都江堰、卧龙、漩口、平武、绵竹、羌寨……一连串熟悉的地方，如今满目疮痍，美丽不再依旧。这使我感到自责：当初在赞叹这里风光壮美时，为什么不能联想到如今的恐怖？说到底是因为知识的缺乏。我对大自然力量的体验和知识积累，主要还是经过做九寨沟和长江三峡规划，我的许多学生都听过我在规划课程上大讲各种地质灾害，虽然这对广东学生似乎是很遥远的事情。去年在参与广东与意大利的高校合作活动时，我对国外某大学用对大自然知识甚少的二、三年级景观学生去做一块高山地区的规划就颇有质疑。我们学科教育的这一重大缺陷，可能是世界性的。另一个值得引发我们学科思考的是对绿地防灾功能的忽视。1923年日本关东大地震后，很多国家都很重视城市绿地在防灾（特别是地震）方面的作用，20世纪60年代初期我在读本科时，好几个老师都在课堂上讲到过绿地的防灾功能。不知何因，现行的《城市绿地系统规划编制纲要（试行）》和《公园设计规范》没有这方面的内容，实在令人有些匪夷所思，看来进一步强化我们行业的法制建设是到了非抓不可的时候了。幸好北京因为办奥运，开始抓起这项工作，并有了一些经验积累，可供制定全国性法规的参考。我想，我们应该切实做好这项工作，作为以实际行动对汶川大地震逝者的悼念。

Phase 6, 2008

Topic of This Issue: LA History Research

When reading the paper of this issue for the first time last month, I could not control my excitement, since the contents are very substantial and the papers are of elegant styles and graceful writings. As the chief editor for two years, I know this is a rare opportunity. At first I just would not like to recommend any articles and let our readers to choose, which was not good after consideration. Then I decide to put my special recommendation on Borrowing the Landscape—Theory in China's Yuan Ye (1634) and Practice in Japanese Garden Art of the Seventeenth Century, Chinoiserie and the Migration of the Chinese Garden Pavilion to the West, and Chinese Influences on American Design. Although not easy to read and the arguments are open to discussion, they can give us a lot of inspirations. I think the most important thing is that you will draw others' attention as long as you really make something personally and sound while you must not always imitate others awkwardly. I believe that those readers who are interested in theories and history can have their own thinking and feelings after reading the three papers.

My special gratitude goes to Mr. YANG Hongxun. The three papers by foreigners on Chinese landscape architecture research all came from the 2006 Weifang International Symposium on Chinese Classic Gardens organized by Mr. Yang. They are published in this issue with Mr. Yang's concession and the authors' agreement, and some of them are proof-read by Mr. Yang.

The Wenchuan Earthquake is the biggest thing of May in China. It will have a significant impact on China's social development and national spirit, which would certainly go down in history permanently. I have been working and living in Chengdu for nearly 20 years, and a series of familiar places like Dujiangyan, Wolong, Xuankou, Wuping, Mianzhu, and Qiangzhai are now devastated, no longer beautiful. This brings me self-accusation that why I had not thought of the terrors today when praising the magnificent sceneries then. After all it is because of the lack of knowledge. My experience and knowledge of natural forces mainly came from the planning for Jiuzhaigou and Yangtze River Three Gorges, and even my students have heard that I talked about geological disasters in the planning courses although these are quite far from Guangdong students. In my participation in the cooperative activities between Guangdong and Italian colleges, I queried that certain foreign university let the sophomores and juniors majored in landscape architecture with little knowledge of the nature to do mountain landscape planning. The defect of the education of this discipline may be universal. Another aspect which we should think about is the neglect of the disaster prevention function of green space. After the Great Kanto Earthquake in 1923 in Japan, many countries have attached importance to the function of green space in disaster prevention (especially the earthquake). And some teachers talked about the disaster prevention function of green space when I was in my undergraduate years in early 1960s. I do not know why this content is missing in the present Urban Green Space Planning Outline (Trial) and Park Design Specifications, which is unimaginably queer, and it appears that the legal system of our industry has to be further strengthened. Thanks to the Beijing Olympics, it has been started and some experience has been accumulated, which can be reference for national regulations. I think we should conscientiously do this work well, as a practical action to mourn for the deceased in the Wenchuan earthquake.

2008 年 7 期

本期主题：植物多样性

　　城市生物多样性比纯自然界生物多样性复杂得多，城市的主体是人，人与各种生物的互动是个永远说不完的话题。虽然城市植物多样性只是生物多样性的一部分，需要深入探讨的地方仍然很多：大到城市动植物的生态与城市居民生态的关系，城市植物多样性与人类生活环境质量（包括温度、湿度、大气、水体、食物、害虫、疾病等）的关系，对（由于能量、物质、人工、设备等大量输入造成的）高度城市化地区城市植物多样性的评价，城市植物多样性的量化方法和评比方法等；中到乡土植物的定义，顶级群落在城市中的合理分布及其在绿地中的合理比例，栽培品种特别是观赏品种的评价及其对城市植物多样性的意义等；小到具有生态意义的绿廊宽度及其断点的宽度，绿廊断点的技术处理，植物多样性与人的室外生活的关系，植物多样性与社会安全的关系，或者更具体一些，假如一块林地紧靠着生物量几乎为零的铺装广场，它们边界上的物种是多还是少？等等。我想，这些关系中，简单的线性关系一定很少，多数应该是复杂的曲线关系，然而，即使从全世界来看，对此的研究也很少，多数报告都是一些适用范围相当狭窄的线性结论，更加遗憾的是引用这些结论时大多不管其适用的条件。当然，要我们这个学科的人弄出来很复杂的方程是很困难的，但是摆脱单线联系和线性发展的思维，搞出思维结构框架还是可能的。所以，这是我国在这方面研究走在世界前面的一个机遇。

　　其实需要多样化的不仅是物种，也包括景观、文化、制度、思想等。在人类没有掌握真理之前，我始终对当前过早地全球化感到担忧：被资本掌控的社会体制果真是社会形态的最高形态吗？万一人类一起犯同样的错误，不是太可怕了吗？这是我主张学术多样化的根本心态。最近参加IFLA大会刚刚借道北欧回来，稍微仔细观察，粗粗看来十分相似的几个北欧国家之间确实有着深厚的风土人情差异，于是更增强了我走自己民族道路的信心。对文化景观异质性研究动向感兴趣的读者，可以看看介绍乡土景观的文章。其实很多问题不需要读大部头，看看好的评介文章就可以了，黄昕珮的文章就属于这种。

Phase 7, 2008

Topic of This Issue: Plant Diversity

The bio-diversity in city is much more complicated than the bio-diversity in the pure nature. Human is the main body of the city, and the interaction between human and various living things is always a never-ending topic. Although the urban bio-diversity is only a part of the bio-diversity, there are still a lot things that could be deeply discussed, such as the big ones like the relationship between the urban bio-ecology and the citizens, the relationship between the urban plant bio-diversity and the quality of human living environment (including temperature, humidity, atmosphere, water, food, pests, diseases, etc.), the evaluation of the bio-diversity of the highly urbanization regions (due to large scale importation of energy, material, labor, and equipment), and the quantitative and evaluation methods of urban bio-diversity, the medium ones like the definition of native plant, the reasonable distribution of top communities in the city and in green space, and the evaluation of cultivars or ornamental varieties and their significance to urban bio-diversity, and the small ones like the width of ecological green corridor and breakpoints, the technical processing of the breakpoints, and the relationship between plant diversity and human outdoor activities and social security, or more specifically, the quantity of bordering plants between the woods and its neighboring pavement square. I think there are few linear ones among those relationships and more of them are complex curve relationships. However, even in a global view, there are little study on this. Most of the reports are rather narrow linear conclusions, and the more regrettable thing is that these conclusions are cited regardless of the applicable conditions. Of course, it is very difficult for the professionals of this field to develop a very complicated equation, but it is quite possible to break through from the one-way links and linear thinking and build a framework of thinking. Therefore, it is an opportunity for China to take a leading position in the research of this field.

In fact, the need for diversity is not only in species, but also in landscape, culture, system, and thinking. Before humans having the truth, I am always worried about the current early globalization: the social system controlled by capital is really the highest form of social pattern? It would be too terrible if humans make the same mistake. This is my fundamental mentality of why I advocate academic diversity. During my way back from the IFLA General Assembly via north Europe, the careful observation showed me that the seemingly similar Nordic countries have profound differences in local customs and features, which enhanced my confidence of adopting our national way. Readers who are interested in the research trend of the heterogeneity of cultural landscape can take a look at the papers introducing local landscape. Actually many problems do not need to seek solutions from volumes of books, and good review papers are enough, and Huang Xinpei's paper is one of them.

2008 年 8 期

本期主题：乡村景观建设研究

　　我国大陆的风景园林学科对农村建设的主要关注集中在景观特性、文化遗存、生态保护等方面，本期的主题栏里我特别推荐《生产·生态·生活——"三生"一体的台湾休闲农业园区规划与建设》一文，对于乡村建设这样的实践主题研究来说，台湾从实际出发并不断修正的观光农业发展道路的做法更加切实可行，尊重实际的学术方法比从理念出发的学术思维也更有实际价值。

　　刘叙杰先生的《纪父亲刘敦桢对传统古典园林的研究和实践》，详叙了刘敦桢前辈对中国园林理论体系奠基的重大贡献，读之令人肃然起敬。当下在一般的青年学子中民族虚无主义十分流行，一些名人也在其中推波助澜，我在《评所谓"中国迷园"》（发表在《风景园林》2007年第3期）一文中曾写道："虚无主义（nihilism）本意是全盘否定传统价值，凡是持有这种思想的人必然是只知道一些所谓的结论而对人类文化成果积累的艰辛毫无感觉的人。"所以，读一读介绍梁思成、刘敦桢（20世纪中叶有"南刘北梁"之说）、童寯等先辈的工作态度、探索经历和爱国热情的文章，对许多本行业的新从业者应该是大有裨益的。

　　近来听到越来越多对于我国风景园林界缺乏深度理论探讨和切实作品评论的批评，我觉得这是好现象，说明我们行业正在摆脱只重视埋头实践（或者挣钱）的状态，意味着具有中国特色的新时代的风景园林理论体系的曙光已在地平线升起。本期有几篇理论探讨的文章：《在城市中营造野态环境的途径》介绍了当前欧美势头正旺的野态造景理论与国内的若干实践经验，国际上这也是个讨论激烈的题目。《景观设计的语言学分析方法》也许有些难读，但我们学科的水平要获得社会承认，这种工作是必须要做的。我们不必被海德格尔、德里达、索绪尔、胡塞尔……这些哲学和语言学的大人物吓倒，语言是出于交流的需要，景观或风景园林是人类生境的一个部分，生境的艺术方面有交流的内涵，但生境建设的基础是直接为人服务的环境功能，依据的是肯定性的科学。此外，中国园林中的"此处无声胜有声"、"禅"和"悟"，提出了当前西方语言学理论难以解决的问题。《当代景观营建方法的类型学研究》是我刊首次介绍类型学的研究方法，不善于分类（细分）是中国传统思想方法的弱点，也是古代我国科学进步缓慢的重要原因，但对于园林这样的综合事物，用什么方法细分，细分到什么程度才是适度的，细分的同时如何与综合考虑相结合，还是大有继续探讨的空间。《设计作品的形成及其参考点》的作者显然是年轻新锐，有一种非我其谁的气势，发表此文，并不是说文章已经十全十美，而是为着鼓励年轻学子积极投身到中国风景园林的理论建设中来。

Phase 8, 2008

Topic of This Issue: Village Landscape Construction Research

The concerns of the landscape architecture subject of mainland China to countryside construction mainly focus on the aspects like landscape features, cultural relics, and ecological protection. In the theme column of this issue, I especially recommend Production, Ecology, and Living — The Planning and Construction of Taiwan's Leisure Agriculture Parks. For the practical theme research of countryside construction, China Taiwan's tourism agriculture development approach which proceeds from reality and under continuous revision is more practical, and the academic method in respect of the reality is of more practical value than the academic thinking out of conception.

Mr. LIU Xujie's paper On My Father LIU Dunzhen's Study and Practice of Traditional Classical Gardens, which recounted in detail the senior LIU Dunzhen's major contribution to the foundation of the theory system of Chinese gardens, arouses my respect after reading. The national nihilism is very popular in the young students currently and some celebrities are also adding fuel to the flames. In my paper On the So-called Chinese Fascination Garden (published in Chinese magazine of Landscape Architecture, Issue 3, 2007) I wrote that, "the purpose of nihilism is a total rejection of traditional values, and all holders of such thinking must be those who only know some of the so-called conclusions and know nothing about the painstaking accumulation of human cultural achievements." So, reading the papers on the work attitude, exploration experience and patriotic enthusiasm of seniors like LIANG Sicheng, LIU Dunzhen, and TONG Jun should be of great benefit for the new comers of this profession.

Recently, there are more and more criticism that China's landscape architecture profession lacks in-depth theoretical studies and comments on practical works, which I think is a good thing. It shows that our industry is breaking off the situation of simple focus on practice (or earning money) and means that the dawning of landscape architecture theory system of the new era with Chinese characteristics is coming. There are several papers on theoretical discussions. Approaches to Constructing Wild Environment in Cities introduces the wild environment creation theory, currently popular in the European and American countries, and some practical experience in China. That is also a topic with fierce discussion around the world. A Linguistic Analytic Approach to Landscape Design might be difficult to read, but this work must be done if the level of our discipline wants to be recognized by the society. We do not have to be scared by the big names of the philosophers and linguists like Heidegger, Derrida, Saussure, Husserl and so on. Language is just out of the need of communication, and landscape or landscape architecture are part of human habitat. The artistic aspect of habitat has the content of communication, but the basis of habitat construction is the environmental function of direct service for people and its foundation is the affirmative science. In addition, the "silence speaks more eloquently than words", "Buddhist meditation", and "enlightenment" in the Chinese garden have solved the three problems that the current western linguistics theories find no solutions. The Typology Study on Approaches of Contemporary Landscape Architecture is the introduction of the typology research method in our journal for the first time. Being not adept in classification (subdivision) is the weakness of the traditional Chinese way of thinking and is also an important reason why the scientific progress of ancient China was very slow. But for the garden as a comprehensive thing, discussions are still needed on the classification methods, the appropriate levels of classification, and the combination of classification with comprehensive considerations. The author of The Formation of Design Works and Their Reference Points is obviously the how there is obviously the brilliant one of the younger generation, and has an imposing manner that no one else but him can do it. Publishing this paper does not necessarily mean it has been perfect, but to encourage young students to actively participate in the theoretical establishment of Chinese landscape architecture.

2008年9期

本期主题：城市避灾绿地

　　本刊2007年7月推出的一组以"防灾避险绿地"为主题的文章，当时我写过下面一段话："城市园林绿地的一个大用途，是避灾。不知是否受到单纯景观（顶多加上自然生态）观念的影响，对此已经多年不怎么提到了，甚至我们的城市绿地系统规划编制办法中都没有将其列入必须做的工作内容。其实所有城市都有避灾问题，即使真的没有自然灾害的城市，人为灾害（战争、恐怖主义行动等）也是城市规划和建设者应该考虑的。"

　　不想，仅仅10个月后就发生了汶川大地震，惨痛的经历使我们这一期又推出了同样的主题，所不同的是去年以介绍外国经验为主，本期则主要是国人自己的论述。其间更大的不同在于社会背景发生了根本性的变化，在百度上输入"防灾绿地"，可以检索到90 000多条，说明绿地对于城市安全的作用已经深入人心，而这点其实是最最重要的，因为建设城市避灾绿地在技术上其实没有什么难度，问题主要在于思想重视，而有了思想重视，才有政策和措施的跟进。

　　这里，我特别推荐《城市安全与我国城市绿地规划建设》一文，文章最大的贡献是通过灾后调查研究指出我国城市绿地指标严重不足。依我看，这是我国占统治地位长达近半个世纪的一种"集体无意识"思想的最严重后果，那就是认为我国城市建设占地太多。实际上，城市占地增加是因为人口增多，只要有了人，他就要占地，不管他是在农村还是城市，仅就生活用地而言在农村占地也比在城市还多，对此全国人大几年前有过调查。拼命压缩城市用地的严重后果还表现在城市交通堵塞、空气污染和对城市化发自内心的恐惧，正由于这种恐惧，很多政策在阻碍着城市化，结果是近亿人口在城乡两头用地，实际上浪费土地更多。

　　经过深入考虑，本期主题采用"城市避灾绿地"的说法，因为城市绿地对于地震、海啸、洪水、战争等灾害的预防作用是不大的，主要作用在于灾后的避难和救护，甚至像这次北京奥运期间一样可通过提供示威场所增加城市的整体安全。防灾绿地的说法很可能源于日本，希望有能力的学者予以考证，20世纪60年代初我的老师讲课时就是用的这个词，当时距东京大地震才30多年。不准确的汉语词汇对中国的危害已经够多的了，比如建筑、景观、美学等，为了避免今后有人在"防灾"二字上做文章贬低城市绿地，希望这次我们不再重蹈覆辙。

　　最后，我衷心推荐本期谭玛丽、M.欧伯雷瑟—芬柯和罗纳德·韩德森3位外国友人撰写的3篇文章，它们可以让我们深入体会"以人为本"的真实内涵。

Phase 9, 2008

Topic of This Issue: Urban Disaster Shelter Green Space

The July Issue 2007 of our journal had "emergency and disaster- prevention green space" as its theme. I wrote the following words then, "one of the major use of urban landscape green space is disaster shelter. I wonder if it is because of the influence of the pure landscape (plus natural ecology at the most) concept that this aspect has not been mentioned for many years. Even the urban green space system planning has not included it into the necessary work. Actually all cities have the disaster shelter problem. Even if there is the city without natural disasters, the man-made disasters (war, terrorist, etc.) should also be considered by city planners and constructors".

I had never thought that Wenchuan Earthquake happened ten month later. The painful experience make us launch the same topic, and the difference lies in that the issue of last year is mainly about the foreign experiences while this issue is the discussion of our own experience. During this period the bigger difference is the radical change of social background. The search of the key word "disaster prevention background" at baidu.com can return over 90 000 items, which means that the role of green space in city security has enjoyed popular support. This is the most important thing, because the construction of urban disaster shelter green space is not difficult technically and the problem mainly lies in the ideological attention. Policies and measures will follow up only if there is ideological attention.

Here, I particularly recommend the paper of Urban Green Space Planning and Construction in Consider of City Safety in Case of Disasters. Its biggest contribution is to point out the serious shortage of green space index through post-disaster investigation and studies. In my view, this is the most severe consequence of the over-half-a-century overwhelmingly "collective unconsciousness" that our urban construction has occupied too much land. In fact, the increase of urban land is because of the growth of population. As long as there is a person, he will need to occupy land, no matter he is in rural area or city. Only in terms of land for living use, more land is occupied in rural area than in city, which was investigated by the National People's Congress a few years ago. Desperately compression of urban land brings other serious consequences of urban traffic congestion, air pollution and the fear of urbanization from inner heart. It is because of the fear that many policies are impeding the urbanization, with the result that nearly 100 million people use land both in rural area and city, which wastes more land.

After in-depth consideration, the current theme adopts the wording of "urban disaster shelter green space", because the urban green space has little prevention function for earthquakes, tsunamis, floods, wars and other disasters and its main function is for post-disaster shelter and rescue, or enhancing the overall security of city through such as providing demonstration sites during the Beijing Olympic this time. Disaster prevention green space might originate from Japan, which I hope the capable scholars can research. My teacher used this term in lectures in early 1960s, which is only over 30 years after Tokyo Grand Earthquake. Inaccurate Chinese terms have brought us too much trouble, such as architecture, landscape, and aesthetics. In order to avoid that the urban green space be depreciated by making a fuss about the term of "disaster prevention", I hope we will not repeat the same mistake this time.

Finally, I sincerely commend the three papers by Trudy Maria Tertilt (Tan Mali), M. Oblasser-Finke and Ronald Henderson, which will give us an in-depth experience of the true meaning of "human-oriented" idea.

2008年10期

本期主题：风景园林评论

越来越多的人认识到我国很久以来只能听到由作者自言自语，缺乏对照言论的风气对于我们行业的发展没有好处，呼吁《中国园林》应该带头开展风景园林批评。从上半年到现在，在几位年轻学术精英的勇敢支持下，我们终于可以组织起一期"风景园林评论"这样的主题，奉献给读者。

按照《辞海》，评论即批评，此处之批，可理解为眉批、夹批之意。可能由于"文革"前的时代里，批评与批判、批斗几成同义语，批评在一般的理解中成为大不敬的词汇。我想，还是还其批评的本意为好。

开展学术批评和艺术批评，不是一件容易把握的事，除了我们必须有在战略上藐视的决心，还必须有在战术上重视的心态和方法。

首先是要把握住大局，从大局出发，就不容易发生大的偏差。恰好，国际风景园林师联合会（IFLA）主席戴安妮女士应我在2008年IFLA大会时的请求为我刊写来了专稿，她站在全球的高度归纳了LA行业的概貌，预见未来它的发展，提出了对未来从业者在知识、态度和工作方法等方面的要求。短短9 000多个英文字符（约3 000汉字）就把这样的大事阐述得清楚明了，戴安妮女士强大的观察、思维和语言能力令人折服。

其次是把握住批评的目的，即批评是为了我们事业的健康发展，一定要避免个人意气的发泄。从科学角度，批评还好掌握，总算有实践检验的标准；从艺术角度，批评很难说有绝对的是非对错，因此，抱着商榷的态度和有容乃大的心境，可能是开展好东方式艺术批评的要诀。

当前国际经济形势由次贷危机到资源涨价到金融海啸的剧烈震荡，依我看在示意着一个任由以流动金融资本为代表的资本掠夺全球的时代的衰落的开始，伴随着全球的资源紧张和气候变化，这一切意味着人类正处在一个最重大的历史转折点上，即人类正从可以乐天甚至胡闹的青少年时代转向必须成熟的时代。在未来的时代里，当下流行的绝大部分观念都会产生翻天覆地的变化。正因此，及早开展批评才更为重要，因为我们这个行业本身从事的就是百年大计的事业。作为刊物的态度，我们不以对错论英雄，提倡对事不对人，凡言之有理，依据确凿的文论，我们皆双手欢迎。

齐康院士在2007年的教育分会年会和2008年第3届青年风景园林师沙龙上都对我国青年风景园林师发表了热情洋溢的讲话，齐先生为本刊写作的《建筑·风景》一文也在本期发表，代表着一位建筑大师长久以来对风景园林事业的观察和理解，文章可以说与戴安妮女士的文章双璧生辉。相似的，还有闫水玉等阐述已故黄光宇先生用十数年精力思考我国城市绿地系统规划的文章《"交互校正"的城市绿地系统规划模式研究——以陕西安康城市绿地系统规划为例》，不容错过。

每年的IFLA大学生设计竞赛都是青年人比较关心的，本期请来3位2008年的获奖者专门撰文介绍其作品，文章显然比展板上的简要文字让人更容易对他们的作品进行理解。

Phase 10, 2008

Topic of This Issue: Landscape Architecture Criticism

More and more people realize that only the author's monologue can be heard in our country and the lack of contrast views will not benefit the development of our industry, and appeal to Chinese Landscape Architecture to take the lead of landscape architecture criticism. From the early half of the year up to now, with the courageous support of several young scholars, we finally organize this issue themed landscape architecture review to our readers.

According to Ci Hai, review is namely criticism, and criticism contains the meaning of notes and commentaries. Perhaps because in the period of the "Cultural Revolution" the word criticism was nearly the synonym of criticizing and denouncing, criticism becomes the word the great disrespect in the ordinary understanding. I think it is better to return its original intention.

It is not an easy thing to handle to carry out academic criticism and art criticism. Besides the determination to despise strategically, we should also have the attitude and method to attach importance tactically.

First of all, taking the whole situation into account will not be prone to have large deviations. Precisely, IFLA President Dr. Diane MENZIES wrote a special article to our journal at my request at the IFLA General Assembly this year, and in that article she summed up the general situation of landscape architecture industry with the global perspective, foresaw its development and set the requirements of knowledge, attitude and working method for future practitioners in this industry. Only about 9 000 words (about 3 000 Chinese characters) put such a major thing in very clear understanding, which reflected her powerful and impressive observation, thinking and language skills.

This was followed by the grasp of the purpose of criticism that criticism is for the healthy development of our cause and the personal emotional outlet should be avoided. From the scientific point of view, criticism can be handled better, for there is practical standard to check, while in the artistic perspective there is seldom absolutely right or wrong criticism. Therefore, the attitude of discussion, tolerance and capacity may be the tips to carry out the oriental art criticism.

The current severe shock of international economic situation from the subprime lending crisis to resource price increase and financial tsunami, as I think, shows the start of the decline of the time of capital plundering globally characterized by the free flow of financial capital. Together with the global resource tension and climate change, all these mean that the mankind is standing at a significant historical turning point, namely the mankind is growing from the carefree or even mischief adolescence to the mature age. In the future time, most of the current popular concepts will undergo enormous changes. For this reason, it is more important to carry out criticism earlier, since what we do in this industry is crucial for generations to come. As the attitude of our journal, we do not judge people from them being right or wrong, advocate that business is business rather than getting personal, and welcome papers with reason and solid foundation.

Professor QI Kang, member of Chinese Academy of Sciences, delivered warm speeches for young landscape architects of our country in both the annual meeting of the education sub-conference last year and the 3rd Youth Landscape Architects Salon. His special article for our journal Architecture and Scenery is also published in this issue. The paper represents the architecture master's long-term observation and understanding of the landscape architecture cause and shines with Dr. Menzies by reflected glory. Similarly, readers should not miss the article Study on the Mutual-correction Model of Urban Green Space Planning— Case of Shannxi Ankang Urban Green Space Planning, in which the late Mr. HUANG Guangyu's thinking over the urban green space system planning for more than a decade is elaborated.

The annual IFLA student design competition is much concerned by the young people. In this issue three winners of this year are invited to introduce their works, and their papers obviously make it easier to understand their works than the brief text on the panels.

2008 年 11 期

本期主题：风景名胜与旅游规划

业界盼望已久的中国风景园林学会第四次全国会员代表大会成功召开并完成了其历史使命，本期用较大篇幅对此作了报道，我们和广大会员一样热盼着学会的新领导班子能够引导行业在业务发展、机制完善和学科建设等诸多方面取得长足进步。

毋庸讳言，当前金融海啸必将使整个地球面临一个长期紧缩的年代，我们行业面临的形势也将是严峻的。资本的自私本性决定了它不会有良心去为别人分忧，除非预见到会波及自己。中国巨大的城市化市场将成为世界行业沙漠中少有的绿洲，所以资本肯定会在我们行业大举入侵，抢夺材料、设计和工程的市场，对此，我们最好有所准备。

最近常有人问我：你上期说"人类正从可以乐天甚至胡闹的青少年时代转向必须成熟的时代。在未来的时代里，当下流行的绝大部分观念都会产生翻天覆地的变化"。乐天，胡闹到底指的是什么？我的回答是：以为人人都可以过上美国人那样的日子就是一种天真，耗费巨大的资源去专门建设某种刺激视觉的景观就是一种胡闹，皆因地球已经没有那么多资源了。其实，资源约束还使自由资本主义（即信奉市场经济准则）也成为幻想，因为掌握资源的人肯定与其他人处在不平等的地位上。虽然资源问题仅是当前经济危机的原因之一，但从时代的眼光来看，它将是人类历史大转折的根本原因。

在新形势下，我们行业能够为国家做出贡献的方面主要有两个：一个是积极投入城市化工作；另一个是做好风景名胜区的旅游工作；总之是配合国家拉动内需。本期的主题"风景名胜与旅游规划"与过去有些不同，主要是研究如何搞好旅游工作，也可看作是配合当前的经济大形势。当然，我们不是旅游部门，为了子孙后代保护国家资源，始终是我们做规划的首要指导思想。我特别推荐刘滨谊教授等人写的《建设慢节奏、生态化的风景地区旅游小城镇——以华阴市为例》，因为它提出的慢节奏、低消耗、理性发展，与当前时髦而且无节制的快节奏、高投入、强烈感觉刺激，形成鲜明对比。另外，《密歇根州旅游地的历史档案信息资源研究——以费耶特历史州立公园为例》一文，可以帮助我们改进风景名胜区的档案工作和信息收集工作，提高其详细度和可信度。通过对照美国人文资源浅薄的情况，我们还可以更加了解中国的优势，不要在美国历史资源很少时跟着他们藐视历史，等到美国积累了一定可以自豪的历史资源时又跟着他们反过来批评我们这一代人忽视历史。

2008年年初的南方大冻害是研究园林植物的天赐良机，可惜至今这方面的成果很少。广州市主管城建的苏泽群常务副市长亲自抓住这个课题，并将成果交予本刊发表，应对各地具有很强的启示意义。

Phase 11, 2008

Topic of This Issue: Scenic Spot and Tourism Planning

The industry's long-awaited 4th National Member's Congress of the Chinese Society of Landscape Architecture was successfully held and completed its historic mission, which is reported with large space in this issue. We and the general members all look forward that the new leadership will make substantial progress in guiding business development of the industry, improving the mechanism, building the discipline and so on.

No need for reticence, the current financial tsunami is bound to make the whole world facing an era of long-term austerity, and our industry will also face a severe situation. The selfish nature of capital decided that it lacks the conscience to share others' worries, unless it is foreseen to affect itself. The enormous urbanized market will become the rare oasis in the desert of the industry around the world, so certainly the capital will mount a large-scale invasion into our industry to seize the market of materials, designs and engineering projects, which we should better be prepared.

Recently people often ask me: You mentioned in last issue that "the mankind is growing from the carefree or even mischief adolescence to the mature age. In the future time, most of the current popular concepts will undergo enormous changes". What the carefree and mischief refer to? My answer is: thinking that everyone can live a life like Americans is a kind of naive, and spending enormous resources on some visually stimulating landscapes is a kind of mischief, both of which are because of the limited resources on the earth. Actually resources constraints also make free capitalism (namely the belief in the market economy norms) become a fantasy, since the one having the resources is certainly in the unequal status upon others. Although the issue of resources is only one of the reasons for the current economic crisis, it will be the root cause of the major turning point in human history from the perspective of the times.

Under the new situation, our industry can contribute to the country mainly in two aspects: one is to be actively involved in the work of urbanization, and the other is to do a good job in tourism of scenic zones, all in all to help the country to increase domestic demand. The theme of this issue "scenic spot and tourism planning" is different from the past and is mainly to study how to improve tourism, which can also be seen as taking coordinated action on the current general economic situation. Of course, we are not in the tourism sector, and the most important guiding ideology for our planning is to protect national resources for future generations. My special recommendation is Professor LIU Bin-yi and his co-authors' *Building Small Slow-rhythm and Ecological Tourism Towns in Scenic Areas – A Case Study of Huayin City*, since it puts forward the slow-rhythm, low-consumption and rational development, in sharp contrast with the current fashionable and incontinent fast pace, high investment and strong sensational excitement. In addition, the paper of *Historic Archived Information Resources For Tourism Sites In Michigan: A Fayette Historic State Park Case Study* can help us improve archives and information collecting efforts of scenic zones to enhance the details and credibility. Through comparing with the limited human and cultural resources of the United States, we can also have a better understanding of the advantages of China, and we should not follow the Americans to contempt the history with their limited historical resources or criticize our generation of neglecting the history when they have accumulated a certain amount of historical resources with pride.

The big freeze disaster early 2008 is the heaven-sent opportunity to study landscape plants, but so far there are few results in this regard. Mr. SU Ze-qun, Deputy Mayor of Guangzhou for urban development, takes charge of this research topic and hands over the results to us for publishing. It should be of strong inspiration significance for other places.

2008年12期

本期主题：观赏草

　　历史上极不平凡的2008年即将过去，想到年初《主编心语》对今年的美好期望，似乎有点失望。真所谓人算不如天算，人类还是不要太天真，更不应胡闹。但令我惊喜的是，我此生竟然能够目睹美国式资本主义真正走向衰败的转折点，以及中国如何在如此严酷的逆境中成为世界瞩目的中流砥柱，我是因为担心美国式的全球化是否会将人类迅速带向灭亡，才从一个世界大同主义者走向文化多样性拥护者。也许数百年以后的人会感觉自己的日子太平淡，不知他们将如何羡慕我们这些经历过2008年的人。

　　年轻时候在北京儿艺看过一个话剧，里面有一句话曾一度成为名言："忘记过去就意味着背叛"，不幸我当时正是个"没有过去"（王蒙语）的狂妄者，曾私下将此话猛批了一番。现在看来，此言确实点出了现实众生相之一种。我国的风景园林事业，既有数千年的祖先遗产积淀，更有近百年来前辈的辛勤劳动积累，忘记他们当然是一种背叛。孟兆祯院士的《中国风景园林师的天职——继往开来，与时俱进》，从继承和发展两方面分析中国风景园林师的责任，其中涉及许多对中国文化的深层修养，正是所谓"生长在新社会"的我辈所严重缺乏的，期望年轻人中能有人奋起直追，不要使太多的中国文化积累在我们这个世纪被断送。谈及近代史，我们不能忘记的第一人当是陈植先生，我殷切希望广大年轻读者读一读《中国造园学的倡导者和奠基人——陈植先生》这篇文章，特别是在这个"以钱为荣"的大环境中能够继承陈先生求真唯实的精神。

　　周干峙院士为本刊撰写的文章《对生态城市的几点基本认识》深刻地指出城市生态"应包含自然和社会两部分"。我们行业中的人重视生态和可持续发展一般是没有问题的，但有些同仁只想到自然生态，没有想到人，我曾经和一位中山大学生物学教授在讨论广州市生态规划时发生过争论，我问："是否广州大街上有了狮子老虎才算生态了？"这种思想，甚至有意无意地影响到一些对城市环境研究的实验设计，这就成了大问题。此外，城市生态理当是我们这个学科的核心之一，当前，至少应占一半的权重，但现在业界中很多人仅仅满足于理念、图形、结构、格局、系统之类的虚空概念，不去扎实地分析城市里影响人类生活的各种因素和了解各种过程的机理，如此下去，本行业的城市生态科学是否能获得真正的进步，我对此颇为担忧。

　　本期的主题是"观赏草"。虽然细究起来我们很早就有应用沿阶草等观赏草的传统，但应该承认当前世界上正在流行的观赏草风有着自己的起源，很可能与湿地风的勃兴有一定的关系。观赏草的应用在我国沿海的一些大城市开始兴起，在广大内地还很少被波及，而内陆的草资源十分丰富，有可能为新景观创作提供一种途径。

Phase 12, 2008

Topic of This Issue: Ornamental Grasses

The historically extraordinary year of 2008 is about to end. Thinking of the fine expectations in the *Editor's Words* at the beginning of the year, it seems a bit disappointed and really unexpected. Human beings should not be naive or even be mischievous. However, what surprises me is that I was able to witness the turning point of American-style capitalism on the real decline, as well as how China has become the mainstay with the world's attention in this hard adversity. It is because I am worried that the American-style globalization would take human beings to faster distinction that I turned from a world peace and harmony believer into an advocate for cultural diversity. Perhaps people of hundreds of years later will feel their days are dull and we do not know how they will envy us who have experienced the year of 2008.

In a drama I watched in Beijing Children's Art Theater when I was young, a statement had once become a famous saying:"Forgetting the past means betrayal."It was unfortunate that I was an arrogant guy"without the past" (as Wang Meng said) at that time and I criticized the words in private. It now appears that this statement does point out a certain kind appearance of our living creatures. China's cause of landscape architecture has both the heritage of thousands of years from ancestors and the accumulation of hard work of the older generation in recent one hundred years, and forgetting them, of course, is a kind of betrayal. Academician Meng Zhaozhen's paper of *Carrying on Past Brilliance and Making Timely Progress – The Bounden Duty of Chinese Landscape Architects* analyzed Chinese landscape architects' responsibilities from the two aspects of inheritance and development. It is about considerable accomplishments of Chinese culture, which is lacking in us who are called"the ones grew up in the new society."Hope the younger generation will be able to catch up, not to make too much of the accumulation of Chinese culture ruined in this century. Turning to the recent history, the first person that we must not forget is Mr. Chen Zhi. I earnestly hope that our young readers can read the article of *Mr. Chen Zhi, the Advocate and Founder of Chinese Garden Making* and in particular carry forward his truth-seeking spirit in this general environment that"money brings glory."

Academician Zhou Ganshi's paper for our journal, *Some Thinking about the Eco-city*, profoundly pointed out that the eco-city"should include both nature and society." It is all right for us the ones in this industry to pay attention to ecology and sustainable development, but there are still some colleagues only thinking of natural ecology rather than human beings. I once argued with a professor of biology of Zhongshan University when discussing the ecological planning of Guangzhou city, and I asked if the city could be called ecological only with lions and tigers in streets. This thinking somehow impacts the experimental design for environmental studies, which has become a big problem. Furthermore, Urban ecology should be one of the key topics of our discipline and accounts for at least the half currently, but now many people in the industry are just satisfied with abstract concepts of ideas, graphics, structures, layout and system and do not conduct solid analysis of the urban factors that affect human life or understanding of the mechanism of various processes. If it goes on like this, I am quite worried if the science of urban ecology of our industry will achieve real progress.

The topic of this issue is "ornamental grasses." Although with careful study we have had long tradition of using ornamental grass such as *Ophiopogon japonicus*, it should be recognized that the current popular trend of ornamental grass has its own origin, most probably of certain relation with the rise of trend of wetlands. The application of ornamental grass began to rise in some of China's big coastal cities but has not spread to the vast inland. The grass resource of the inland is quite abundant and is possible to provide a way of new landscape creation.

2009年

2007年　2008年　　2010年　2011年　2012年　2013年　2014年　2015年　2016年

2009年1期

本期主题：风景园林教育

本期的主题"风景园林教育"是为配合2008年11月在上海同济大学召开的教育年会。每年一次的学会教育年会，虽然仅仅开了三届，已经显现出其强大的生命力和对学科发展的重要作用。

孟兆祯院士的文章讲了自己对风景园林专业本科教育的体会，除了重点谈到"人工与自然的关系"和"承前启后，与时俱进"这两个众所关心的问题，特别突出了对本科生中国文化教养的培育的关注，我的体会是这是我们学科"自立于世界民族之林"的根本，也是为了更好地承担起"提高含本学科在内的综合国力"的需要的重任。欧百钢调研员等人的文章从全国的高度系统地研究了我国风景园林高层次人才培养体系的建立问题，对学科的整合与发展必将起到重要作用。李雄教授则从北京林业大学的经验探讨了硕士研究生的培养，很值得设有硕士点的教学研究单位参考。

中国香港的做法更趋向与国际接轨，从香港的情况也可看到我们与国际发展的差距。陈弘志教授《专业精神与行业竞争力：以香港为例谈如何确立风景园林行业之专业地位》一文，从制订园境师专业操守、教学评估制度、园境师注册考试制度和加强行业规范等全方位入手，直指提高行业的国际竞争力。

近读玄奘法师译经时有"五不翻"，其中第二是"含多义故，如'薄伽'，梵具六义"；第五是"生善故，如'般若'尊重，'智慧'轻浅"（参见宋·周敦义《翻译名义集·序》）。故而"薄伽"，"般若"皆用音译，所谓"不翻"。大师高见，令我浮想联翩，"五不翻"如果推广到世界，很多文化交融的障碍可以消弭。比如landscape明明是多义的，非要翻译成汉字语义十分片面的"景观"，带来了很多荒唐和苦恼；中国"园林"是富有极深厚精神和文化内涵的，非要翻译成轻浅的garden，何苦来？又比如，kungfu（功夫）就音译得很好，如果译成boxing，甚至误译成skill，不是索然无味了吗？所以我建议园林就翻译成yuanlin。由此想到，香港将LA翻译成"园境"，摆脱了日译藩篱，其精神很值得学习。最近看到一篇文稿，研究"中国古典园林声景观"，"声景观"这种说法十分唐突，在咬文嚼字的古代是不会出现的，显然是这个浮躁社会的产物，于是我向作者建议译成"声境"。

Phase 1, 2009

Topic of This Issue: LA Education

The theme of this month, "LA Education" is organized to coincide with the annual education conference held in Tongji University, Shanghai, in November 2008. Although it was only held for three times, the annual education conference of CHSLA has already demonstrated its strong vitality and the important role in the development of this discipline.

Academician MENG Zhaozhen's article is about his own experience of LA undergraduate education. In addition to the focus on the two widely-concerned issues of "the relationship between artificiality and nature" and "bridging the past and the future and advancing with the times", the paper also emphasizes his concern to the undergraduate education of Chinese culture in particular, and my experience is that this is the basis of our discipline to "be self-reliant up the nations of the world"and also the necessity to better shoulder the task of"raising the overall national strength including this discipline." Director OU Baigang and others' papers systematically study the establishment of high-level talent training system of China from the national perspective and are bound to play an important role in the integration and development of the discipline. Professor LI Xiong explores the education of Master degree candidates from the experience of Beijing Forestry University, which is of reference for those teaching and research institutions with Master degree training programs.

The approach of Hong Kong is more internationalized, and the gap between us and the world can be seen from the Hong Kong conditions. Professor Leslie H. C. Chen's paper of *Professionalism and Competency: Hong Kong's Model on Establishing the Profession of Landscape Architecture* starts from the full scope of drawing up the code of professional conduct, education accreditation system, and professional registration and examination system and strengthening professional norms, and aims to improve the internationral competence of the industry.

Recently I read that Master Monk Xuanzang did not translate five kinds of things in his scriptural books and documents translation. The second one is"no translation of multiple meanings" and the fifth one is "no translation for respect". So some words and terms are in transliterating, namely no translation. The Master's wisdom makes my mind thronged with thoughts that many intercultural barriers would be removed if the "five no translations" principle is extended to the world. For example, the word "landscape" is obviously of multiple meanings, and the very partial Chinese translation "jingguan (景观)" brings ridiculousness and troubles. It is also boring to translate"yuanlin (园林)", the Chinese word with profound spiritual and cultural content into the shallow word of "garden". Another good transliterating example is "kungfu (功夫)", and it would be unpalatable to translate it into"boxing" or even "skill" wrongly. So I suggest " 园林 (yuanlin)" is transliterated into "yuanlin". LA is translated into "yuanjing (园境)" in Hong Kong, which breaks through the hedge of Japanese translation, and the spirit is worth learning. Lately I read a paper of research on the "sound landscape (声景观) of Chinese classical gardens". The term of "sound landscape (声景观)" is very abrupt, which would not be in the wording ancient time and is obviously a product of the impetuous society, so I propose to the author to translate it into"soundscape (声境)".

2009年2期

本期主题：风景园林与气候变化

本期主题是"风景园林与气候变化"。赵彩君博士的《气候变化——当代风景园林面临的挑战与变革机遇》对此做了比较全面的阐述。最近奥巴马成为新闻主角，巧的是，他就职那天我在电视上看到一则新闻，说《伦敦邮报》发表社论说：奥巴马只有4年的时间来拯救地球了，意思是全球的气候变化已经到了临界点。我们不是气象学家，对地球总体状态的了解也不多，难以判断这则新闻是不是过于夸张，但有一点可以肯定，这个地球的确已经到了难以容忍人类再胡闹下去的时刻。

Alan Barber博士的文章《绿色基础设施在气候变化中的作用》，可为我们提供很多思考的空间。我们知道，居住用地大概占城市用地的一半左右，我国低些，国外高些。文章透露，英国政府对大都市的居住用地指标近来从每公顷22户紧缩到30户，紧缩后大概可折合成每平方千米10 000人，而我国城市的居住用地平均是每平方千米20 000人以上！可见，有些人宣称我国城市用地超过发达国家，因此要减少城市绿地是没有依据的。文章后半部对英国的绿地灌溉政策提出质疑。由于英国自然条件与我国差别很大，该质疑本身仅供我国个别地区参考，但作者的深入分析精神值得我们借鉴。如果我们的学者也秉承这种精神对待问题，就不会有对中国城市用地特别是绿地多了的荒诞看法到处流传。

2008年在同济大学召开的教育年会，产生了很好的效果。本期我们继续刊登一些相关的文章。华中农业大学高翅教授于2008年第1期曾将《国际风景园林师联合会—联合国教科文组织风景园林教育宪章》译成中文交本刊发表，今年又将《国际风景园林师联合会——风景园林专业教育认证指南》译文发来。正巧香港大学的陈弘志教授将IFLA这2个文件的英文版发给我们，于是我们将二者同时发表，供我国教育界研读、参考，以改进我们的专业教育工作。我觉得，从中可以体会到IFLA平衡行业基本要求与地域差别的良苦用心，也可看到IFLA对于地域文化差别的尊重。

观察近年的论文，可以发现一个倾向，凡是讲一个问题，即使是常识，也必引用一段从别的书籍或文章上抄下来的话，好像这样才是追赶着科技发展潮流，才能成为真理。由于信息泛滥，其实发表出来的东西不一定就正确，但是这种风气一流行，很多荒谬的东西经过不断引用就似乎成为真理，于是谬种流传。在中国，还特别喜欢引用最近的洋人的话，即使是对于几百几千年前哲人们早已看清楚了的问题，或者中国人早已处理得很好的问题。我的老师们50年以前就一贯强调关注气候和地形，强调适地适树，强调景以境出。忽视气候、土壤、地形等的失误是所谓现代景观兴起以后才普遍出现的倾向，不信？可以看看那些几十几百年前建造的优秀园林。如果我们不谈这些，就会给后代们造成我们以及我们的祖先原先很笨的误解。现代科学技术的发展当然带来了很多新的东西，我们也不会看不见这些功绩。但不要将一切都归功于现代的西方人。其实现代的全球气候和资源危机的根源本来就出自西方。

Phase 2, 2009

Topic of This Issue: Landscape Architecture and Climate Change

The topic of this issue is "Landscape Architecture and Climate Change". Dr. ZHAO Caijun's paper of *Climate Change – The Challenge and Transformation Opportunity for Contemporary Landscape Architecture* has made a relatively full presentation on this. Barack Obama has been the news topic recently, and coincidentally I saw on Phoenix TV at the day of his inauguration that the *London Post* editorialized that he would only have four years to save the earth, which means the global climate change has reached a critical point. We are not meteorologists and do not know much about the overall state of the earth, and it is difficult to judge if this news is much exaggerated. But one thing is definite that it is now the moment that the earth could no longer tolerate human nonsense.

Dr. Alan Barber's article of *The Role of Green Infrastructure in Climate Change* can provide us much space for thinking. We know that the residential space roughly accounts for half of urban land use, and the percentage is lower in China and higher in foreign countries. The article reveals that the UK government recently decreased the residential space quota from 22 households per hectare to 30 households per hectare, namely about 10 000 people per square kilometer now, while the number in China is 20 000 people per square kilometer or higher. Obviously there is no basis for those who claim that the urban land use of China is more than the developed countries and the urban green space should be reduced. The later part of the article questions the UK green land irrigation policies. As a result of the great differences in natural conditions between China and UK, the query is only for the reference of a few districts of China. But the author's spirit of in-depth analysis is worth our reference. If our scholars treat problems in this spirit, the absurd views that China's urban land use and green space in particular are too much would not spread everywhere.

The annual education conference held in Tongji University in 2008 had very good results. In this issue we continue to publish some of the relevant papers. Professor GAO Chi from Huazhong Agricultural University published his Chinese translation of *IFLA-UNESCO Charter for Landscape Architectural Education* in the first issue of our journal last year, and he sent the translation of *Guidance Document for Recognition or Accreditation – Professional Education Programmes in Landscape Architecture* this year. Professor Leslie H.C. Chen from the University of Hong Kong happened to send us the English version of the two documents, so we publish both for the study and reference of the educational circle of China so as to improve our professional education. I think our readers will understand the much thought of IFLA really given to the balance of basic professional requirements and geographical differences and the respect of IFLA to regional cultural differences.

In review of recent papers, a tendency can be found that the words are usually quoted or copied from other books or articles when talking about an issue, even if it is common sense, as if this would catch up with the development trend of science and technology and become the truth. Because of the proliferation of information, in fact the published things are not necessarily correct. However, when the tendency becomes rife, many ridiculous things seem to be truth after constant citations and thus error is disseminated. In China people are particularly fond of quoting recent foreigners, even if it is the problem that the sages of hundreds or thousands of years ago had made clear or the one Chinese had well dealt long ago. Fifty years ago, my teachers had always stressed the concern of climate and terrain, emphasized suitable tree for suitable place, and focused on the scenery from conception and environment. The mistake of ignoring climate, soil, topography and other matters is the tendency emerging generally only after the rise of the so-called modern landscape. Do not believe? You can look at the outstanding landscapes built tens or hundreds of years ago. If we talk nothing about these, it will make our descendants misunderstand that our ancestor and us were originally stupid. Of course the development of modern science and technology brings a lot of new things and we would not be blind to these achievements. But do not attribute all these to modern westerners. In fact, modern global climate and resources crisis originated from the West.

2009年3期

本期主题：墓园

  长时间执行计划生育国策的中国，正在加速进入老龄化社会，当前突出的问题是老龄保障，下一步的突出问题将包括丧葬高峰，有鉴于此，我们策划了"墓园"这一期主题。年轻时，曾经为但丁超越生死的名言"活着都不怕，难道还怕死吗？"而感动，作为比较现实的中国人，实际上我更倾向于遵从着孔子的话语"未知生，焉知死"来对待死亡。由此，我对所谓丧葬风水和厚葬之风大体上是嗤之以鼻，以为它们的好处不过是提供了一些聊天资料和保存了一些文物。在墓葬文化的东西方对比上，我更倾向于源于西方的墓园。更好的方式有没有呢？当然有，那就是以最节俭的方式哪里来哪里去，彻底回归大自然。至于纪念，当人类文明史从现在的数千年变成数万年后，我们可能真的不知道该如何纪念，能留下一本书或一件物品，也就够了。

  对我国的城市绿地系统，特别是市域大范围的绿地规划与管理，近年来有了不少新发展和新探讨，本期"城市绿地系统"栏目对此做了一些介绍，包括郊野公园、城市"生态线"以及"其他绿地"等。而《欧盟国家城市绿色空间综合评价体系》一文介绍了欧盟评价城市绿地的一种比较科学严谨的做法，值得我国参考。

  早在2000年前后，就听说了美国亨廷顿植物园要搞中国园林的事，一晃8年，此事终于成真，流芳园一期于一年前完工并开放。《海外最大苏州园林"流芳园"规划设计回顾》和"Liu Fang Yuan at The Huntington——An Overview"分别从中、美双方角度介绍了该园的情况。我建议凡是具有四级以上英文水平的读者都通读一遍"Liu Fang Yuan at The Huntington——An Overview"，第一，它出自美国人之手，绝不是一篇Chinglish，读了对于提高英文专业水平很有好处；第二，让我们知道真金不怕火炼，即使在西方世界，要想贬低中国园林其市场也是很有限的。

  文化的快餐化和平面化与文化的精致化和深厚化是有矛盾的，前者更适合当前大多数民众快节奏式的社会生活，后者则有贵族化的嫌疑。如果再深一步想一想，大多数民众快节奏式生活果真是为了自己的利益吗？在80%的人群仅仅掌握20%财富的时代，答案必然是否定的。而那20%的人不但掌握着大量财富也有闲享受着大部分人类文化财产。所以，提倡文化的精致化，最终将引导人类放弃整体的快节奏，导致多数人从少数人那里分享得到自己应得的那份文化成果。快餐文化的出现只是文化从贵族那里解放出来的第一步，高雅文化的普及才是真正的文化解放。这就是"以为批评中国古典园林是贵族化就能打倒中国园林"最终不能得逞的根本原因，也是本刊坚持继承发扬中国园林优秀传统方针的信心所在。

Phase 3, 2009

Topic of This Issue: Cemetery

China with its long-running family planning policy is accelerating into the aging society, and the current key problem is the aging protection while the next will include the peak of funerals. For this, we take "Cemetery" as the topic of this issue. I was once moved as a young man by the famous saying beyond life and death of Dante that "we are not afraid of being alive, and are we afraid of death?", but as a practical Chinese I am actually more inclined to comply with the words of Confucius that "not knowing life, you cannot know death" when facing death. Because of this, I usually scoff at the so-called funeral geomancy or elaborate funerals and I think their benefits are just providing some chatting information and preserving a number of cultural relics. In the comparison of eastern and western funeral cultures, I am more inclined to the cemeteries of the west. Is there any better ways? The answer is affirmative, and it is the most frugal way to return to nature completely. As to memory, we may not know how to commemorate when the history of human civilization changes from present thousands of years into tens of thousands of years in the future, and it is enough if a book or an item can be left.

For the urban green space system of China, particularly large-scale urban green space planning and management, there are many new developments and explorations. The "urban green space system" section of this issue makes some introductions to this, including country park, urban "ecological line", "other green space" and so on. The paper of *Comprehensive Assessment System of Urban Green Spaces in the European Union* introduces a relatively scientific approach of European Union assessment of urban green space, which could be of reference for China.

As early as around the year of 2000, I heard that the Huntington Botanical Gardens would have a Chinese garden. It finally came true after eight years that the first phase project of Liu Fang Yuan was completed and opened last year. The paper of *Review of the Planning and Design of the Garden of Flowing Fragrance (Liu Fang Yuan), the Biggest Overseas Suzhou Garden* and the paper of *Liu Fang Yuan at the Huntington – An Overview* introduce the garden from the Chinese and American perspectives separately. I suggest those readers with English levels above CET-4 read through the paper of *Liu Fang Yuan at the Huntington — An Overview*. The reasons lie in that it was written by American people rather than a paper of Chinglish and helps improve the level of professional English, and it lets us know that true gold fears no fire and those who abuse Chinese garden are very limited even in the Western world.

The fast-food and plane culture contradicts the refined and profound culture, and the former is more suitable for the majority of people in the current fast-paced social life while the latter has aristocratic suspects. If thinking a little deeper, is the fast-paced life of the majority of people really for their own benefits? In the time that 80% populations only control 20% wealth, the answer is definitely negative. And the rest 20% populations not only grasp substantial wealth but also have leisure enjoying the majority of human cultural property. Therefore, to promote the fine culture will eventually lead humanity to give up the overall fast pace and result in the majority of people share their deserved cultural achievements from a small number of people. The emergence of fast-food culture is the first step of cultural liberation from the aristocracy while the popularization of elegant culture is the real cultural liberation. This is the fundamental reason the thought that criticism of Chinese classical gardens being aristocratic will defeat Chinese landscape architecture could not succeed, and it is also the confidence that our journal adheres to the principle of inheriting and carrying forward the fine traditional Chinese garden.

2009年4期

本期主题：园林绿色废弃物循环利用

　　本期主题是"园林绿色废弃物循环利用"。废弃物这个概念本是人提出的，无机界本无所谓废弃物，有了会吐故纳新的生物，才有了所谓废弃物。大自然的神奇在于，一个物种的废弃物可能是其他物种的需要物，所以我们在"原生态"的自然中，几乎找不到所谓的废弃物。可见，当前大量人造废弃物的出现，是人类行为已经严重违背天理的证据。绿地占据城市用地的1/3以上，用枯枝烂叶、人畜排泄物等制造堆肥是我国农民的传统技能（当然有待改进），并不困难，然而眼下却成为难题，问题在哪？在于所谓的"现代"价值观和"现代"管理体系。

　　用以金钱为核心的价值观来看，根据广州的经验，将$1m^3$绿色废弃物填埋需要60~80元（这笔钱出自环卫和市政），制作绿肥并推广需要40~60元（但需要园林部门补贴20元），而付诸一炬顶多只需要一个工人的1小时工资。像这样的一个部门做了好事，得益的是另一个部门，就需要上级政府出面协调。而很多领导，喜爱的是直接插手部门内部事务，比如哪条路种什么树，却放弃麻烦得多的对部门间关系的调控。在西方，存在着众多立法的或非政府机构监督着或调解着部门间的关系，我国的体制在这方面本来就较弱，如果不能建立强大的政府内部协调机制，恐怕麻烦会越来越大。

　　这种做法还源于所谓的科学分类和评价打分，以及由此推衍出来的管理方法。看看千层饼、评定量表、优教评价等，不都是这样一种思想方法吗？其实分类和评价只产生于人类头脑，或只对具体物种或个体有效，大自然并不进行分类与评价，对大自然来说，只有过程和规律。用大量的分类（树立新概念）和评价（价值判断）替代艰苦地对具体过程和规律的研究，是当前所谓学术的一大问题。

　　许自力先生《珠江三角洲流域水系景观特征及结构性问题》所述之"景观"确系地理学的景观原意。这些客观的实际调查分析和描述，反映了人为的干预对自然的影响，说明无科学发展的远见，势必自食其果。这种实际的工作结果对各专业规划和发展决策都是有价值的基础资料，也让我们想想如果想让我们学科向真正的科学发展，应该怎么做。

　　吴承照教授等人的《生态链设计与风景环境修复研究》是一篇用我们专业科技支持三农发展的好文章。抓住农村产业的转型和提升的科学可持续发展的基础，才会有中国特色的农村风景园林的繁荣，还可以为生态文明发展做出更大的贡献。这不是只做些自我陶醉式"景观"赐予的表面文章可比拟的。

Phase 4, 2009

Topic of This Issue: Garden Waste Recycling

The topic of this issue is "garden waste recycling". The concept of waste is made by people, and in the inorganic world there is nothing to be called waste and the so-called waste comes with the living things that exhale the old and inhale the new. The nature's magic lies in the fact that the waste of a species might be the need of another species, so we can seldom find so-called waste in the "original ecology"of nature. It can be seen that the current emergence of a great quantity of man-made waste is the evidence that human behaviors have seriously violated the course of nature. The green space occupies more than one third of urban land use, and composting with rotten leaves and branches of trees and human and animal waste, a traditional skill of Chinese farmers (of course it need to be improved), is not difficult, but it has become a problem at the moment. Why? It lies in the so-called"modern"values and"modern"management system.

In the view of money-concerned values, the experience of Guangzhou shows that $1m^3$ green waste landfill will require 60 to 80 Yuan (the money from the public sanitation department and municipal administration), and the production of green manure and promotion will require 40 to 60 Yuan (but the landscape department should subsidize 20 Yuan), while burning down need only 1 hour wage of a worker at the most. Those things that are done by one department but benefit another one need the coordination of the government of a higher level. A lot of leaders like to intervene in internal affairs directly, such as the species of trees along a specific road, but they give up the more troublesome regulation and control of inter-department relations. In the west, there are many legislative and non-governmental bodies to monitor or mediate the inter-department relations. The Chinese system is weak in this aspect, and the trouble will grow bigger if strong internal coordination mechanisms within the government could not be built.

This approach also stems from the so-called scientific classification and scoring evaluation, and the management method derived from it. Take a look at the multi-layered pie chart, the rating scale, the evaluation of excellent education and so on, are they not in this way of thinking? In fact, the classification and evaluation are only from human mind or only effective on specific species or individual, while the nature does not classify or evaluate. For nature, there are only processes and laws. Substituting many classifications (to establish new concepts) and evaluations (judgment of value) for arduous research of the concrete processes and laws is a big problem of the current so-called academic circle.

The"landscape"referred in Mr. Xu Zili's paper of *The River System Landscape and Structural Problems of the Pearl River Delta* is exactly the original geographical meaning of landscape. Those objective analysis and description of the actual survey reflect the impact of man-made intervention on the nature and shows that without the vision of scientific development, it is bound to suffer. This kind of actual work results are valuable fundamental information for the planning development decision-making of various majors, and also make us think how to develop our discipline into the real science.

Professor Wu Chengzhao and his co-authors' *Research of Eco-Chain Design and Landscape Recovery* is a good paper that uses our professional science and technology to support the development of agriculture, village and farmers. Only the basis of the scientific sustainable development with transformation and enhancing of rural industries will bring the prosperity of Chinese characteristic rural landscape architecture, which can also make greater contribution to the development of ecological civilization. This cannot be compared with the ostentation from self-satisfactory"landscape".

2009年5期

本期主题：遗产与风景名胜

本期主题"遗产与风景名胜"有好几篇值得我国学术界注意的文章。

《关于"世界混合遗产"概念的若干研究》提醒我们，在文化景观纳入世界遗产并开展广泛研究的十余年间，中国并没有积极地参与这种研究，致使中国很多遗产的文化景观属性没有被识别出来。文章将混合遗产分为分立型、融合型、化合型，并强调中国以融合型为突出特征，很有意义。

《世界遗产"完整性"原则的再思考——基于〈实施世界遗产公约的操作指南〉中4个概念的辨析》指出，对文化遗产的意义的认识近年来有着重大的转变，"最重要的变化当属由国家遗产向世界遗产的转变"。维护世界遗产的完整性迫切需要"不同区域的管理机构或政府间的协作，目的不仅仅是所谓的成功申报，更应首先树立的是遵从世界遗产的基本理念（和谐）"。切中当前我国申遗工作的要害。

《丽江古城旅游商业人口和空间分布的关系研究》实际在讲述旅游经济对文化遗产的入侵，依我看这种入侵不过是全球化的一种表现，它引发我们更深入地思考另一个问题：全球化与文化多样性之间果真构成一种二律背反的关系？我看不一定。根源还在于人心的贪婪。如果我们能够把人类对财富的疯狂追求转化为对安身立命的合理追求，恢复天下大同的大爱无疆的本质，这一组"二律背反"假象将不再存在。

本期出版欣逢纪念"五四运动"90周年，我从小所受教育整个建立在西方二元对立思想体系之上，没有背过四书五经，与中国传统思维完全脱节。当前世界似乎渐入一个大乱的前期，此时的我渐能体会到王国维先生投昆明湖前的悲凉心境。可以说我对中国传统文化的认识全赖于对中国园林的学习，每思及此，更加庆幸老天爷迫使我从事了这个专业。

2009年中国风景园林学会将召开首届年会，并举办论文和设计竞赛，这是推动学科学术进步的重要措施，本期刊登了有关通知，希望大家注意。

本期《规划设计学中的调查方法7——KJ法》一文介绍了一种整理混乱意见的方法，对广大技术工作者很有用，也很容易学，值得大力推广。不过学术界应该清楚，机械操作式的KJ法不可以完全替代归纳思维。

《桑塔基洛试验》可能读起来有点枯燥，其意义是回复了对地形的注意，特别是强调了微地形和水流的关系，这对景观生产模式推广以来盛行把地块当作若干平面作构成设计是一个批判。其实，在我读书时，孙筱祥、孟兆祯、杨赉丽等老师就教给我们如何重视竖向设计和水流分析，当时所缺乏的是电脑和多学科"开会法"，全部工作都是凭个人能力进行手绘和手工计算，工作量很大，令人视之畏途。试问如今还有几个老师能够教学生真正用等高线做竖向设计？当然，孙先生讲授的方法本是洋人发明的，可见忘记历史已成为现代世界的流行病。

Phase 5, 2009

Topic of This Issue: Heritage and Famous Scenic Sites

On the theme of this issue"Heritage and Famous Scenic Sites", there are quite a few papers that worth the attention of Chinese academic circle.

Study on the Concept of World Mixed Heritage reminds us that in over ten years when cultural heritage was included in world heritage and widely studied, China did not play an active role in such research, which resulted in that the cultural landscape properties of many Chinese heritages have not been identified. It is of great significance that the paper divides mixed heritages into the discrete type, fusion type and compound type, and stresses that China has a prominent feature of the fusion type.

The paper of *Retrospection on the "Integrity" Principle of World Heritage — Review of the 4 Concepts of the Operational Guidelines* for the Implementation of the World Heritage Convention points out that there is major changes in the understanding of the significance of cultural heritage in recent years and"the most important one should be the change from national heritage to world heritage". Maintaining the integrity of world heritage urgently needs"the cooperation between regulatory agencies and governments in different regions, and the purpose is not only the so-called success of declaration while the basic concept of world heritage (harmony) should be complied first."It really hits the crucial point of current Chinese declarations of world heritage.

Study on Tourism Commerce Population and Spatial Distribution in Old Town of Lijiang is actually about the invasion of tourism economy in cultural heritage. In my view this invasion is just a manifestation of globalization, which triggers our deeper thinking of another question: do globalization and cultural diversity really constitute an antinomy relationship? Not necessarily so, as I think. The root is still the greed of people. If we are able to transform human frenzied pursuit of wealth into the reasonable pursuit of living in peace and restore the boundless love nature of universal harmony in the world, this illusion of"antinomy"will cease to exist.

The publishing of this issue happens to come upon the commemoration of the 90th anniversary of the May 4th Movement. My education from an early age was wholly established on the Western dualistic contrary ideology, and I have never recited *the Four Books and Five Classics* and have been totally separated from the traditional Chinese way of thinking. The present world seems to enter the early stage of big chaos, and I gradually realize Mr. Wang Guowei's desolate state of mind before he drowned himself in the Kunming Lake. It can be said that my understanding of Chinese traditional culture fully depends on the study of Chinese gardens. Every time when thinking of this, I feel more fortunate that God compels me to engage in the profession.

The first annual meeting of Chinese Society of Landscape Architecture will be held this year, and the paper and design competitions will be organized, which is an important measure to promote academic progress of this discipline. The related notices are put in this issue, and hope all will pay attention.

The paper of *The Survey Methods in Planning and Design 7 — KJ Method* introduces a way to collate disorder views, which is very useful for the great majority of technical personnel. It is also very easy to learn and worth vigorous promoting. However, the academic circle should be clear that the mechanical operation KJ method cannot completely replace inductive thinking.

The reading of *The Santa Gilla Experiment* might be a bit boring, and its meaning is a reply to the attention of the terrain, with particular emphasis on the relationship between micro-topography and water flow, which criticizes the prevailing of the land being regarded as a number of planes to design since the promotion of landscape production mode. In fact, when I was a student, Professors Sun Xiaoxiang, Meng Zhaozhen, Yang Laili and other teachers taught us how to pay attention to vertical design and water flow analysis. There were no computers or inter-disciplinary cooperation at that time, all work depended on personal abilities to have hand drawing and manual calculation, and the heavy workload was regarded as a dangerous road to take. Today, may we ask if there are still a few teachers truly knowing to teach students to do vertical design with contour line? Of course, the method that Mr. Sun taught was the invention of foreigners, and it is thus clear that forgetting history has become epidemic of the modern world.

2009年6期

本期主题：郊野公园

"郊野公园"（country park），可能翻译为"乡野公园"更得当些，这个词在百度百科中有清晰的解释："'郊野公园'这个名称源于英国。在英格兰和威尔士地区，有大约250个郊野公园。这些郊野公园大部分都是英国政府在20世纪70年代根据《Countryside Act 1968》划分出来的。近年来因为缺乏经济上的支持，少了新的郊野公园。而除了小部分的私营郊野公园之外，大部分都是由地方部门所管理。……郊野公园的存在是为了使市民在邻近市区的地方可以享受到郊野的康乐和教育设施，而无须深入广大的国家公园。至于辟设特别地区的目的，则主要是为了保护大自然。"由此看来，我们有些地方把郊野公园当作是国家公园的一种，或是位于郊区的城市公园的看法和做法，是一种误解。

近来，"雨水花园"这个词的出现率越来越高。前不久我在一个湿地公园的规划上，见到湖中一个岛被命名为"雨水花园"，这种现象让我想起更早几年我见到过的另一个设计，它在一个几公顷的大草坪上故意围出若干数十平方米的小空间，称之为"口袋公园"，这种事情既令人发笑，又几近荒唐。

我们喜欢讲我们的学科是一门科学，可是科学是要求概念严谨的。过去大学里教学，出现一个新词总是先讲清概念的定义，包括内涵与外延。（顺便讲一句跑题的话，有人以为外延越大，内涵就越多，那是不对的。）我们的教育已经有很长一段时间弱化了有关形式逻辑、辩证逻辑以及真理与实践关系的综合素质培训，也就是科学论证的基本功训练，得宠的东西是西方新观念，包括那些欺骗了全世界的金融新产品，这对我国的学术发展很不利。一个缺乏深刻思想的民族是不可能引导全球的。

眼前这个社会，与现在七八十岁的老先生们当初的成长环境相比，已经发生了巨大的变化。即使面对神圣的科学，《辞海》1979年版的解释是："科学是关于自然界、社会和思维的知识体系，它是适应人们生产斗争和阶级斗争的需要而产生和发展的，它是人们实践经验的结晶。"而当下十分流行的伏尔科夫的定义则说："科学本身不是知识，而是产生知识的社会活动，是一种科学生产。"（夏禹龙，《科学学基础》第45页，科学出版社，1983）。两者当然都有问题，但其间的差异，足以说明在一些新锐的人看来，科学已经从对真理的追求堕落为对知识的生产。生产背后的目的是什么？是产生利润。当前盛行的各种廉价易学的机械、孤立、片面、静止的思想方法和对形式化、浅薄化、个人的主观意识的崇拜，不过是为了制造大量个人成功的假象，其实是服务于资本利益的扩张。

Phase 6, 2009

Topic of This Issue: Country Park

"Country park" might be more appropriate to be translated as "乡野公园", and it has a clear explanation in Baidu encyclopedia that: "The name of 'country park' originated from the United Kingdom. There are around 250 Country Parks in England and Wales. Most Country Parks were designated by British government in the 1970s under the *Countryside Act 1968*. In recent years there has been no specific financial support for country parks and fewer have been designated. Most are managed by local authorities, although other organisations and private individuals can also run them... The purpose of a country park is to provide a place that has a natural, rural atmosphere and recreational and educational facilities for citizens who do not necessarily want to go out into the wider national park. The reason why this kind of special region is set is mainly to protect the nature." So it is a misunderstanding that in some places country parks are regarded and treated as a kind of national park or city park in suburbs.

Recently, the term "rainwater garden" appeared more and more frequently. Not long ago in a wetland planning I noticed that the island in the lake is named "rainwater garden". This phenomenon reminds me of another design I saw in a few years earlier. In that design, several small spaces of tens of square meters are deliberately enclosed in a big lawn of several hectares and are called "pocket gardens", which is both funny and almost ridiculous.

We like to say that our discipline is a science, but science requires strict concepts. In past university teaching, if there is a new term, the definition of the concept including connotation and denotation should be made clear first. (By the way, some words running off of the topic: it is wrong to think that the greater the extension is, the more the connotation.) For a long time our education has weakened the overall quality training of the formal logic, dialectical logic, and the relationship between truth and practice, which is also the basic training of scientific proof. Western new concepts, including the new financial products that fool the whole world, are favoured ones, which is negative to the academic development of our country. A people lacking of deep thinking can never take leadership around the world.

In the current society there have been dramatic changes comparing to the former growing environment of the elderly. Even about the sacred "science", the *Ci Hai* (1979 edition) had the following explanation: "Science is a knowledge system of nature, society and thought. It emerges and develops to meet the needs of production and class struggle of people, and it is the valuable results of human practical experience." But the current popular Volkov definition is that: "Science is not knowledge, but social activities to produce knowledge, and it is scientific production." (Xia Yu-long, *Fundamentals of the Science of Science*, p45, Science Press, 1983). Of course they both have problems, but the differences between them are sufficient proof of that in some cutting-edge people's view, science has degenerated from the pursuit of truth to the production of knowledge. What is the purpose behind production? Generating profit. The present prevailing low-cost easy-to-learn rigid, isolating, one-sided, and static thinking ways, and the worship to formal, shallow and personal subjective consciousness are just for creating the illusion of success, and they actually serve the expansion of capital interests.

2009年7期

本期主题：医疗花园

我们策划的本期主题"医疗花园"，获得了国内外作者们的热烈支持，高质量的稿子源源不断，以致估计需要用2期来刊登，大大出乎本刊意料，看来这个主题是选对了。

通过环境的综合效应来医治人体的某些疾病，之所以近年来在世界上开始流行，主要是因为它的确见效。然而从西医主流来看还是属于非正规疗法，是偏门左道，因为它是用的什么枪弹，打的什么靶子，全然不清楚。于是走非常规西医道路的人只有另行组织一个国际组织，当然也难免有一些骗子混迹其中。国内仍然盛行不衰的中、西医之争实际上是东、西方2种思想体系之矛盾的集中体现。通过分工和深研不断获得的成功，助长了西方科学的骄傲，而东方从整体关系上来把握宇宙、万物和人的本质的思维体系，在典型的西方体系看来实属原始甚至迷信。但是，当所有的人们都在把对世界的认知拆分为数十万、数百万、甚至数千万个课题，然后分别钻研，如何保证人类离真理不至于越来越远？不是还要回到整体地、联系地、辩证地甚至某种程度上玄幻地思考问题的东方方式上来吗？

于是，我们能否由此得到一个形似悖论的启迪：表面看起来原始的东方思维其实更能保护人类避免过分偏离真理？所以人类需要的是西东结合，而不是唯西是从。我深信，利用天然材料并遵从大自然规律的中医，比起偏执于对抗自然的（特别是那些跟随医药资本家利益跳舞的）西医，更有利于人类整体的延年益寿。

袁晓梅博士的《中国古典园林声景思想的形成及演进》从声境角度揭示了中国园林的又一优先之处。我们知道，西方古典园林虽然有过"水风琴"、"水剧场"之类，但并不直指心灵。文末所言的中国园林"不在纯粹物理的声音感知经验，以及生理学、心理学舒适度把握，而在如何透过声音表达心灵内容，从而引人超越，实现园林环境对人生命的颐养与培育"，其实深入到还甚少被探究的中西文化的一大根本差别。

前几天，一位创造了唱片销量吉尼斯纪录的当代西方文化的巨星迈克尔·杰克逊的去世，依我看象征着一种时代文化开始衰落，即以资本运作为后台的艺术工业文化开始衰落。这里使用开始衰落，暗示着两层意思：一层是这个衰落期可能很长，因为要伴随着资本主义体制的衰落；另一层是作为一种艺术，一种人类历史的记忆，它不会完全消亡。当代看似迅速发展的西方景观，从其理论体系和生产流程来分析，其实是批量生产的艺术工业的一种，这在以金砖四国为代表的迅速发展中国家表现得更为鲜明。这种思想，对于偏爱玩弄资本增值、广告推销之术的人来说是绝对不能接受的，或是没有时间做这样的深入思考，或是打死也不愿意这样想。由此，我不得不佩服列宁讲过但我年轻时很不能接受的一句话：如果几何定理违背了人们的利益，人们也会否定它。

Phase 7, 2009

Topic of This Issue: Healing Garden

The theme of this issue, "Healing Garden", is warmly supported by domestic and foreign authors, and high-quality papers come in a steady flow and are estimated to be published in our journal in two issues. This is really unexpected, and it appears that we made the right choice.

The reason why it becomes popular to treat certain human diseases through the combined effect of environment lies in that it is indeed effective. However, it seems non-formal therapy and heresy from the viewpoint of mainstream Western medicine, since the bullet and target are totally unclear. Then the ones in non-Western medicine have to set up another international organization, and of course a number of swindlers are bound to occupy a place in it. The prevailing contending between Chinese medicine and Western medicine is in fact the concentrated expression of the contradiction between Eastern and Western philosophies. The continuing success of the division of labor and in-depth research encourages the pride of Western science, while the Eastern philosophy of understanding the natures of universe, things and human beings based on their overall relationships appears to be primitive or even superstitious in the typical Western system. However, when all the people are engaged in the perception of the world through the split hundreds of thousands of, millions of, and even tens of millions of research topics, how to ensure that human beings are not getting farther and farther away from the truth? It still needs to return to the whole, integrated, dialectical or even fantastic Eastern way of thinking.

Thus, can we get a similarly paradox enlightenment that: the seemingly primitive Eastern way of thinking in fact provides more protection for human beings to avoid too much deviation from the truth? So what we need is the integration of the Western and the Eastern, not following after the Western. I truly believe that the Chinese medicine using natural materials and following the rules of nature is more conducive to the overall longevity of mankind compared with the paranoid, anti-natural Western medicine (especially those who follow the interests of pharmaceutical capitalists).

Dr. YUAN Xiaomei's paper of *The Formation and Evolvement of Soundscape Thought of Chinese Classical Garden* unveils another advantage of Chinese garden from the point of soundscape. We know that Western classical gardens once had sorts of "water organ" and "water theater", but they were not directed at the soul. As said at the end of the paper, Chinese garden "is not a purely physical experience of sound perception, or physiological or psychological comfort assurance, but is the expression of soul through sound to lead human beings go beyond and realize the landscape environments' care and nurturing of human lives", which actually explores one of the fundamental differences between the Eastern and Western cultures that has seldom been researched.

A few days ago, the superstar Michael Jackson, who had the world's best-selling records and represented contemporary Western culture, died. I think his death symbolizes the beginning of the cultural decline of an era, namely the decline of the art industry culture based on capital operation. The beginning of decline suggests the two-tier meaning: first, the decline period may be very long, since it goes along with the decline of the capitalist system; and second, as an art, and a memory of human history, it will not die altogether. The seemingly rapidly-developing Western landscape is actually a kind of mass-production art industries by analyzing the theoretical system and production process, which is more obvious in the rapidly developing countries represented by the BRIC countries. This thinking is absolutely unacceptable for those preferring the play of capital appreciation and marketing operation, since they do not have time to do this kind of deep thinking or they would rather not to think so. Because of this, I cannot help but admire Lenin for his saying that I could not accept when I was young: if the geometric theorem violates people's interests, people will deny it.

2009年8期

本期主题：医疗花园

医疗花园为我们行业的许多人提供了发展机会，我们除了可尽情发挥自己在景观与植物方面的综合能力，主动地与大量教育、社保和医疗单位进行联系外，还可以广开渠道，与多方面专家进行合作，共创新局面，例如与食疗专家合作筹建食疗园，与草药专家合作建设草药园，与易学专家合作筹划疗养园，与社会学家合作策划和谐社区园和养老园等。这件事情如果我们做得成功，可以成为推展中国园林的新抓手。

这一波金融危机后，资源、环境、低碳等肯定会成为世人瞩目的焦点，城市对水资源的关注，将会持续很长一段时间，本期的"城市水环境"专题，反映了这种倾向。以后本刊也将继续关注这种动态。

丁新权先生的《城市生态公园特征及其规划布局》比较系统和完整地介绍了源于西方的城市生态公园的历史、概念、原则和做法，思路清晰，表达精略，文中所提到的绝大部分原则和做法是必要和可行的。对于城市这种特殊的生态系统，多数学者同意将其归纳为"复合生态系统"，亦即自然系统、人工系统、社会系统的复合，而这3种系统之间的复杂关系，使得人们对城市的认知也变得复杂起来。在我国处于快速城市化的时期，我们必须把握好这3种系统的关系，才能保障我国城市化的基本成功，这是本文的基本价值所在。

（《论传统的概念及继承传统的方法——以中国和韩国为中心》作者博士生任光淳，教授[韩]金太京），特别强调了概念准确化的意义。文章收集的文献资料比较广泛，解释"传统"的含义和分析继承传统的方法也相当详尽而细致，这种研究对于本身拥有巨大传统资源的民族是很有意义的。

"景观"一词已成为我们学科的关键词汇，既然如此重要，就不能任凭每个人自己的主观臆断来界定这个概念的内涵与外延。陈烨副教授的《城市景观的语境及研究溯源》是我看到的对于"景观"词义最全面的阐述，我甚至觉得它有可能成为当前中文语境下解释城市景观概念的权威文章。读者如对本文有意见或建议，欢迎继续来稿。

为什么老城会让人感觉到处充满了文化与人情？为什么大多数统一规划的新城会让人感到单调枯燥？最近我经常在思考这个问题。我想，老城的每一座房子的房主都在为城市的风景和文化做贡献，而现在的城市基本是由一个或少数设计师来决定面貌，可能是其根本原因。城市景观应否仅由少数人（设计师、官员、资本家）决定？如果不应，在景观民主化过程中，作为专业人员的设计师是应该提供优质的专业素质保证，还是一味表现自己的灵感和特性？如果答案是后者，那是应该否定的，对其否定的主要阻力来自哪里，不值得我们思考吗？

Phase 8, 2009

Topic of This Issue: Healing Garden

Healing garden has provided development opportunities for many of our people in this industry. In addition to that we can bring our comprehensive abilities in landscape and plants into full play and actively contact a large number of education, social security and health institutions and units, we can also open up more channels to cooperate with experts in various fields for innovation, for example, dietary healing garden with dietary experts, herbal medicine garden with herbal medicine expert, convalescence garden with experts of yi studies, and harmonious community garden and garden for the elderly with sociologists. This would be a new approach to promote Chinese landscape architecture if we do it successfully.

After this wave of financial crisis, resources, environment, and low-carbon would certainly become the focus of world attention, and the concerns of city of water resources will continue for a long time, and this tendency is reflected by the theme of "urban water environment" in this issue. Our journal will continue to pay attention to the developments.

Mr. DING Xin-quan's paper of *Characteristics and Planning Layout of Urban Ecological Park* systematically and comprehensively introduces the history, concept, principles and practices of Western-originated urban ecological garden, and the paper has clear lines of thought and concise expression. Most of the principles and practices mentioned in this paper are necessary and feasible. As to the special ecological system of city, most scholars agree to categorize it as "complex ecosystem", namely the complex of natural, artificial and social systems, and the complicated relationships between the three systems also make people's cognition of city complicated. During this period of fast urbanization in China, we must grasp the relationships among the three systems to ensure the basic success of the urbanization in China, and this is the fundamental value of this paper.

The paper *The Concept of Tradition and the Methodology Succeeding to Tradition—A Case Study of China and South Korea* (Ph.D. candidate REN Guang-chun and Professor KIM Tae-kyung, who is from South Korea), especially emphasizes the significance of concept accuracy. The documents and literature the paper collects are of a wide range, and the way to explain the meaning of "tradition" and to analyze the succession of tradition is much detailed and meticulous. This kind of study is very significant for the people with huge tradition resources.

"Landscape" has become a key term of our discipline. Being so important, its connotation and denotation should not be defined by our own assumptions. Associate Professor CHEN Ye's paper of *The Context of Urban Landscape and the Origin and Advances of Its Research* is the most comprehensive exposition on the meaning of "landscape" I have ever seen, and I even think that it would possibly become the authoritative paper to explain the concept of urban landscape in present Chinese context. Different views or suggestions for this paper are welcome, if there are any.

Why do old cities make people feel rich culture and sensibilities? Why do most unified-planned new cities make people feel monotony? This is what I am thinking about this period. I think it is the fundamental reason that every house owner of old cities contribute to the city landscape and culture while the outlooks of modern cities are usually determined by one or several designers. Should the city landscape be determined by only a small number of people (designers, officials, and capitalists)? If not, in the process of landscape democratization, should the designers as professionals provide best quality assurance or try hard to perform their own inspirations and characters? If the latter should be negative, where does the main resistance come from? Is it not worthy of our thinking?

2009年9期

本期主题：绿色基础设施

根据学会的安排，本期刊登了中国风景园林学会2009年优秀论文评奖获奖文章获得一、二、三等奖的论文。相信这批高质量的文章将为本期增色不少。

本期的主题"绿色基础设施"对自然和人类生态环境建设具有重要意义。据我所知，绿色基础设施这个概念起源于20个世纪90年代的美国，可惜我不知道这么重要的概念是谁最早提出来的。对国土进行绿色基础设施规划，马里兰州也许是始作俑者，至今仍处于世界领先地位，付喜娥博士等人的《绿色基础设施评价（GIA）方法介述——以美国马里兰州为例》紧紧抓住问题的关键与核心对马里兰的工作做了推介，很值得一读。刚开始的时候，绿色基础设施或许自觉弱小，常常借助于势力强大的城规和地景来阐述自己，但由于这个词抓住了事物的本质，仅仅10多年，它已得到迅猛发展，俨然自立，一些敏感的规划师和景观师反而向它靠拢。今年的IFLA大会主题是"绿色基础设施——高水平展示的景观"（Green Infrastructure: High-performance Landscapes）就有些令人捉摸不透，infra的本意是"在下面""在基层"，怎么反而High-performance起来？另一方面，我们从张红卫博士等人的《运用绿色基础设施理论，指导"绿色城市"建设》可以发现，绿色基础设施的基本语汇（核心、绿道、链接、洼地、雨水、屋顶、渗透、乡土等），都紧紧围绕着生态功能，已与视觉意义上的景观渐行渐远。所以景观如果不把绿色紧紧抓在自己手中，长远来看很难不被绿色基础设施抛弃。但是如果我们定义学科的内涵是人类（室外）活动空间，则绿色基础设施当然是我们工作领域的基础层面。

上海市绿化和市容管理局副总工张浪博士结合近年来的工作，写了一本书《特大型城市绿地系统布局结构及其构建研究》，基本内容实际是围绕着城市绿色基础设施的理论、规划、设计、建设展开，孟兆祯院士和程绪珂先生分别为其写了"序言"，正好结合本期主题一起发表，以飨读者。

本期的一个专题是"公共空间标识系统"。Sign system在国外已经发展成一个学科，国内对其的认识才刚刚起步，名称也很乱。开车的人都对我国道路标牌感到头痛，若对其实行系统改造，从全国看是至少需要数十亿的大型工程，还不算数量更大的其他标识的改造，也就是说，如果我国对标识系统认识得早，这笔钱本来是可以省下的。我以为，作为人类活动空间重要元素的标识系统，当然是我们学科的内容。在实践方面，北京动物园开了个好头。曾振伟副教授的《公共空间标识系统设计的基本理论与评价方法》，比较系统地介绍了这门学科在国际上的发展态势，以及学科的基本概念和工作方法。我国若能有高校在开辟这片领域中起到先锋作用，应该还为时不太晚。

本期的另一个专题"藤本植物"，是很多位于实践第一线的同仁比较关心的问题，在小空间塑造，特别在垂直绿化方面，藤本植物具有重要的作用。希望这一组文章能对这些同仁有所助益。

Phase 9, 2009

Topic of This Issue: Green Infrastructure

According to the arrangement of CHSLA, the first, second and third prize winning papers of Award-winning Papers of CHSLA 2009 Annual Meeting Best Papers Award are published in this issue. I believe this issue will enriched by these high-quality articles.

The theme of this issue, "green infrastructure", is of great significance for nature and human ecological environment construction. As far as I know, the concept of green infrastructure took place in the United States in 1990s, but unfortunately I do not know who is the first one to put forward this important concept. Maryland might be the first one to make green infrastructure planning of land, and so far remains the world leader. Dr. Fu Xi'e and her co-authors' paper of *Introduction to Green Infrastructure Assessment(GIA) – A Case Study of Maryland*, USA introduces the practices and experiences of Maryland by focusing on the keys and essentials of this issue, and it is worth reading. Just at the beginning, green infrastructure might be consciously weak and often articulates itself with the help of the powerful urban planning and landscape. However, since this term captures the essence of things, it has been developing rapidly over just a decade, it has been developing rapidly, as if self-sustaining, and a number of sensitive planners and landscapers turn to it instead. The theme of this year's IFLA General Assembly, "Green Infrastructure: High-performance Landscapes", is somewhat enigmatic. Infra means "below" or "basic", and how it can be "high-performance"? On the other hand, we can learn from Dr. Zhang Hongwei and his co-authors' paper of *Guiding the Construction of Green City with the Theory of Green Infrastructure* that the basic vocabulary of green infrastructure (such as core, green corridor, link, marsh land, rainwater, roof, infiltration, and localness) is closely connected to the ecological functions and has been gradually getting far away from landscape in the visual sense. So if landscape does not integrate infrastructure into itself, it would be possibly be discarded by green infrastructure in the long run. However, if we define that the content of our LA discipline is human (outdoor) activity space, green infrastructure is of course the basis of our working field.

Dr. Zhang Lang, Chief Engineer of Shanghai Landscaping and City Appearance Administrative Bureau, combined his work in recent years and wrote a book *The Research of the Great Space System's Layout Structure and Construction of Hyper-megapolis*, and its contents are actually about the theory, planning, design and construction of urban green infrastructure. Academician MENG Zhaozhen's and Mr. CHENG Xuke's prefaces for this book are just published together with the theme of this issue for our readers.

One of the special topics of this issue is "public space sign system". Sign system has developed into a discipline in foreign countries, while in China the knowledge of it has just started and the terms are also chaotic. The drivers are headache about road signs in China, and their transformation throughout the country will cost at least several billions of money, not including a greater number of other signs, that is, if the sign system is recognized earlier in China, the money could have been saved. I think that the sign system, as the important element of human activity space, is of course part of our discipline. In practice, the Beijing Zoo opened a good start. Professor Zeng Zhenwei's paper of *The Basic Theory and Evaluation Method of Public Space Sign System* systematically introduces the international development trend of this discipline, as well as basic concepts and working methods of the discipline. If there are colleges and universities playing a pioneering role to open up this field in our country, it should not be too late.

Another special topic of this issue, "vine plants", is the concerns of our colleagues in the practical field. Vine plants are important in small space shaping, especially in vertical greening. Hope this set of articles can be helpful for these colleagues.

2009年10期

本期主题：新中国城市园林60年

　　在新中国成立60周年之际，中国风景园林学会成功地举办了一届年会，同时举行了优秀论文评奖和大学生设计竞赛，从获得的结果分析，中国的风景园林学术与教育早已跨越了中度发达阶段，我们再经过若干年努力，应该完全有信心步入领袖群伦的境界。为此，我们不妨将前沿阵线分为2支队伍：在经过了一段近乎狂热地学习西方景观思潮之后，一支继续努力追赶世界的发展潮流并通过冷静地分析吸收其优点以为我用；一支努力学习和复兴似乎已被忘怀的中国传统，并以充满关爱之心从中汲取具有普世价值的营养。2支队伍一起发力，中国大地上必将形成"百家争鸣"式的理论繁荣，与城乡建设中光耀夺目"百花齐放"式的实践兴旺形成日月同辉之景象，进而影响世界，造就一段历史上的辉煌。

　　达到这个目标的主要难点在哪里？虽然在人员队伍、学术气氛、活动架构、研究经费、教育体系等方面还都有些问题，但对学术来讲这些都不是关键，想来想去，关键应该在于语言能力：外语和古汉语的能力，而其背后，是学者的文化科学修养。近10多年来，小到园林，中到人居环境，大到文、史、哲领域，大量含有对古文和外文错误理解的文章充斥各种学术场合，这类肤浅的东西是经不起历史考验的。当前的历史机遇对中国人来说也就这么一次，如果由于各种鼓励浮躁的机制耽误了这次机遇，后人对这段历史的评价只能是"辉煌的泡沫"，我们将真的无颜面对后人。

　　近来看配合新中国成立60周年的电视连续剧《解放》，演绎到1948年夏时，蒋介石说他的全国战略是"三角、四线、十三点"，我猜这种机械式思维大概是他的西点军校毕业的部下提出来的，显然不是中国式的思维，遭到毛泽东、周恩来等人理所当然的嘲笑。中国人的思维方式是顺着事物内在的阴阳互动演变来捋顺思想，并时刻照顾着左右八方的关系，并由此推衍未来，当连则连，当变则变，时刻不忘抓紧关键（或曰"抓住主要矛盾"、"抓住牛鼻子"）。毛泽东的《实践论》和《矛盾论》虽然在论述上仍有可改进之处，但他通过强调"实事求是"这4个字对中国人的这种尊重实践、尊重规律，紧抓重点、及时调整和整体式思维的特征做了很好的概括。写到这里，电视正演到毛泽东在西柏坡院子里打太极拳，太极拳和广播体操，是用形体动作解释东西方思维差异的极好例子。

　　为了对付家里出现的老鼠，我买了一只小猫。我惊奇地发现，小猫对于镜子里和电视里的形象似乎毫无反应。联想到小孩在听不懂话以前对这些平面形象的反应也是如此，我想到对于动物天性来讲，本来就是三维的形象才有意义，二维形象只是实体的一张皮，只有人类才自我训练得那么重视二维的东西，才会误将二维的东西等同于三维的实体和空间。我在这里说的上述2点，是想请我们行业的从业者从中得到一点什么启发，目的是坚信中国人在学科理论上一定会对人类做出重大的变革与贡献。

Phase 10, 2009

Topic of This Issue: Sixty Years of Urban Landscape in New China

At the occasion of the 60th anniversary of the founding of New China, the Chinese Society of Landscape Architecture has successfully held the Annual Meeting and also the excellent paper awards and the college students design competition. It is analyzed from the results that the academic level and education of landscape architecture in China has long crossed the mid-degree developed stage and we should be have full confidence to step into the leadership group after several-year efforts. To this end, in the front lines we may be divided into two teams: after a period of near-frenzy learning the Western landscape thoughts, the one team continues their efforts to catch up with the trend of world development and absorbs their advantages through cool-headed analysis, while the other studies and rehabilitates the nearly-forgotten Chinese traditions and draws the universal value nutrition with the heart full of love. With the joint efforts of the two teams, there will be an "a hundred schools of thought contend" type of theoretical prosperity in China and it will shine by reflected glory with the "a hundred flowers in bloom" type of urban and rural construction practice, which will then influence the world and create brilliance in history.

What are the main difficulties to achieve this goal? Although there are problems in professional personnel, academic atmosphere, activity structure, research funding, and education system and so on, those are not key to the academic. After much deliberation, the key actually lies in language ability: the capacity of foreign languages and classical Chinese, but behind it is the cultural and scientific cultivation of scholars. In recent over ten years, among us, in human habitat or in the large areas of literature, history and philosophy, a variety of academic occasions have been glutted with a large number of papers containing erroneous understanding of classical Chinese and foreign languages, and such superficial things cannot stand the test of time. For Chinese people the current historic opportunity is only once. If it is delayed by various kinds of impulsiveness-encouraging mechanisms, the descendants' evaluation of this period of history can only be a glorious bubble, and we will really have no face for the future generations.

In the recent popular TV series *"Liberation"*, Chiang Kai-shek said in the summer of 1948 that his national strategy is "a triangle, four lines, and thirteen points", which I guess this mechanical thinking was probably put forward a West Point graduate under his command. It was obviously not the Chinese style of thinking, and was regarded naturally ridiculous by Mao Zedong, Zhou Enlai. The Chinese way of thinking is to smooth down thoughts along with the yin and yang interaction and evolution within things, always care about the relationships with all directions, deduce the future, and always pay close attention to the key (or "seize the main contradiction", or "lead an ox by the halter") through links or changes. Although there is still room for improvement in the exposition of Mao Zedong's *On Practice and On Contradiction*, through the emphasis on "seeking truth from facts" he has made a very good summary of the thinking characteristics of Chinese people: respect for the practice, respect for laws, clutching focuses, timely adjustment and overall consideration. When I wrote this, Mao was playing taijiquan in the courtyard in Xibaipo in the TV series. Taijiquan and broadcast calisthenics, they are an excellent example of physical movements to explain the differences between the Eastern and Western thinking.

To deal with the rats at home, I bought a kitten. I was surprised to find that the kitten seems to have no reaction to the images in television or mirror. Associating that the kids have similar reaction to these planar images before they understand what people are saying, I think, for the nature of animals, inherently the three-dimensional image is meaningful while the two-dimensional image is only the skin the entity. Only human beings are self-trained to pay so much attention to two-dimensional things resulting that they mistakenly equate the two-dimensional things with three-dimensional entity and space. With the above-mentioned two points, I want to ask the practitioners in our industry to get a little inspiration from them, and the purpose is that we should believe that Chinese people will certainly make significant changes and contributions to the theories of this discipline for human beings.

2009年11期

本期主题：陈植先生纪念

  人类的发展是靠积累的，所谓的飞跃，也要在量变积累的基础上，所以传统是现代的基石。当前有人鼓吹"忘记过去"，最终将导致重复人类过去的错误，否认前人已经达到的成就，为荒诞甚至谬论的东西得以传播创造社会基础。因而，在现代中国风景园林的发展中，回顾老一辈造园学家的思想与贡献，深入发掘中国古典园林的精华十分重要。陈植先生毕生致力于中国造园教育和古典园林研究，他的学术成就一方面领先介绍了先生在世之时世界造园学术的发展，为创立中国现代造园学立下筚路蓝缕之功，另一方面收集出版并严谨地注释了很多基本的古代园林史料，为后人研究中国古典园林奠定了坚实的基础。为纪念陈植先生的杰出学术贡献，弘扬陈植先生育人不悔、严谨求实的治学精神，值陈植先生诞辰110周年、逝世20周年之际，南京林业大学风景园林学院于2009年11月13～15日举办陈植造园思想国际研讨会。本刊借此机会开设"陈植先生纪念"专栏，以表敬仰之情。

  澳大利亚冯仕达先生《自我、景致与行动：〈园冶〉借景篇》对"意境"在明末的突然勃兴，作出了很好的阐释，本文的译文也很有水平。冯仕达先生的哲学探讨，已经开始影响国内理论界。我觉得文中引用吴光明先生的"施为式思维（performative thinking）"来解释《园冶》的思维方式，说得很准确。计成的造园理论是从中小型空间建造山水园的实践中总结出来的，并不依赖设计图，而是以不断在基地或工地的现场上"游走"并体验，由此而产生"理想意向"（想来是用水墨画来表达的）并以此来指导建设或改造的施工。此种游走式思维不受所谓理论或逻辑体系的制约，体现出一种艺术性的自由，但此种中国式的游走性并非任意胡想，它"严格"地受制于地理、经济、自然等现实条件和自然规律的约束，表现出中国文化"实事求是"的一面。

  学科的本科教育应该适应学生未来的就职要求，应该是必要条件，所以美国景观建筑师注册委员会对注册风景园林设计师的基本要求有3个方面，即知识背景K（knowledge），必备能力A（ability），经验技巧S（skill），所谓KAS模式。但如果考虑到本专业对学子在环境道德甚至人格素质上的要求，浙江大学以素质Q（quality），同济大学以人格P（personality）来替换skill，即提出KAQ或KAP模式，似乎更接近充分条件。建议感兴趣的读者研读一下同济大学李瑞冬老师《论基于KAQP人才培养模式的风景园林本科教育专业教学目标体系的建构》，如能就此展开讨论，将对制定我国本科教育的基本要求作出贡献。

Phase 11, 2009

Topic of This Issue: Commemorating Professor CHEN Zhi

Human development depends on the accumulation, and the so-called leap should also have the quantitative accumulation as its basis, so tradition is the cornerstone of the modern. At present some people advocate "forget the past", which will ultimately lead to repeating the past mistakes of human beings, denying the achievements of the predecessors and creating social foundation for the spread of absurd things or even fallacies. Thus, in development of modern Chinese landscape architecture, it is very important to recall the gardening ideas and contributions of the older generation professionals and make in-depth exploration of the essence of Chinese classical garden. Professor Chen Zhi devoted himself life-long to the Chinese gardening education and the study of classical garden. His academic achievements include his introduction to the academic development of world gardening in his period, which is initiative contributions to the establishment of Chinese modern gardening study, as well as collecting, publishing and annotating volumes of fundamental historical literature of ancient garden, which set a solid foundation for later generations' research on Chinese classical garden. To cherish Professor Chen's outstanding academic contribution and promote his academic spirit of untiring education of others and seeking for the truth, and on the occasion of the 110th anniversary of his birth and the 20th anniversary of his death, the International Symposium on Professor Chen Zhi's Gardening Thoughts will be held by the School of Landscape Architecture of Nanjing Forestry University in November 13-15. The special topic of "Commemorating Professor Chen Zhi" is set up in our journal to show our respect.

The paper *Self, Scene and Action: The Final Chapter of Yuan Ye*(《园冶》) of Mr. Stanislaus Fung (Australia) makes a very explanation for the vigorous growth of "artistic conception" in late Ming Dynasty, and the translation of this article is also of a high standard. Mr. Stanislaus Fung's philosophical exploration has already started to influence the theoretical circle of China. I think in this paper it is very accurate to quote Mr. Wu Guangming's "performative thinking" to explain the thinking way of *Yuan Ye* (《园冶》). Ji Cheng's gardening theories were summarized from the practice of medium and small scale landscape gardens, not relying on the design plan, and the "ideal intention" (presumably expressed with Chinese ink and wash) coming out of the on-site "walk" and experience was used to guide the implementation of construction or transformation. Such walk-type of thinking is not constrained by the so-called theoretical or logical system and shows a kind of artistic freedom. But it is not wild flights of fancy, and it is "strictly" subject to geographical economic, natural and other realistic conditions and laws of nature, showing the "truth-seeking" aspect of Chinese culture.

It is the necessary condition that the undergraduate education should meet the future career requirements of students. Therefore, the basic requirements of the American Society of Landscape Architects Registration Board for registered landscape architects are three aspects, namely K (knowledge), A (ability), and S (skill), the so-called KAS model. However, if taking into account the requirements of the discipline in environmental ethics or even personality quality of students, the KAQ model of Zhejiang University and KAP model of Tongji University (Quality and Personality instead of Skill respectively) seem closer to the sufficient condition. I recommend that interested readers read and study the paper *Research on the Professional Teaching Objective System of Landscape Architecture Undergraduate Education Based on "KAQP" Model* of Mr. Li Rui-dong of Tongji University, and discussion of this topic will contribute to establishing the basic requirements for undergraduate education in China.

2009年12期

本期主题：风景名胜与遗产

在东莞市政府、东莞市城市管理局、东莞市岭南园林建设有限公司的大力支持下，《中国园林》第四届编委会于2009年11月21~22日在东莞召开。此届编委涵盖了我国6个大区和港台地区，基本囊括了行业所涉及的各种专业和部门，近98%的中青年，近80%的博士，约3/4的教授，约1/4的"海归"，高端的学术地位、广泛的代表性和生气勃勃的精神面貌，是本届编委会的突出特点。编委会是科技刊物的学术领导组织，我坚信在新编委会的领导下，《中国园林》必有更高的成就。

将近年底，风景园林学术活动频繁，本期择要简介了IFLA大会，中国风景园林学会第一届单位会员沙龙活动，南京林业大学举办的纪念陈植先生的国际研讨会，并刊登了"中国风景园林学会北京宣言"。一期刊物里能介绍这么多大型活动，本身就预示着我国风景园林学术的必然繁荣。

本期的主题是"风景名胜与遗产"，这个题目不是罗列，世界遗产这个概念是西方人提出来的，风景名胜是典型的汉语词汇，将二者并列就含有比较的意思。

我一直对由西洋人主控的世界遗产事务不满，一是他们主客体分离的思维方法无法适应中国的风景名胜区；二是他们制订了一套实质上主要在限制中国申报遗产数量的规则；三是他们用对待石结构的观点对待木结构的绝对保护主义的态度。当然，中国目前的体制，让一些没有修养又崇洋的官吏在对待遗产上有了过大的决定权，造成了很多问题，但这与西方文化本身的缺陷没有关系。

最近我一直在思考一个词，"资源紧缺"。这里的资源是广义的，包括环境质量、气候资源等。大家知道，当前的主流思潮还是生长在亚当·斯密的自由交换和价值规律上的一系列思想，但很少人注意到，这些思想都隐含着2个条件——资源无限和无限发展。一旦资源紧缺，当前的主流思潮必然土崩瓦解。老实说，这是我敢于对当前似乎在主导着世界的西方主流思潮提出挑战的根本原因。

但是，西方也不是一直生活在资源无限的梦境中，就我所知，可能最早还是美国人认识到风景资源的珍贵，并于1872年建立了世界上第一个国家公园——黄石公园。此后，逐渐发展到现在的世界遗产制度。因此，对风景名胜和遗产的研究，隐含着对生活在资源有限环境中的未来世界体系进行思考的无数种子。对此，迄今在世界上最为成功的经验还是中国古代的名山大川。中国因为人口和资源的严重矛盾是中国不能仅仅追随所谓国际"先进"潮流的基本原因，也是中国在人与自然关系上必将走向世界前列的基本原因。低碳、节约、多文化、少追求刺激，应该是中国风景园林的现实前途，也是未来能够引领世界的前途。

Phase 12, 2009

Topic of This Issue: Famous Scenic Sites & Heritage

With the strong support of Dongguan Municipal Government, Dongguan Urban Administration Bureau, and Dongguan Lingnan Landscape Construction Company, the meeting of the 4th Editorial Committee of Chinese Landscape Architecture was held in Dongguan in November 21-22, 2009. The committee members come from six major regions of mainland China and Hong Kong and Taiwan regions, from almost all the related fields and departments. Among them, 98% are young and middle-aged, 80% hold doctoral degrees, around 3/4 are professors, and around 1/4 are returnees from overseas. High academic position, wide range of representation and vibrant spiritual outlook are the prominent features of the present Editorial Committee. The editorial committee is the academic leading organization, and I fully believe that with the leadership of the new Editorial Committee our journal will obtain higher achievements.

Academic activities on landscape architecture are frequently held near the end of the year. In this issue, the IFLA General Assembly, the Salon for the 1st Institutional Members of Chinese Society of Landscape Architecture, and International Symposium in Honour of the Scholarship of Prof. Chen Zhi in Landscape Architecture are briefly introduced, and the "Beijing Declaration of the Chinese Society of Landscape Architecture" are published. So many major events introduced in one issue indicate the inevitable academic prosperity of landscape architecture in China.

The topic of this issue is "Famous Scenic Sites & Heritage". The topic is not listed. The concept of world heritage is proposed by the Western, while scenic site is typical Chinese word, and putting those two side by side shows the meaning of comparison.

I have always been dissatisfied that the Westerners dominate world heritage matters, which comes from three points: first, their thinking way with the subject and object separated does not fit in with Chinese scenic sites; second, they developed a set of rules which actually restrict mainly the number of world heritage nominations from China; and third, they hold the absolutely protectionist attitude in the way that they treat wooden structures with the stone structure perspective. Of course, China's current system gives those officials who have no self-cultivation but a servile attitude to foreign things too much discretion on heritage, which has resulted in many problems, but this has nothing to do with the shortcomings of Western culture itself.

Recently I have been thinking about a phrase, "the scarcity of resources". The resources here are in a broad sense, including environmental quality, climate resources, etc. As we all know, the current mainstream thinking still grows up in Adam Smith's ideas of free exchange and value laws, but few people notice that these ideas all have two implicit conditions — unlimited resources and infinite development. Once the resources are scarce, the inevitable collapse of the current mainstream thinking comes. To be honest, this is the root cause of my courage to challenge the mainstream Western thinking that seems to dominating the world.

However, the Westerners have not been always in the dream of unlimited resources. The earliest awareness of this issue, as far as I know, is that the Americans realized the valuable scenic resources and in 1872 established the world's first national park — Yellowstone National Park. Since then, the system of world heritage gradually developed into the present one. Therefore, scenic sites and heritage studies, implies the numerous seeds of thinking of future world system in the living in the resource-limited environment. In this regard, the world's most successful experience so far is still the famous mountains and rivers of ancient China. The serious contradiction between population and resources in China is the underlying cause that China should not simply follow the so-called international "advanced" trend, and is also the underlying cause that China is bound to move to the front of the world in the relationship between human and nature. Low carbon, economizing, multi-culture, and less pursuit of stimulation should be the real prospect of landscape architecture in China and also the prospect to the world in the future.

2010年

2007年　2008年　2009年　　2011年　2012年　2013年　2014年　2015年　2016年

2010年1期

年年辞旧迎新，难免一些套话，今年第一期的内容很充实，套话就免了。

本期的主题是"植物园"，重点在宏观上探讨了数百年来植物园的内涵演变，本刊首任主编余树勋先生结合中国实情，对此讲得比较系统。上海辰山植物园应是近年来世界上最重要的植物园建设之一，我觉得这次规划很有水平，当然规划阶段落实到施工图过程，很多实际问题需要我国人员深入参与，二者结合得好，互相补充调节，则辰山植物园完全有可能作为改革开放的纪念物，成为世界上最著名的植物园之一。

在设计方面，本期还发表了日本的来稿《CAPITAL MARK TOWER景观》，将其与瓦伦丁先生规划的辰山植物园联系起来看，我们会发现国外在追求新奇、刺激、炫耀的作风之外，还有一大批设计师坚持走细心、求实、平和、淡雅的道路，值得我们的青年设计师和学生深思。

吴良镛先生的《关于园林学重组与专业教育的思考》，听说是花了近一个星期时间思考的结晶。在探讨人居环境问题上，吴先生当之无愧是宏观第一，我们学科一定要更积极、更主动地把建设和维护人居生态环境的任务担当起来。此外，我特别为吴先生"园林涉及人们的日常基本生活需要，学科的综合性和实践性很强，较易普及，其享用者是人民大众。所以，从守边大吏、植树将军，以至知名者或不知名者，都有发言权，是大家的共同创造，这值得我们治此专业者深思"的这段话所感动。相较某些动辄"天地神"之类的高论，越感吴先生情系民众，心系国家之可尊。

在理论方面，本期有2篇年轻人的文章可以连读：邬东璠、庄岳的《从文化共通性看中国古典园林文化》和陈鹭的《继承中国古代园林传统的探讨与思考》。前者讲要真正了解中国古典园林，通晓中国文化，我对此颇有同感，常常为自己在诗词、戏曲、国画、典籍等方面的浅薄而懊悔；后者实质上在讲世界大同的基本前提是民族文化之间的通融，包括消除误解。老实讲，二者之间是有矛盾的。出于政治生境的原因，我从小就对那些偏狭地宣扬民族文化优越性感到厌恶，最初发现俄国人有此偏好，后来发现一些中国人也有这种毛病，比如电脑的汉字输入法解决以后，有些人就大吹汉字的高效率，完全忘记了中国人要花10多年时间认字的低效率。因此，只要有可能，我总会把"中国——世界园林之母"修改成"中国——世界花园之母"，也从来不说《园冶》是世界上第一部造园专著。我觉得20世纪20年代前后留学，特别是留学美欧的人对这种矛盾解决得比较好，他们有比较深厚的国学底子，在亲身感受到西方文明优越性的同时，能冷静地分析以实物和利益为中心社会的潜在危险，因此，他们回国后几乎都成为彬彬君子和热烈的爱国者。

Phase 1, 2010

Conventional words are hard to avoid upon the coming of the new years. With the rich content of this issue of our journal, let us dispense with those words.

"Botanic Gardens" is the topic of this issue, with its emphasis on the macroscopic content evolution of botanic gardens over hundreds of years. Mr. YU Shu-xun, the first editor-in-chief of our journal, talked about this systematically in connection with China's reality. Shanghai Chenshan Botanic Gardens should be most important botanic garden construction project around the world recently. I think the plan is of a high standard, while the practical problems in the process of planning into implementation need Chinese professionals be deeply involved. With the combination and complementation of these two aspects, it is entirely possible that this garden, as one of the monuments of reform and opening up, would become one of the best botanic gardens around the world.

As to design, the paper from Landscape of *CAPITAL MARK TOWER* is also published in this issue. Linking this paper with Valentine's plan of Chenshan Botanic Gardens, we can find that many foreign designers adhere to the careful, realistic, calm and elegant approach apart from the pursuit of novelty, stimulation, and showing off styles, which is worth our youth and students' deep thinking.

Mr. WU Liang-yong's *Thinking of the Reorganization of Landscape Architecture and Professional Education* is said to be the valuable result of thinking in nearly one week. Mr. Wu is definitely the master in human habitat research. Our discipline should actively and positively shoulder the task of building and maintaining the ecological environment of human habitat. In addition, I am especially moved by his words in the paper that "Garden deals with people's basic daily living needs. The discipline is highly comprehensive and practical and easy to spread, and the public are the users. Therefore people from governors and generals as well as the well-known and unknown persons all have a voice, and it is everyone's co-creation, which is worthy of the thinking of our professionals." Compared to some frequently "Heaven and Earth and God" sort of high theories, Mr. Wu's passion for the people and the country is more respectable.

As to theory, there are two young people's papers to read: WU Dong-fan and ZHUANG Yue' *Researching the Culture of Traditional Chinese Garden from the Nation Cultural Universality*, and CHEN Lu's *Discussion and Thinking of Inheriting the Ancient Chinese Garden Traditions*. The former says that Chinese culture is the key to true understanding of Chinese classical gardens. I quite agree on this, and I am often regretful for my superficial knowledge of poetry, drama, Chinese painting, and literature. The latter actually says that the basic premise of universal brotherhood is the accommodation of cultures, including the elimination of misunderstanding. Honestly speaking, there is a contradiction between the two. For political reasons, I have been disgusted with preaching the superiority of national culture since my childhood. First I discovered the Russians have this preference, and later I found that some Chinese people also have such problems. For example, some people brag about the high efficiency of Chinese characters after the computer input of Chinese characters is solved, completely forgetting about the low efficiency that it will take Chinese over ten years to learn the characters. Therefore, as long as possible, I always put the "China—the mother of world landscape" be amended to "China—the mother of world gardens", and I never say that Yuan Ye was the world's first monograph on gardening. I think those people studying abroad around 1920s, especially those who study in the United States and Europe, solved this better. They have profound studies of Chinese ancient civilization, and they were able to calmly analyze the potential dangers of the material and profit based society while feeling first-hand the superiority of Western civilizations, so almost all of them became gentlemanly and hearty patriots after they returned to China.

2010年2期

本期特别奉献"纪念钱学森先生"专题。从我们学科的角度来讲，钱学森先生的去世是极大的损失，如果有钱老的继续护佑，我们将更顺利地摆脱一些顽固守旧思想的羁绊，在广阔的新领域上纵马驰骋。作为世界级理工学者精英，又在西方文化特别是音乐文化的氛围中浸泡了20年的钱老，一接触中国园林便倾心于她，可给出的解释是太多了。前不久，在苏州聊到一个话题：为什么20世纪二三十年代出去留学的人绝大多数都非常热爱中国文化？钱老一篇《不到园林，怎知春色如许——谈园林学》留给我们的启示，岂止如此。我向我们的70后、80后、90后推荐这篇文章，读了以后再分析分析某些自诩旧世界反叛者究竟是深刻还是肤浅。而钱老另一篇《园林艺术是我国创立的独特艺术部门》，则已经开始探讨我国园林与城市建设的关系，后来更发展为"山水城市"的思想。在我看来，当下流行的景观都市主义，其思路本质上和"山水城市"差不过，只不过具有更时髦的条件罢了。

我对本期《北京公园规划修编工作实践体会》一文很感兴趣。北京这次对公园规划的修编很重要，特别是以资质作为抓手，从园林本质出发，没有空谈生态，而是抓住了文物保护，市民对文化生活、停车、必要建筑等的需要，体现了实事求是的精神。我觉得新中国成立后建设的北京植物园、玉渊潭公园、紫竹院公园、陶然亭公园、北京动物园以及全国建设的一些好公园，是我国抵制前苏联影响、未受西方影响、自主探索独立发展中国公园从理论到实践成就的见证，是打破当前对西方盲目崇拜的极好例证和教材，是中国园林未来重新占据世界顶峰的基础，最多再过一二十年就应属于文物。认真总结这一历史时期，我们并不输于西方，甚至很多方面高于西方，缺的是由于我们的体制问题，没有制造名家，没有制造理论体系。当前有些地方的匆忙"旧园改造"，不一定是好事，需要仔细分析。

本期"名家名作"向读者推介杭州市的老领导余森文，余先生毫无疑问是现代西湖的奠基者。在中国，有深厚文化修养的地方领导人物，如杭州历史上的白居易和苏轼，为中国风景园林做出巨大贡献，这在西方历史上基本是看不到的。我经常痴想：如果我国不曾孤立于世界30年，岂有某些西方所谓大师的成名机会！

在1月底召开的达沃斯世界经济论坛上，创始人施瓦布明确提出对当前的世界体制需要重新思考，重新设计，重新构筑（rethinking, redesigning, rebuilding）。一贯善于表演的法国总统萨科奇令人惊讶地提出：自由贸易和市场经济不是目标，只是手段。他们当即遭到与会金融资本家的冷脸，并引起全球资本主义原教旨主义者的群起攻之。在我看来，如果没有主流思潮的怀疑倒是奇怪的，但对一个新世界的探讨，如东方地平线上的初日，正不可遏止地喷薄欲出。

Phase 2, 2010

In this issue the special dedication is the topic of "Commemoration Mr. Qian Xuesen". From the perspective of our discipline, Mr. Qian's death is a great loss. If he continues to protect us, we would be able to shake off some stubborn old-fashion thinking more successfully and gallop in the vast new field. As a world-class scholar of engineering and technology, and with twenty years experience in Western culture, especially music culture, he was attracted by Chinese garden once he entered this field, which could be explained in many ways. Not long ago we talked about a topic in Suzhou: why most of those studying abroad in 1920s and 1930s were passionate about Chinese culture? Mr. Qian's paper of *You Can't Know How Beautiful Spring Is unless You Go into Gardens – Talking about Landscape Architecture* gives us more than some inspirations. I recommend this article to those born in 1970s, 1980s, and 1990s, and after reading this we can analysis whether those claiming to be rebels against the old world are profound, or superficial. And Mr. Qian's another paper *Yuanlin Art (Chinese Landscape Architecture) as a Unique Art Department Created by China* had begun to discuss the relationship between landscape and urban construction in China, which developed into the "shan-shui city" thought later. In my opinion, the basic thinking of current popular landscape urbanism is not much different from "shan-shui city" essentially, while the former has more stylish conditions.

I am quite interested in the paper *Practical Experiences of Park Planning Revision Work in Beijing*. The park planning revision of Beijing is very important. It emphasizes qualification, with the nature of park as the starting point, without empty talk of ecology, and it stresses the protection of cultural relics and citizens' needs of cultural life, parking and necessary buildings, which reflects the spirit of seeking truth from facts. I think that Beijing Botanical Garden, Yuyuantan Park, Zizhuyuan Park, Taoranting Park, Beijing Zoo and other good parks all over the country built after the establishment of New China are the proof of achievements from theory to practice that we resisted the impact of the Soviet Union, had no Western influence, and independently explored and developed Chinese parks. It is an excellent example and teaching material to break the blind worship of the West and the basis for Chinese gardening to re-hold the acme of the world in the future, and it would be cultural relics in at most another decade or two. Conscientiously summing up this historical period, we do not lose to the West, and in many respects we are even better than the West. What is missing is that due to our systemic problems we have not made famous scholars or theoretical system. At present in some places there is a rush of "old garden transformation", and it is not necessarily a good thing and requires careful analysis.

The former leader of Hangzhou Mr. Yu Senwen is presented to our readers in this month's "Masterpieces of Celebrities". Mr. Yu is definitely the founder of modern West Lake. Under the conditions in China, the local leaders with profound cultural cultivation, like Bai Juyi and Su Shi in the history of Hangzhou, make significant contributions to Chinese landscape architecture, which is seldom seen in Western history. I often afford to wishful thinking: if China had not isolated in the world for 30 years, there would be no chance for some so-called Western masters to acquired fame.

At the World Economic Forum at Davos at end of Jannuary, the founder Klaus Schwab clearly stated that the current world system needs rethinking, redesigning, and rebuilding. French President Nicolas Sarkozy, consistently with good performances, surprisingly proposed that: free trade and market economy are not the goals, but the means. They were immediately given the cold shoulder by the participating financial capitalists and caused the rallying attack of global capitalism fundamentalists. In my view, it would be strange if there is no doubt about the mainstream thinking, but the discussion of a new world is irresistibly emerging in all its splendour, like the rising sun above the eastern horizon.

2010年3期

2009年末，在杂志社编委会换届时，我们表达了要更充分发挥副主编和编委作用的想法，得到大家的认同。作为启动的第一步，本期的主题部分"城镇绿地生态构建与管控技术研究"是由同济大学策划和组织的，今后我们还将朝着发挥更多著名院校和设计科研单位作用的方向去努力，使本刊更好地代表我国和世界在本学科的先进水平。

最近，海地和智利的特大地震再次提醒我们需要注意避灾绿地的问题。《我国城市防灾避险绿地系统规划体系》较好地对汶川大地震后我国防灾研究的进展做了综述，并指出在当前我国还没有立法的情况下，可以从制定"城市防灾避险绿地规划方面的设计规范和行业标准"着手，以利于推进绿地避灾工作。

《对铁路选线涉及的景观保护区法规的解读》第一次从第三者的角度观察我国风景资源多头管理的问题，参考价值很大，是篇很好的文章。

《城市自然遗留地景观美学评价的方法研究——心理物理学方法的理论与应用》是本刊迄今为止对心理物理学方法最系统的介绍，从中可以看到它如何尽量地减少误差，使结果更加接近客观。作者收集了国内大量的心理物理学派的评价研究，发现很多研究对心理物理学派评价的方法并没有完全掌握，评价的中间环节模糊，由调查数据无法推出相应的结果。这里凝聚了大量学者的心血，特别是那些真正明白并崇拜科学方法的人的心血。我个人对此的看法是两分的：既有一定合理性，又没有必要迷信它。30多年前，中国完全否定人类的共性（包括自然美），是一种片面。现在，这种方法又走到另一极端，百分百地追逐共性美（这里表现为自然美）。无论如何，改进我国学科的科学水平仍是当前的主要矛盾，所以，我们再次发表了这方面的文章。

《绿道网络的生态廊道功能及其规划原则》给我们最重要的新信息应该是"环带是否在城市发展中产生负面影响，由于城市发展的复杂性、不可逆性和多因素性，很难有确论"。另外，"绿道边缘的网格密度是600m，人均城郊自然保护绿地占有量的建议标准是每千人1hm^2，也就是人均10m^2"的数据也是很有参考价值的。

《居住区生态服务功能补偿研究——以扬州海德公园居住区为例》提供了许多分析方法和基础资料，对其他人继续开展工作很有好处，因之必然推动我国在这一方面的研究。但在一些具体问题上，还有值得商榷的地方，可引起讨论。本案例之所以能够成功，皆因原来用地大都是农田，按32%来折算，差不多有38%的绿地（其中70%多的林地）就可"补偿"了。设想一下，假如在一片林地上开发居住小区，如何做到完全补偿？

Phase 3, 2010

On the occasion of the editorial committee starting its new term last year, we expressed our idea that the roles of the deputy chief editor and the members of the editorial committee should be brought into full play, which gained support from all. As a first step to start, this month's theme section, "Research on the Ecological Construction and Management Technology of Urban Green Space", is planned and organized by Tongji University, and in the future we will put more efforts in giving full scope to more famous universities and research institutions, so as to make our journal better represent the advanced level of this discipline in China and around the world.

The recent big earthquake brought the problem of disaster prevention green space to our attention again. *The Planning Work System for Urban Disaster Prevention and Avoidance Green Space of China* reviews pretty well the progress of disaster prevention research in China after Wenchuan Earthquake, and points out that in the current situation of no legislation for this in China, the establishment of "the design specifications and industry standards of urban disaster prevention and avoidance green space planning" could be a starting point in order to facilitate the work of promoting disaster prevention of green space.

Explication of the Regulations for Landscape Protection Area Related to Railway Route Selection views the issue of multi-headed supervision of scenic resources of China for the first time from the viewpoint of a third party. This is a good article with great reference value.

Study on Landscape Assessment of Urban Remnant Natural Area – Theory and Application of Psychophysical Method is the most systematic introduction of psychophysical method in our journal to date, and you can see how it tries as much as possible to reduce errors and make the results closer to the objective. The authors collected a large number of evaluation studies of domestic psychophysical faction and found that a lot of research did not fully grasp the psychophysical faction evaluation method, and the middle part of evaluation was ambiguous and the appropriate outcome could not be deduced from the survey data. In this paper is embodied the painstaking efforts of a lot of scholars, especially those who really understand and worship the scientific method. My own view is with aspects: admitting its rationality, but no necessity to have superstition of it. Over thirty years ago, China completely denied the common humanity (including the natural beauty), which was one-sided. This approach also goes the other extreme, an absolute pursuit of common beauty (here expressed as the natural beauty). In any case, to improve the scientific level of our discipline in China is still the principal is the current scientific level is still the principal work, so we publish again the article on this topic.

The most important new information from *The Greenway Network as Ecological Corridors and the Associated Planning Principles* should be that "whether the ring belt has a negative impact on urban development is difficult to have a affirmative conclusion due to the complexity, irreversibility and multi-factor nature of urban development." In addition, the data that "the grid density of the edge of greenway is 600 meters, and the proposed standard of per capita share of suburban green space conservation is one hectare per thousand people, namely 10 square meters per capita" is also of much reference value.

Study on Ecological Service Compensation of Residential Communities – With Yangzhou Haide Park as the Example provides a lot of analytical methods and basic data, very good for others to continue the work, which will necessarily promote the research on this aspect in China. However, in some specific issues, there may be open to question and a matter for debate. The success of this case lies in that the original site is generally farmland, and with the conversion rate of 32% there is around 38% green space (over 70% is woodland) for "compensation". Just imagine, if a residential community is developed on woodland, how could full compensation be fully achieved?

2010 年 4 期

2010年3月恰逢陈从周先生仙逝10周年。感谢同济大学为本刊组织了一组纪念文章。对于中国园林美学来说,陈先生无疑是一座高峰。我觉得从周先生对中国园林的感情,似乎比计成更胜一筹。计成毫不隐讳其造园乃为求生的窘境,而陈先生则不计得失地醉心于园林。也许,对园林是"爱"还是"醉",蕴含着中西两种文化的差别。"醉心"这种状态,我们在陈从周先生和孟兆祯院士身上都曾看到,这是不是与他们都热爱戏曲有关呢?或者与他们都身陷中国文化的整体氤境有关呢?其中又有什么深意呢?实在是值得深入探讨的问题。至少,置人于醉心状态,几何型园林做不到,牧场式园林做不到,推崇理性的现代主义园林更做不到。陈先生的突出特点,是他使用充满灵性的诗化语言来展开他对造园理论的探讨,不强调理论结构的整齐和概念定义的确定,以散文的面貌出现,指摘事象以阐明园林理论,同样给读者留下了充裕的想象空间并保持了理论的张力。我在想,陈先生是否会成为连接诗与园的最后一人?倘若如此,那将是历史的悲剧。

王舒扬等的《诗话园林——用中国山水诗解码私家园林的视觉审美特征和审美追求》,也触及同样的问题。"诗意栖居"本是中国人的知识产权,不知为何现代许多国内学者却偏偏不断引用荷尔德林的本来与建筑无关的诗句来装点自己文章的门面。仅仅用传统《画论》研究中国园林终究有两点不足:一是偏于静观,二是偏于视觉。山水诗则没有这些缺陷,而且更容易进入心灵和哲理。所以将园林与诗词相联系来解析中国园林应该很有前途。

本期的另一组专题是介绍国外新出现的地理设计(GeoDisign),GeoDisign其实就是汪菊渊、孙筱祥等先生提倡的大地景物规划(只不过当时没有以3S+CAD为代表的现代高科技手段),而且它可能成为比人居环境还高一级的学科,因为它已经超出人居的领域。我们也许应推崇它为一个新的学科门类,并力争成为这一工作领域的带头人。

我推荐《景观形态之理性建构思维》这篇文章,是因为当前在感性和独创被强调到了天上的这个历史阶段,能与之唱出反调,反映出作者坚强的社会责任心和良好的独立思考能力。

我还要特别感谢韩国造景学会会长、韩国汉阳大学教授曹世焕先生为本刊撰写了专稿《风景园林与城市的融合:对未来公园城市的提议》。加强地球上各民族的交流,是世界走向大同的基本条件,我们期望未来有更多的国外专家的文章在本刊出现。

Phase 4, 2010

This March coincides with the 10th anniversary of Professor Chen Congzhou's death. I would like to express my gratitude to Tongji University for organizing a set of commemorative articles. For the aesthetics of Chinese gardens, Professor Chen has undoubtedly been on the peak. I think that Professor Chen's affection for Chinese gardens was much more than that of JI Cheng. JI Cheng never avoided mentioning that gardening was actually for the awkward situation for living, while Professor Chen was deeply engrossed in gardens without any thought to personal gain or loss. Perhaps the difference between the Eastern and Western cultures is contained in the "love" or "being engrossed" in gardens. We have seen the status of "being engrossed" from Professor Chen Congzhou and Professor MENG Zhaozhen. Is it because that both of them loved traditional Chinese opera very much? Or is it because that they were deeply soaked in the enshrouding Chinese culture? Is there any profound meaning in that? This is really a question to be further discussed. At least, to make human being engrossed, geometric gardens cannot do so, nor the ranch-style gardens or modernism gardens that praise highly rationalism. Professor Chen's outstanding characteristic is that he used the spiritual poetic language to expand his exploration of gardening theories, and did not stress the regularity of theoretical structure and the determination of concept definition, and his papers took the essay style and picked actual things to clarify landscape theories, which left readers ample room for imagination and kept the stretching force of theories. I wonder whether Professor Chen would be the last one to link poetry with gardens. If being so, it would be a historical tragedy.

WANG Shuyang and her co-authors' paper of *Poetically-presented Gardens — The Script Decipher of Visual Aesthetic Pursuit and Aesthetic Feature in Chinese Private Garden by Chinese Landscape Poetry* also gets to the same question. "Poetic dwelling" is originally of Chinese intellectual property, but I do not know why many domestic scholars constantly quote Holderlin's poem that has nothing to do with architecture to decorate their articles. After all there are two shortcomings in simply using the traditional *On Painting* to study Chinese gardens: first, too much emphasis on contemplation, and the second, too much emphasis on vision. However, landscape poetry has no such defects and can enter the spiritual and philosophical fields more easily. So it should be promising to associate landscape with poetry to analyze Chinese gardens.

Another special topic of this month is the introduction of GeoDesign newly emerging in foreign countries. GeoDesign is in fact land scenery planning advocated by Professor WANG Juyuan and Professor SUN Xiaoxiang (but without modern high-tech means represented by 3S+ CAD at that time), and it might become an discipline even higher than the human living environment discipline, since it has gone beyond the field of human habitat. Maybe we should promote it to be a new subject category and try to become the leader in this working field.

The reason I recommend The *Thinking for the Rational Structuring of Landscape Form* is because that to strike a reverse tune in this historical period when emotional and creative things are stressed too much reflects the author's keen sense of social responsibility and excellent capability of independent thinking.

I would also like give special thanks to Mr. Cao Se Hwan, President of Korean Institute of Landscape Architecture and professor of Hanyang University for his special article for our journal, *Landscape Architecture and Integration with City: The Proposal for Future Park City*. Strengthening the exchange of various nationalities on Earth is the basic condition for the universal harmony in the world, and we look forward to more foreign experts' papers in our journal in the future.

2010 年 5 期

电视上转播的上海世博会开幕式让我看得心潮澎湃，或许后人回忆起来，它和2008年北京奥运开幕式一起将成为世界史翻开新的一页的标志！

其实引起我深思的还不止这一点：

在景观大行其道的今天，那巨硕无比的LED大屏幕，那直刺天空锐利的激光剑，那绚烂瑰丽的火树银花，那摇曳多姿的水上喷泉，甚至那由江中的河灯或悬空的白球组成的变换图案和曼妙女郎排列的舞蹈队形，谁敢说它们不是景观？应不应该成为景观学的研究对象呢？地球应该分配多少资源来支持人类的这类"需要"呢？

就以焰火来说，品种数量比起我们使用的植物材料和建筑材料肯定少多了，可人家把它用起来，用好了，就成为大师！差别在哪里？开创性是答案之一，但也许还有其他的解答。

上海世博会的园林能否成为新时代的宣言，还有待时间检验。无论如何，上海的同行们已经花费了巨大的心智，应用了许多新思想、新理念、新技术、新材料，值得各地的同仁学习。上海市绿化和市容管理局为本刊组织了本期的主题文章，相信对全国业界都有助益，我在此表示感谢。

在探讨"当代中式"设计问题这方面，我们已明显落后于建筑界，这是很不应该的。是到了大规模探讨"当代中式"风景园林的时候了。陈跃中先生《"当代中式"景观的探索：上海世博中国园"亩中山水"设计》一文能否成为这个运动的正式开端，还要看过程的发展，但作为历史脚印和探索样板之一，当无疑问。

为了配合IFLA世界大会在苏州的召开，本期我们特别推出中国最值得珍重的自然风景遗产介绍给世界：无论和谁比，三江并流和九寨沟都无愧是绝世珍品。

本期的"名家名作"专栏，由朱育帆教授执笔代表孟兆祯院士的几位高徒介绍了孟院士。遍览今世全球的活跃IFLA人士，孟院士无疑是以深入中国传统文化（特别是高度的综合性表演艺术——戏曲）的人士之身份孑然独立的。在陈从周、朱有玠、孙筱祥等少数几位前辈的中国传统园林文人之后，孟院士几乎单线独传地承担起了传承道地中国园林文化的任务。我无法想象，缺少了孟院士，中国园林文化会变味成什么样子。

关于中国古代园林，有两块是世人所知不多的，一块是岭南，另一块是巴蜀。《清末广州十三行行商伍氏浩官造园史录》和《诗意栖居的成都杜甫草堂园林空间解析》二文可以让我们开阔眼界，破除迷信。

本刊的台湾编委郭琼莹教授亲自撰写了《台湾的另一波无形空间革命——城乡风貌改造运动的意义与效益》一文，以示对本刊的支持。该文对于我们的新农村建设很有参考价值，建议从事这方面工作的人士一阅。

Phase 5, 2010

The opening celebration of the Shanghai World Expo via TV broadcasting made me full of excitement. Perhaps when future generations recall the big event, it, together with the opening celebration of Beijing Olympic Games in 2008, will be the milestone into an new era in the history of the world.

In fact, what evoke my deep thinking are more than this:

At the present time when landscape is prevailing, the unparalleled giant LED screen, the sharp laser sword piercing the sky, the display of fireworks and a sea of lanterns, the swaying waterfront fountains, and even the transforming patterns composed by lanterns in the river or floating white balls, and the dancing formation of graceful girls, who can say that these are not landscape? Should these be studied by the landscape discipline? How many resources should Earth allocate to support such human "needs"?

Taking the fireworks for example, the varieties and quantities are much less than the plant and building materials we use, but they use them and use them well and become masters. What is the difference? Innovation is one answer, but perhaps there are others.

Whether the Shanghai World Expo garden can become a declaration of a new era remains to be seen through time. In any case, our Shanghai peers have spent a great mind, and applied a number of new concepts, new ideas, new technologies and new materials, which deserve the visits and learning of our colleagues from all over the country. Shanghai Landscaping and City Appearance Administrative Bureau organizes the papers of this topic, which is believed to be helpful for the whole industry of China, and I would also like to express my gratitude to the Bureau.

In the exploration of "contemporary Chinese style" design, we have obviously lagged behind the architecture circle, which is very wrong. It is the time for a large-scale discussion of "contemporary Chinese style" landscape architecture. Whether Mr. Chen Yuezhong's paper of *Modernizing Classical Garden Designs: The Chinese Garden for 2010 Shanghai Expo* would become the formal start of this movement depends on the development of the process, but it could be undoubtedly regarded as one of the historical footprints and exploration models.

In line with the IFLA World Congress to be held in Suzhou, we specially introduce the most cherished natural heritage of China to the world in this issue: no matter compared with anyone, Jiuzhaigou and the Three Parallel Rivers are absolutely matchless wonders.

In the "Masterpieces of Celebrities" column, Professor Zhu Yufan write on behalf of Meng Zhaozhen's students to introduce Academician Meng. Among all the active IFLA people around the world, Academician Meng is no doubt the special independent one with deep traditional Chinese culture (especially the highly comprehensive performance art—Beijing opera). After the very few seniors of Chinese traditional gardens such as Chen Congzhou, Zhu Youjie and Sun Xiaoxiang, Academician Meng has shouldered the responsibility of inheriting and passing on the Chinese garden culture almost by himself. I cannot imagine what the Chinese garden culture will change into without Academician Meng.

As to ancient Chinese gardens, there are two parts less known to the world: one is Lingnan (South China), and the other is Bashu (Sichuan). The paper of *Review of Howqua's Gardens at Canton in Late Qing Dynasty* and the paper of *Research on Poetic Landscape Space of Dufu Caotang in Chengdu* would open up horizons for us and break down superstitions.

Professor Monica Kuo, the Taiwan member of the editorial committee of our journal, personally writes an article of *Taiwan's Next Invisible Spatial Revolution—The Significance and Impact of the Townscape Reform Movement* to show her support for our journal. The paper is of great reference value for the new countryside construction in the mainland, and I suggest those in this field to read.

2010 年 6 期

首先，感谢清华大学组织了本期的"低碳与风景园林"主题。在与疯狂消耗自然与社会资源的资本扩张做斗争的过程中，作为节约型社会的一个方面的低碳，正合我们行业的胃口，我们有什么理由不抓住它？

上期本刊刚刚轰轰烈烈地介绍了上海世博会的室外空间规划设计与工程，当下，我看着本期敲定的目录竟有"低碳""IFLA世界大会""西安世园会"三大板块，有点发呆，怎么像三国鼎立那么热闹！在一期杂志里有这么多本行业的热点，再加上最近"东方园林"股价一度成为中国第一，"棕榈园林"也高调上市，"南棕榈，北东方"一时成为网络热语，风景园林行业现在真是风光得很。

过去，因为我国园林企业的门槛很低，我总以为园林不可能搞成大企业，除了有国家市政投入做后盾的国有公司，私有企业做到拥有一亿元左右的资产可能就到顶了。现在看来，我的设想不对，其中的经济学道理，值得有人去研究，以助中国园林开创另一个世界第一。如果我们拥有了世界第一强经济实力的公司，开拓国外市场，当也不远了。那时，我们的海外学子，除了向中国传播西方的学问，还可以从事向全球推销中国风园林的工作。

最近收到台湾的中国文化大学环境设计学院院长，本刊编委郭琼莹女士来信，阅后颇为感动，摘录如下：

"拜读了您在五月期刊里的主编心语，确有很多感触。

现代化或当代中式风景园林该有以新科技、新工法、新材料作为创新之行动了。欧美在面对气候变迁、人力短缺，以及能源之局限时，也逐渐发展出在有限资源条件下，设计出另类新型之风景园林价值观。

此次到苏州参加IFLA大会，更花了两天的时间再度拜访许多园林，也到了木渎古镇之严家花园与启园。对于中国园林擅用空间转折、借景、布局等智慧更有深切之体悟。个人以为中国园林手法应可再转化为现代化垂直都市之空间营造，尤其在土地昂贵、高密度发展之城市人居空间中，自公寓到小区到都市公园，乃至大尺度之广场开放空间，甚或都会公园、国家公园等。

中国园林在有限空间中开创多元、多层次之空间体验与景致之变化，确实有再演化之崇高价值。另，现代化之中式园林更可借助计算机3D仿真，让空间转化、光影变化……乃至四季变迁，作为设计方法之辅助与设计内涵之深化。

……期待更多年轻学子能将古人之智慧再转化为现代人生活中之珍贵空间文化资产。"

郭女士的话语，应该对所有中华文化圈中的风景园林人都有启发。还是那句老话，如果再过20年，当中国即将走完城市化的道路之时，回头一看，满眼是西洋景观，生活在这个时代的中国风景园林人，是无颜以对后人的。

根据中国风景园林学会的布置，本社承担编辑了《中国风景园林名家》一书，通过本次苏州IFLA世界大会上的发放，向全世界更多的人介绍对中国风景园林事业做出巨大贡献的前辈。其中介绍本刊首任主编余树勋先生的文章，本刊尚未刊登过，故本期再次将其刊发，以利未参加苏州大会的读者了解余先生。余先生也写了一篇随笔《"写点东西"交卷》，首次发表，以飨读者。

Phase 6, 2010

First of all, I would like to express my gratitude to Tsinghua University for organizing the theme of "Low Carbon and Landscape Architecture" for the June edition. During the process of battling the capital expansion that crazily consume natural and social resources, low carbon, as an aspect of conservation-oriented society, is exactly to our industry's liking, and we have no reason not to grasp it.

In last month's edition, we just had a spectacular introduction to the outdoor space planning and design and engineering of Shanghai World Expo. At this moment looking at the three columns of "Low Carbon", "IFLA World Congress" and "Xi'an IHE" finalized for this month, I am in a daze and feel that it is bustling a bit like the tripartite confrontation of the Three Kingdoms. With so many hot spots in just one issue, the stock price of "Oriental LA" once ranking top in China recently, "Palm LA" also being listed in the stock market with high profile, and "South Palm and North Oriental" becoming an Internet hot word for a while, the landscape architecture industry is indeed a wonderful sight now.

In the past, because the threshold of garden business is very low in China, I always thought that it was impossible for gardening enterprises to grow big. Apart from state-owned enterprises backed by national municipal investment, private companies will reach the peak with assets around one billion yuan. It now seems that my idea is wrong, and the economic reason in that is worth studying to help Chinese landscape architecture create another number one in the world. If we have companies with strongest economic power around the world, it is also not far away to open up foreign markets. At that moment, our students and scholars overseas can promote Chinese garden to the world as well as diffusing Western knowledge to China.

Recently I received a letter from Ms. Monica Kuo, Dean of College of Environmental Design of Taiwan's Chinese Culture University and member of the editorial committee of our journal, and I am greatly moved after reading the letter, which is excerpted as below: "I really have a lot of thoughts and feelings after reading the Chief Editor's Words in the May edition.

"There should be actions of innovations with new technologies, new construction methods and new materials in modernized or contemporary Chinese style garden. Values of landscape architecture to design new and unique patterns under the conditions of limited resources when facing the problems of climate change, manpower shortage and energy limitation.

"During my stay in Suzhou for the IFLA World Congress, I spent two days to revisit many gardens, and I also went to Yan's Family Garden and Qi Garden in the ancient city of Mudu. I had deeper understanding of the wisdom of Chinese garden in using spatial turns, borrowing landscape and layout skillfully. In my view, Chinese gardening skills could be transformed for the creation of modern vertical urban space, especially in the urban human habitat space with expensive land and high-density development, from apartment to community to city park, even to the open space of large-scale square, or even metropolitan park, national park and so on.

"Chinese garden creates a pluralistic, multi-level experience and scenic changes in a limited space, which is of great value to be further transformed. In addition, with the help of 3D computer simulation, modernized Chinese garden could also make spatial transformations, lighting and shadow changes, or even seasonal changes as a support to the design method and the deepening of the design content.

" …I hope that more young students could convert the ancients' wisdom into valuable cultural assets in modern people's living."

Ms. Kuo's words should be inspiring for all landscape architecture professionals in the Chinese cultural circle.

As the old saying goes, if there are all Western landscapes after twenty years when China has finished its urbanization, our landscape architecture professionals living in this era will be discredited when facing later generations.

As the arrangement of Chinese Society of Landscape Architecture, our group shouldered the editorial work of the book of *Masters of Landscape Architecture in China*, which strives to introduce the predecessors who had made great contribution to Chinese landscape architecture undertakings to more people of the world through delivering to the IFLA World Congress in Suzhou. In the book the paper introducing Mr. Yu Shuxun, the first chief editor of our journal, has never been published in our journal before, and now we publish it to help readers not attending the conference in Suzhou to know about Mr. Yu. We also offer to the readers Mr. Yu's essay of *"Write Something" to Fulfill My Assignment*, which is published for the first time.

2010年7期

本期的重头戏是2010年第47届IFLA世界大会的主旨报告。

仇保兴副部长的报告比较全面地向世界同行介绍了我国风景园林的成就和经验，读者可以从中领悟到中国的工作不但规模巨大，而且循章有序，并不像他们过去在某些IFLA会议上得到的印象是一片混乱和粉饰。为此，我们发表了该报告的英译稿。

或许我们之中很多人没有想到过，他们在文章或说明书中常用的将"以人为本"和"生态优先"并列的所谓设计原则，本身是矛盾的。科克伍德先生是睿智的，他的《弹性景观——未来风景园林实践的走向》直面了这个难题。从遣词造句看，科克伍德显然是在基督教文化的语境下成长的，我深信：在人类的共同命运面前，除了少数极端主义者，人们肯定拥有共同语言。但这并不代表这个地球当前必须拥有"全球景观"，因为这不是一个美学问题，也不是一个科学问题，更不是一个地域或民族观念问题，而是因为在其上面还有一个更高层级的问题：现代城市，特别是特大城市，已经成为金融资本吞噬全人类创造力（先激发你的创造力，后再吞噬之，是资本比先前的各种统治手段高明之处）的基地。此外，城市也不可能一统全人类，因为人类的基本生活资料和实体经济原料不能全靠城市供应，除非科技发展真的能解决这个问题，但到那时也就没有什么生态问题了，现有的许多"先进"理念也就作废了。我们从进士五十八先生的主旨报告《从多样性展开的风景论》可以看出，进士五十八先生也不相信这种科学至上的"远景"，宁可服从佛教"众生平等"的信念。实际上科克伍德先生通过他的"弹性景观"和提倡"保守的手术"，特别是指出"风景园林已从建筑领域独立出来，相比之下，建筑学则拒绝了这些积极的、建设性的社会参与，原因在于其理论是关于几何的组织结构、构造和艺术生产等"，直接否定了当前在中国似乎很春风得意的景观国际主义和建筑化倾向。

进士五十八先生的报告提醒我们，日本的同行是在极其认真地对待生态问题，所以他们正在深入地调查研究日本传统第二自然（里地里山）的"人和自然的关系——生产和生活技术中的智慧和知识"，并想由此创建某种具有国际性意义的模式。对比起来，我们很多人仍在喊口号，玩理念，比如对传统园林和风景区的生态过程，至今未见深入研究，更不用谈从传统的生产和生活技术中汲取智慧和知识了。对乡土景观之美（从文字到视像）的深入开掘，我们也大大落后于港台。又如，在20世纪五六十年代国家贫困时期，我们很多公园的种植设计也是十分重视植物生长规律和种群发展规律的，可惜也没人研究总结。

如果说科克伍德重点讲未来，进士五十八重点讲现在，孟兆祯先生的《认识苏州古代园林》则重点讲过去。在IFLA大会在苏州召开之际，向世界的同行们介绍作为世界文化遗产的苏州园林达到的高度，自然是应有之义。我觉得，一个平衡的社会，必须过去、现世、未来三者皆具，正如很多大雄宝殿里供养着三世佛。我不明白为何在一些人的头脑里只有自己，而对数千年来数以亿计先辈的智慧积累皆视为粪土。其实如果将每个具体的人（当然也包括我）的全部思维活动仔细地、"科学"地、追踪溯源地剖析一下，真正属于他自己的可能不到万分之一。

值此之际，向读者汇报了一下我在本次IFLA大会上的一些收获。

Phase 7, 2010

This month's highlights are the keynote speeches of IFLA 47th World Congress.

Vice Minister Qiu Baoxing's report more comprehensively presented to our counterparts from all over the world the achievements and experience of China in landscape architecture, from which the audience could get to comprehend that China's efforts is not only big in scale but also in orders, not like the impression of chaos and whitewash they got in some IFLA meetings. To this end, we published the report with English translation.

Perhaps many of us have not thought that the paralleled so-called design principles of "people first" and "ecology in priority" usually used in articles or manuals are contradictory in themselves. Mr. Kirkwood is wise and he faced up to this problem in *Resilient Landscapes – Dimensions of Future Landscape Architectural Practices*. From the wording and phrasing, Mr. Kirkwood's growth is clearly in the context of the Christian culture. I am sure that facing the common destiny of mankind, people must have a common language, except a few extremists. But this does not mean that the Earth must now have a "global landscape", because this is not an aesthetic issue or a scientific problem, nor an issue of geographical or national conception, but because there is a higher level question above it: modern cities, especially extra large cities, have become base of the financial capital swallowing the creativity of all mankind (first sparking your creativity and then swallowing it, which is the smart point of capital better than former means of ruling). In addition, the city cannot dominate all mankind, since it cannot fully supply the basic means of livelihood and the real economy raw materials of mankind, unless the development of technology can really solve this problem, but then there will be no ecological problems and many of present "advanced" concepts will also be void. From Mr. Shinji Isoya's keynote speech of *Landscape Theory Developed from Diversity*, we can see that he does not believe in such "vision" based on the scientific supremacy and would rather obey the Buddhist belief of "being equal". In fact, Mr. Kirkwood has directly negated the landscape internationalism and architecture tendency, which seems to be very popular in China now, through his "resilient landscape" and "conservative surgery" he promotes, and in particular indicating that "landscape architecture has been independent from the field of architecture, and in comparison architecture has rejected those positive and constructive social participation, because its theory is about the organizational structure, structuring and artistic production of geometry."

Mr. Shinji's report remind us that our Japanese counterparts deal with the ecological issues very seriously, and they are doing in-depth investigation and study of "the relationship between human and nature—the wisdom and knowledge in production and living technologies" of Japan's traditional second nature (satoyama, or the border zone between mountain foothills and arable flat land) and would like to create some kind of pattern with international significance from this. In contrast, many of us are still shouting slogans, playing ideas, for example, about the ecological processes of traditional gardens and scenic areas, there has been no in-depth studies, let alone learning wisdom and knowledge from traditional techniques of production and living. As to the in-depth development of the beauty of local landscape (from text to video), we also lag far behind Hong Kong and Taiwan. And another example: in 1950s and 1960s, the poverty period of China, the planting design of our many parks also attached great importance to the laws of plant growth and plant community development, but unfortunately no one studied or summed up that.

If we say that Mr. Kirkwood eyes the future and Mr. Shinji talks about the present, then the focus of Mr. Meng Zhaozhen's *Traditional Chinese Gardens in Suzhou* is on the past. In the occasion of IFLA 47th World Congress held in Suzhou, it is natural and proper to proficiently present the height of Suzhou Gardens as the World Cultural Heritage to our counterparts from all over the world. I think, a balanced society should have all the past, the present and the future, just like the Buddha of the Past, Present and Future in temples. I do not understand why some people only have themselves in their minds while regarding the wisdom accumulation of billions of our ancestors for thousands of years as dirt. In fact, if all the thinking activities of each particular person (including me, of course) are carefully and "scientifically" tracked and analyzed, the person's real own part might be less than one out of ten thousand.

On this occasion, I report my gains from this year's IFLA World Congress as above to our readers.

2010年8期

我想大力推介一下《适应公共生活变化的公共空间》，很多知识分子喜欢北欧，那里的确具有较多的人类高级形态社会的特征，特别是文中的"步行环境品质"表格，真正地表现了对人的细致关怀。我想请至今还把我们行业看作创造视觉效果的行业的人留意一下，在这个表格里，视觉的分量有多少？

同理，我提醒大家注意《极简主义园林与日本传统园林融合的探索》一文指出的："作为西方公司在日本的总部，IBM的庭院空间的设计追求的是给人视觉的欣赏和彰显其企业传统的氛围，而不是将参与性考虑成设计的焦点。"所以在这种环境我们自然有理由忽视"步行环境品质"表格所列举的大多数项目。但这种情况应该成为多数吗？

在本期主题部分，最引起我兴趣的是《可控廊道的城市绿地系统》，这是我第一次见到认真批评混淆地理的和景观的廊道，以及分析廊道崇拜的文章。而《住宅环境的价值——智利圣地亚哥的绿地引入策略（节译）》一文，则将我们带入国情完全不同的一个世界，在这里，一切要通过资本家的利益核算。这在人少地多的南美或许可行，因为大自然并不强迫那里的城市必须留出多少绿地，城市缺少的功能大自然会补上。但在中国不行，我们必须从功能需要（而不是从利润）出发仔细安排每一块土地。而且大自然为人提供的服务基本上是豪爽的，不计价的，难以进入市场的。目前不少人在努力计算大自然的"价值"，这种思维在说服百姓或当权者时颇为有用，但这种价值是真实的吗？再者，岂不是把上天或上帝严重玷污了吗？

公园服务的公平性近来似乎引起了很多人的注意，来稿不少，本期就有2篇关于这个问题的文章。问题虽然不复杂，也的确稍微有点难度，可做多方面的探讨。总的来看，关心公平性是我们社会进步的表现，也是我们学科从注重精英到同样关心普罗大众之转化的一个必然步骤。

相对于"爱情是文学永远的主题"，我们可以说"天人关系是园林永远的主题"。《中国古典园林中人、自然、园林三者关系之探究》重提这个问题，似乎是与当前这个只重实利的社会唱反调，我很高兴，很喜欢。对文中的论点，我倒不全都赞同。我不热衷于把一切归结于"天人合一"，那样的话，这个永恒的主题不是太单调了吗？实质上它与"制天命而用之""天人不相胜"有巨大的差别，与"人与天调"也有似细微实重要的差别。我们恢复了"合二而一"的应有地位，是改革开放的重大成果，但若走向否定"一分为二"，是不是另一个极端呢？

Phase 8, 2010

I would like to strongly recommend *Public Space for a Changing Public Life*. Many intellectuals like Northern Europe, the place with more characteristics of advanced human society, especially the table of "walking environment quality" in the paper, which really shows meticulous care of the people. I would like to ask those who regard our industry as creating visual effects to note that: how much is the proportion of visual elements in this table?

Similarly, I draw everybody's attention to what is pointed out in the paper *Exploration about the Integration of Minimalist Garden and Traditional Japanese Garden*: "As the headquarter of a Western company, IBM's courtyard space design pursued to give the people a visual appreciation and highlight its traditional business environment rather than the focus on participation and interaction in design." So in such an environment we naturally have the reason to ignore most of the items listed in the table of "walking environment quality". But should this be the majority?

In the topic of this month, *Urban Green Space System with Controlled Corridors* catches my interest most, and it is the paper seriously criticizing confusing geographical and landscape corridors and analyzing the worship of corridors, which I read for the first time ever. *The Value of Residential Environment – The Strategy of Green Space Introduction of Santiago, Chile (abridged translation)* takes us into a world with completely different situations, where all will be subject to capitalists' profit accounting. This might be feasible in South America with large population and less land, since the nature does not force the city to set apart green space and the lacking function of city will be filled up by the nature. But it is not workable in China, and we should carefully arrange each piece of land on the basis of functional needs (rather than profits). And the services to people provided by the nature are basically forthright, not priced, and difficult to enter the market. Now many people are trying to calculate the "value" of the nature, and this way of thinking might be quite useful to convince people or those in authority, but is this value real? Furthermore, is the Heaven or God not seriously tarnished by this?

The fairness of park service has recently attracted many people's attention, and we have got a lot of contributions, among which two papers of this month are on this issue. The issue, although not complicated, is indeed a bit difficult, which can be discussed from many aspects. In general, the concern about fairness shows the progress of our society, and it is also the inevitable step of our discipline's conversion from focusing on elites to focusing on the general public as well.

Comparing to "love is the eternal theme of literature", we can say that "the relationship between the universe and human is the eternal theme of landscape architecture". *Exploring the Relationship among Human, Nature and Landscape in Chinese Classical Gardens* brings this issue up again and seems to sing a different tune in this materialistic society, and I am pleased and love it very much. But I do not fully agree on the paper's argument. I am not keen on it all boils down to "human is an integral part of the universe". In that case, is the eternal theme not too monotonous yet? In fact it is greatly different from "understanding and using natural laws" and "the natural power cannot be compared with human power", and also has seemingly slight but actually important difference from "human activities should be coordinated with the nature". We resumed the due position of "combining two into one", which is a major achievement of the reform and opening up, but if we tend to negate "one divided into two", does it no go another extreme?

2010年9期

前两日偶然翻阅到2010年7月27日《中国社会科学报》上武汉大学刘杉的《试析当今西方的"大话中国书籍"》一文,其中讲到"在整个20世纪里,一些西方人对中国的喜好来自他们对中国历史文化和现实发展的求知欲,另一些人的兴趣则出于对'黄祸论'杜撰出来的傅满州博士(Dr. Fu Manchu)之类邪恶故事的恐惧,或源于对传统女人小脚的好奇。随着中国的日益开放,西方人来中国从事商务和旅游的人数越来越多,中国落后愚昧的刻板印象已经不在。西方大众对中国的兴趣很多已经不是一种异域探奇的需要,而被转移到了对中国政治经济和文化发展的深层次了解方面。"这段话对我颇有启发。看来,过去我对在国外宣扬"中国园林女人小脚论"的担心实在是有些杞人忧天。近年来西方大量出版的很多描绘或预测中国的书,"由于这类书籍并非都具有学术研究所推崇的系统、严谨和客观的特性,有西方学者把这类书称为big China books","大话中国书籍"便是对这一词传神的中文翻译。由此看来,"中国园林女人小脚论"对于西方正经的学者应该是不屑一顾的。

近来东方园林和棕榈园林2个公司的先后上市,特别是东方园林创出了中国股市第一高的奇迹,应该记入园林史。它给我的启发是:一、过去以为"园林公司做不大"的想法得改变了。二、园林这个行业,开始由众多小公司自由竞争的态势向企业兼并扩张转变。三、上市公司的出现,特别是创出股市第一高奇迹,让全社会,特别是那些从来不知道这个行业的金融界,迅速开始了解和关注园林,其效果,是学术界不可能做到的。四、巨资涌入园林公司,将为本行业和全社会带来什么,有待观察,也是学术界可以探讨的新事物。五、将金融纳入本行业学术的视野,看来也将是必然趋势。我可以想象,对上述问题的正确认知,在初期的冲动平复之后才有可能。战争和金融,可能是以后的人类最不能理解的前辈作为,因为二者都是掠夺,只不过有暴力和非暴力之别。不管现代的人类如何看高自己,我以为现代只不过是由充满幼稚与幻想的人类青年时代向必须成熟转化的时代。所以,中国就像有军队和将军一样,还必须学习金融,精通金融。

本期主题"风景名胜遗产"讨论的是国家大事。对"风景"的认知,是中国人可以在人与自然关系史和人类美学史上大书特书凭以自傲的事,然而以科学的态度全面分析"风景"这个概念,杨锐副主编的《"风景"释义》还是首次。韩锋的文章《文化景观——填补自然和文化之间的空白》,澄清了一件事:在西方,"自然与文化双遗产"是将自然与文化分开并置的,而"文化景观"才像我们的"风景名胜",将二者连接成一体。贾建中、邓武功的《建立具有中国特色的国家遗产保护体系》,系统清理了我国在遗产保护问题上的混乱,提出了科学合理的解决建议,如能得到实现,将是福及子孙的幸事。

Phase 9, 2010

Recently, I accidentally read over the paper of An Analysis of "Books Talking Big about China" in the Modern West, written by Liu Bin from Wuhan University and published on *China Social Science Today* on July 27th 2010, which said that "in the whole 20th Century, some Westerners' love for China comes out of their curiosity about Chinese history and culture and actual development, while others' interests comes from the fear about the fabricated evil stories like Dr. Fu Manchu under the 'perilous yellow people view', or their curiosity about the bound feet of traditional women. With China's increasingly opening up, the number of Westerners to China for business and tourism grows up, and the stereotype image of China's backwardness and ignorance is gone. The Western public's interest of China has no longer been the need of exotic adventure, but the deep understanding of China's political, economic and cultural development." It is quite inspiring to me. It appears that my fear in the past of spreading abroad "the timid and conservative Chinese garden view" is somewhat unfounded. In recent years, a large number of Western books describing or forecasting China have been published. "Since not all those books have the system, serious and objective characteristics highly praised by academic research, some Western scholars call them 'big China books'." "Books talking big about China" is their vivid Chinese translation. From this, "the timid and conservative Chinese garden view" should beneath serious Western scholars' notice.

Orient Landscape and Palm Landscape being listed in the stock market, and Orient Landscape creating the highest stock value in China's stock market should be recorded in landscape architecture history. It is inspiring to me as the following: First, the old view of "landscape companies never grow big" should be changed. Second, the landscape industry becomes to change from the free competition of many small companies to business annexation and expansion. Third, the emergence of listed companies, and particularly creating the miracle of the highest stock value have made the whole society, especially the financial circle knowing nothing about landscape architecture in the past, to know and pay attention to landscape architecture quickly, and it is impossible for the academic circle to do so. Four, what the big amount of money into landscape companies will bring to the industry and the whole society has yet to be seen, and is also a new thing which could be discussed in the academic circle. Five, making finance part of the academic scope of this industry seems to be inevitable. I can imagine that the correct understanding of these issues could be possible only after the subsiding of the initial impulse. War and finance might be the conducts that could be the most incomprehensible one for the future generations, since those two are both looting, with only difference in violence or non-violence. No matter how highly modern human beings think themselves, I consider that the modern is just a period of human beings transferring from the youth full of childishness and fantasies into the inevitable maturity. Therefore, China should also learn and master finance.

The theme of this issue "scenic site heritage" is discussing about a national affair. The knowledge of "scenery" is the one that Chinese can write a lot and be proud of in the history of human and nature relationship and in the human aesthetic history, but it is the first time to have a scientific and comprehensive analysis of the "scenery" concept in our deputy chief editor Professor Yang Rui's *Meanings of "Feng Jing"*. Han Feng's paper *Cultural Landscape – Filling the Gaps between Nature and Culture* clarifies one thing: in the West, "the natural and cultural dual heritage" separates and juxtaposes nature and culture, while "cultural landscape" combines the two into one, like our "scenic site". Jia Jianzhong and Deng Wugong's *To Build the National Heritage Protection System with Chinese Characteristics* systematically cleans up the confusion in China's heritage protection, and proposes scientific and reasonable solutions, which would benefit the nation and future generations once realized.

2010年10期

中国的风景名胜事业，已经启动了30余年，并积累了丰富的理论和经验，应该建立一门学科了。北大的谢凝高先生，积其近半生的探讨，撰成《风景名胜遗产学要义》，为"风景名胜遗产学"的创立立下了筚路蓝缕之功，也是对中国风景园林学的一大贡献。

最近偶然重温了一次波普尔的证伪理论，感触良多。我们学界似乎满足于近乎死水微澜的状态，老年的大师陶醉于中国传统园林文化正在获得世界上越来越多的人的赞美，而中年的学者则很得意于在理论和教育体系方面已被现代景观体系彻底改造的学科现状，虽然二者皆有漏洞，双方却也平安相处。据说，这样很好，既符合中国传统的宽恕之道，也符合西方现代的民主精神。不过我想，如果自然科学（我们现在不是还被列为自然科学吗）也失去了"较真"精神，人类还有资格追求真理吗？

本期的主题"绿心"，较全面地介绍了绿心式规划思想的历程和现状，有利于想采用这种模式的同仁作出更完整的思考和对方案进行解说。但其中罕有地包括了一篇以批评绿心为主的文章，傅凡和赵彩君的《分布式绿色空间系统：可实施性的绿色基础设施》，应该视作是负责本期主题组稿的北京林业大学做出的表率。此外，对理论感兴趣的读者可以认真地读一读邓毅等人的《基于非权重方法的城市景观规划设计生态性评估探论》，该文批评了"权重法"（即上期我所谓的"分层叠加法"）的问题，介绍了"足迹法"（在我看来这是一种很接近综合平衡，并有可能发展为动态平衡的方法）的优势，值得大力推广。可惜本文在具体做法上的改革并不彻底，没有洗脱通过计算"钱"来折合"足迹"的资本印记，而且设想对列表中的项目的"重要性进行层级标志"，仍然回到权重思维的老路。但是，足迹法仍然是一种革命，它可以估计地球的环境容量，也可揭露很多真相，如美国的人均足迹（资源占有）是世界平均的6倍，就是这样得到的。又如，它揭示了旅游不是"无烟工业"，而是一种比起平常生活消耗资源高数倍的"工业"。

1980年前后我在读研究生期间，最佩服陈志华先生的博学和思维的犀利与深刻，虽然那时我还算年轻，很狂，但还能感觉到在我之上有一批很稀有的"人之龙"，陈志华先生就是其中之一。也许由于陈先生之高，很多人不能认知，反而忽视了陈先生在园林方面的贡献。本期我们特约张振威先生撰写了《陈志华园林史学思想探析》，希冀能够彰显陈先生的成就，并启迪后来者。

在中国现代园林史中，"大地园林化"总是处在忽隐忽现的状态。无情的历史背景为"大地园林化"盖上了摆不脱的烙印，但如果处理得好，"大地园林化"也有可能成为中国在"地球表面规划设计"学科中很有意义的贡献。赵纪军的《对"大地园林化"的历史考察》比较客观中肯地探讨了这个问题，可以一读。无论如何，中国现代园林史是个更值得研究的方向，因为它与我们当前现实的关系实在太紧密了。

Phase 10, 2010

China's scenic sites cause has been launched for around thirty years and has accumulated a wealth of theories and experience, and a discipline on this should be established. Mr. Xie Ninggao of Peking University completes his article of *Essentials of Scenic Sites Heritage Studies* based on the exploration in nearly half of his life, which arduously contributes to the creation of the "scenic sites heritage study" and also contributes greatly to the study of Chinese landscape architecture.

I thought a lot when I accidentally reviewed Popper's falsification theory recently. Our academic circle seems to be satisfied with the almost backwater academic situation. It seems that old masters revel in the world's increasing praises for traditional Chinese gardens, while the middle-aged scholars are quite proud of the current academic situation which has been totally reformed by modern landscape system from the aspects of theories and educational system. Though both of them have loopholes, the two sides are still in peaceful harmony. It is said to be good like this, which is consistent to the traditional Chinese way of forgiveness and the modern Western democratic spirit as well. But I think that if the natural sciences (our discipline is still listed as a natural science, isn't it?) lose the "contentious" spirit, do human beings still have qualifications to seek the truth?

This issue's theme of "Green Heart" makes a more comprehensive introduction to the process and current situation of the green heart planning idea, which is helpful for our colleagues who want to adopt this mode to make a more complete thinking and explanation of the plan. But unusually an article full of criticism of green heart is also included, Fu Fan and Zhao Caijun's *Distributed Green Space System: An Implementable Green Infrastructure for the City*, which should be regarded an example set by Beijing Forestry University, responsible for soliciting contributions for the current theme. In addition, readers interested in theories can seriously read Deng Yi and his coauthors' *Research on the Eco-characteristic Assessment System of Urban Landscape Planning Based on Non-weighting Method*, which criticizes the problems of the "weighting method" (what I called "layered superposition method" in last issue), introduces the advantages of the "footprint method" (in my opinion this is a method very close to the overall balance and probably dynamic balance after development), and it is worth promoting. Unfortunately, the article's reform in practice is incomplete, not away from the capital mark of equivalent "footprint" through "money", and envisaging the "level signs of importance" for the listed projects comes back to the old thinking of weight formula. However, the footprint is still a revolution, and it can estimate the Earth's environmental capacity and reveal a lot of truths as well. The result that the U.S. per capita footprint (resource share) is 6 times of the world average is thus obtained. And it reveals that tourism is not a "smokeless industry", but an "industry" with the resource consumption several times higher than daily life.

During my graduate study around 1980, I most admired Mr. Chen Zhihua's knowledge and sharp and deep thinking. Although I was fairly young, very arrogant, I could still feel that were a group of very rare top-talented people above me and Mr. Chen Zhihua was one of them. Many people do not know Mr. Chen perhaps because of his elderliness, and his contribution to landscape architecture is somehow ignored. In this issue we specially invite Mr. Zhang Zhenwei to write the article of *Analysis of Chen Zhi-hua's Historical Thought of Garden*, hoping to highlight Mr. Chen's achievements and inspire newcomers.

In the history of modern Chinese garden, the "national landscaping and gardening movement" is always in a state of flickering. The relentless historical background stamped an everlasting seal on the "national landscaping and gardening movement", but if handled well, it would be a very meaningful contribution of China to the discipline of "earth surface planning and design". Zhao Jijun's *A Historical Inquiry about 'National Landscaping and Gardening Movement'* takes a more objective and fair view to discuss this topic, which is worth reading. In any case, the history of modern Chinese garden is a more worthy research direction, because it is so closely tied with our current reality.

2010年11期

重视25周年，百年的一个quarter，大概是舶来品。以人作参照，25岁正当青年，但刊物不同于其他东西，它从一诞生就肩负有社会责任。所以刊物至少应从青年起身，那么25年的刊物应该到了壮年，和社会上的中年人一样，一方面是社会的中坚，一方面上有老，下有小，负担最重。当此之时，读到那么多中、外的学界泰斗、名家和本刊的前辈为创刊25周年写来的题词、贺词、叮嘱和回忆录，我除了非常感谢，还真有些受宠若惊、忐忑不安，担心在大家心中分量如此之重的刊物，万一在我的手中出了大纰漏，不知如何向师友和后来人交代。

感谢同济大学，让我们得以重温冯纪忠先生的两篇大作。《组景刍议》我在30年前读研时就拜读过，这是我国首次系统地专论动态景观的文章，是它引导我从动态来观察分析赏景，逐渐开始怀疑用效果图、照片集的静态方法来替代动态实践的创作路线，开始明白国画，特别是长卷的观察方法，并进而逐渐进入中国造园者的内心。特别是文中抓出柳宗元"旷如""奥如"的空间感受，我觉得与"开放空间、闭锁空间"的舶来说法相比高妙多了，从此进入了我最爱用的词汇。《人与自然——从比较园林史看建筑发展趋势》一篇，则更为全面地展示了冯先生的中外文化修养和对事物的剖析能力，可惜发表时间欠佳，正逢20多年前的政治风波，人们的思想被扰乱了，很多人可能和我一样，没有拜读过这篇文章，影响力被大大缩减。文章从中、日、欧的园林发展史，挖掘到园林的"形、情、理、神、意五个层面"，令人耳目一新。如果对现今的欧美大部分景观设计和论文做一统计分析，可以发现他们基本又退到仅有"形、理"两个层面，其中的含义，值得我们深思。这种风气，自然也影响到中国，特别是硕、博士论文，因为此种论文必须有"科学"的外貌。

根植于中国文化的中国赏石，早已超越了国外纯功利的对宝石的追求，《中国赏石的分类与鉴赏》是系统探讨这个方面的论文。和鱼虫花鸟一样，赏石是中国园林的组成部分。过去，各地的园林学会有金鱼、禽鸟、摄影、书画、赏石等分会或小组，不知如今状况如何。若任其败落，是否可作为现代景观对封建园林胜利，或凡事讲"条条系统"行政体制畅行的实例？

2010年IFLA国际大学生设计竞赛获奖者，摆脱了说明书式的写法，更多从创作过程来着墨，应该更受师兄弟们的欢迎。

来自台湾地区华梵大学的《溪石之间——石碇小镇时空依存之美的魅力地方营造》告诉我们，港台地区很有一批学者认真关心和热爱地方文化、景观和资源。现在中国有100多家大学的风景园林本科，其实多数应该走这条路线。

Phase 11, 2010

Attention to the 25th anniversary—a quarter of one hundred years—probably comes from abroad. With people as a reference, being 25-year-old is just youth, but journals and publications are different, since they should shoulder the social responsibility from their birth. So the journal should start at least at its youth, and during the 25 years it should enter its prime of life, like the middle-aged people, being the backbone of society and shouldering the hardest burden by taking care of the old and the young as well. At this time, when reading the congratulation inscriptions, messages, advice and memories from so many domestic and foreign academic authorities, celebrities, and pioneers of our journal, I am very grateful, and also feel uneasy at heart, thinking of never making mistakes in the journal that is valued so much, otherwise I do not know how to justify myself.

We should thank Tongji University for giving us chances to re-read Mr. Feng Jizhong's two masterpieces. *Over "Landscape Organizing"*, which I read 30 years ago, was the first systematic monograph on dynamic landscape, and it was the paper that led me to observe and analyze landscape from the dynamic perspective, gradually begin to suspect the creation routes that use the static method of effect plans and photos instead of dynamic practice, begin to understand the observation methods of traditional Chinese painting—especially the long roll painting, and thus gradually enter Chinese gardeners' heart. Especially Liu Zongyuan's spatial perception of "vastness" and "profoundness" pointed in the paper, which I think is better than the Western words of "open space, closed space", has been part of my favorable vocabulary. *Man and Nature—On the Architectural Development Trend from the Perspective of Comparative History of Landscape Architecture* gives us an even more comprehensive display of Mr. Feng's Chinese and foreign cultural enrichment and his analysis capabilities, but unfortunately the publishing time was bad, coincided with the political turmoil 20 years ago when people's minds were disturbed and many people, like me, have not read through this article, and its influence was significantly reduced. From the landscape histories of China, Japan and Europe, the paper tapped the five dimensions of "shape, emotion, reason, spirit, and conception" of landscape, which is really refreshing. If making a statistical analysis of most of the present Western landscape designs and papers, it could be found that have basically gone back into the only two dimensions of "shape, and reason", and the implication is worth our consideration. This trend, naturally, affects China, especially in Master and Ph.D. degree candidates' dissertations since such papers should have a "scientific" appearance.

Rooted in Chinese culture, Chinese ornamental stones have long been beyond the foreign purely utilitarian pursuit of precious stones. *Classification and Appreciation of Chinese Ornamental Stones* is systematic study of this topic published in our journal. Like fish, insects, flowers and birds, ornamental stones are a composite part of Chinese garden. In the past, local landscape architecture societies had branches or groups for goldfish, birds, photography, painting, ornamental stones and others. I do not know how the situation is now. If allowing them to decline, can it be regarded as an instance of modern landscape architecture's victory over traditional garden, or the unimpeded passing of hierarchical administrative system?

The authors of this year's IFLA Student Design Competition winners cast off the writing style of manuals and write more about the creation process, which should be more welcomed by their fellow students.

Between the Brooks and Stones—The Charming Local Construction of the Space-time Beauty of Shiding Town from Huafan University, Taiwan, tells us that a big group of scholars from Hong Kong and Taiwan are quite seriously concern with and love local cultures, landscape and resources. There are undergraduate programs of landscape architecture in over one hundred Chinese universities, and in fact most of them should follow this route.

2010年12期

广州亚运会焰火的熄灭，或许象征着从1990年北京亚运会奏响序曲，经2008年北京奥运会和2010年上海世博会的高潮，中国重返世界舞台中心的这个历史大仪式的谢幕。张扬过后，我们是否该冷静下认真想想，下一步咱们该怎样踏踏实实地做事了。

如果讲城市生态，就应从基本科研数据入手，在总图上玩弄图形是靠不住的。我很重视本期的主题"绿地效益评价"，类似的主题以后还会做下去，即使不同实验所得的数据有差异，也是追求真理道路上的常态。对此，最怕凭空想象，也怕人云亦云。例如，根据《福州植物园绿量与固碳释氧效益研究》一文来推算，过去被大量使用的每个城市人口需要$10m^2$树林即可平衡其吸入的氧气之说便大有问题，可能需要增大几倍到十来倍；如果考虑到还有汽车和工业的耗氧，则问题更大。至于真相，技术上并不复杂，资金和仪器对现在的中国也不是问题，更不需要寻找洋人的理论或数据做自己的靠山，只要我们踏踏实实地做实验，总会接近真理。一旦中国人老老实实地得到真相，并以此为根据建立了理论系统，无疑也就确定了中国在这个领域的世界领导地位。

目前我们风景园林界面临的基本困境大体上可概括为：学科地位、法规建设、执业制度和行业管理，这4项不是单凭本身努力可以解决的，中国风景园林学会正在努力寻求相关行政管理部门的支持和帮助，逐步逐个地解决问题，但我们希望学界为此做出自己的贡献。《我国城乡园林绿化法规分析》一文，有助于推进我们行业的法规建设。《深圳市综合公园建设标准研究》则从基层层面在踏踏实实地做工作。

光、色，是人居环境的基本元素。夜景照明，近年来也在各地兴起。对此，过去学术界的关注度还跟不上实践的需要。发表《光色变幻——杰基尔造园的浪漫主义绘画特征》、《夜景照明与历史古典园林的保护和发展》、《等级制度在古建筑夜间景观中的应用研究》，可看作我们从美术和实用的角度来涉入这方面的试探。

绿道建设，正在中国大地兴起，也颇具中国特色。凡有兴趣关注并想较全面地了解这方面学术动态的读者，我推荐读一读《中国绿道研究进展》，也许它还能有助于各地更好地开展这一项工作。

21世纪的第一个10年过去了。在走向纷乱的世界面前，我想我们最好还是以更加冷静的心态来面对。

Phase 12, 2010

The extinction of the fireworks of Guangzhou Asian Games might be a symbol of the closing of the historical ceremony of China's return to the center stage of the world starting after the start from Beijing Asian Games in 1990, and the climax of Beijing Olympic Games in 2008 and Shanghai World Expo in 2010. After the blaze, we should calm down and think over how we go next down to earth.

If talking about urban ecology, we should start from basic scientific data, and playing with graphics on the general plan is unreliable. I attach great importance to this issue's theme of "green space benefit evaluation", and we will continue to select similar themes in the future, even if the data obtained from different experiments are different, which is normal on the way of pursuing truth. In this regard, groundless imagination has no use, nor does parrot. For example, calculating according to the paper of *Research on Vegetation Quantity and Carbon-fixing and Oxygen-releasing Effects of Fuzhou Botanical Garden*, the argument that citizens can balance their inhaling oxygen with a $10m^2$ wood for each, a widely used view in the past, is actually a big problem, and you may need to increase several or tens of times, and the problem is even greater if we consider vehicle and industrial oxygen consumption. The truth is not complicated technically, capital and equipment is not a problem for present China, and we do not need to seek foreigners' theories and data as backing either. We will get close to the truth as long as we do experiments down to earth. Once the Chinese get the truth honestly and establish the theoretical system on this basis, China's leading position in this field will no doubt be identified.

Currently the basic dilemma that our landscape architecture circle is facing can be summarized as below as general: the disciplinary status, the legal system, the professional accreditation system, and the industrial management. These four aspects cannot be solved with our own efforts alone, and the Chinese Society of Landscape Architecture is seeking relevant administrative departments' support and help and solving the problems one by one, but we hope that the academic circle make its own contribution. *Analysis of the Urban and Rural Greening Laws in China* helps promote the legal establishment of the industry, while *Study on Comprehensive Park Construction Standards of Shenzhen City* does the job from the grassroots level.

Light and color are the basic elements of living environment. Night lighting also rises throughout the country in recent years. In this aspect, the attention of the academic circle did not keep up with the practical needs in the past. *Publishing Brilliance – Romantic Drawing Characters of Jekyll's Gardening, Integrating Nightscape Lighting with Protecting and Developing Classical Garden*, and *On the Application of Hierarchy in the Nightscape of Ancient Buildings could be* regarded as our attempt into this aspect from the aesthetic and practical perspectives.

Greenway is rising in China and full of Chinese characteristics. For readers who are interested in or want to know more fully about the academic trends of this topic, I recommend *Progress of Greenways Research in China*, which perhaps can help better the work of related regions.

The first decade of the 21st century has passed. Facing the crazy world in the front, I think we should better have a calmer mind.

2011年

2007年　2008年　2009年　2010年　　2012年　2013年　2014年　2015年　2016年

2011年1期

新年伊始，世界已经感到地球正处于发生巨变的转折点，中国也已经醒悟到她将遇到史无前例的机遇和困难。我想，《中国园林》这几年唱出一些与世间主流时尚不同的更具深刻性和前瞻性的歌，子弹已经打出去了，让它飞一会儿，总会看到效果。

请读者注意《论〈全球风景公约〉的重大意义》一文，由IFLA提出的全球第一部《全球风景公约》必将成为风景园林行业发展中的里程碑，对行业各方面的理论和实践都具有重要意义。

本期的主题是"低碳的实践与研究"，华中科技大学为其作出了很大贡献，我在此表示感谢。对低碳，我觉得可以采取"两点论"：学术上自由争论，行动上尽量一致。学术自由与学术民主，二者并不对等，学术自由是为了探求真理，学术民主是为了统一行动。对于非直接实践学科，或基础学科，没有统一行动的必要，当然以学术自由为尊。低碳不一定全是真理，但是大家必须一起行动，所以可以采用学术民主。

与"低碳"主题关系相当密切的是"城市绿地系统"栏目中的3篇文章，有兴趣的读者可以参比阅读。《有关绿色基础设施几个问题的重思》和《基于绿色基础设施理论的城市绿地系统规划——以河北省玉田县为例》探讨了国内绿地规划的一些深层次问题。《国外开放空间研究的近今进展及启示》则基本上反映出国外思想和研究的基本动态，文章给我总的感觉是介绍、追随和仰望，这是当前主流，其实以后还可以从批判和审查出发来撰写这类文章。

《景观意象论——探索当代中国风景园林对传统意境论传承的途径》让我想到：中国文化有自己的一套，和西方艺术的差别是如此之大，是否应有几个中国院校的风景园林专业站出来抓一抓中国传统艺术的传承，把京剧或昆曲、古建、诗词、琴棋书画等列入必修课？特别是我国在风景园林专业即将制定培养标准的时候，千万要注意不要只剩了追逐"国际水平"。这里实际上提出一个两难问题。新近读到台湾林谷芳先生编著的《谛观有情——中国音乐传世经典》，对林先生敢在与西方相比明显处于弱势的音乐领域奋起捍卫中国传统文化，我非常佩服。虽然我对林先生特别强调文化整体性的观点尚有疑虑：如此而言，世界各民族间的理解沟通还剩下多宽的途径呢？但书中实际上提出了一个非常重要的问题：对于从小在西方教育体系下成长的人，能够真正理解中国文化的精髓吗？

2010年我们发表过钱学森先生关于园林的3篇文章，其实还有许多学术大家和文艺巨匠也非常关心和热爱风景园林。前不久齐康院士将其"景园学"讲课记录稿（共16课）全部赠给本刊，以此为契机，本期起我们开辟一个"大师谈园林"栏目，我们以后还将主动邀请我国乃至世界更多的各行各业大家巨匠参与议论风景园林，此举必将有助于广开我们的耳目，启发我们的智慧，促进行业的发展。

Phase 1, 2011

Just upon the beginning of this year, the world has felt the Earth at the turning point of its radical changes, and China has also realized that it will face unprecedented opportunities and difficulties. I think that *Chinese Landscape Architecture* has sung some in-depth and forward-looking songs that were different from the mainstream fashion of the world in recent years. The bullet has been shot playing out, let it fly for a while and we will always see the results.

I suggest that our readers pay attention to the paper of *The Momentous Importance of Global Landscape Convention*. The *Global Landscape Convention* formulated by IFLA for the first time around the world will become a milestone in the development of the landscape architecture industry and is of great significance for the theories and practices of all aspects of the industry.

The theme of this issue is "Practice and Research on Low Carbon", to which Huazhong University of Science and Technology has made a great contribution, and I would like to express our gratitude. For low-carbon, I think we can take "two points": free for academic debate, and consistent as far as possible in action. Academic freedom and academic democracy are not equal, since academic freedom is to seek the truth while academic democracy is for unified action. For the indirect practical or fundamental disciplines, there is no need for unified action, and of course academic freedom is respected. Low carbon is not necessarily the whole truth, but all must act together, so academic democracy could be adopted.

The three papers in the "urban green space system" column are closely connected with the theme of "low carbon", and interested readers can do comparative reading. *Rethinking of Several Issues of Green Infrastructure*, and *Urban Green Space System Planning Based on the Green Infrastructure Theory – A Case Study of Yutian County in Hebei Province* explore some of the deep-level issues of domestic green space planning. *Recent Progress and Inspiration of Research on Open Space Abroad* basically reflects the trends of thoughts and researches abroad, and it gives me the general feeling of introduction, follow and looking up, which is in the current mainstream, and in fact such papers can also be written from the critical and examining perspectives.

Theory of Yixiang (Image) in Landscape – Explore the Method of Chinese Contemporary Landscape Architecture Inheriting Traditional Yijing (Concept) Theory makes me thinking that: Chinese culture has its own set, and the difference from the Western art is so great. Should there be a number of Chinese landscape architecture majors standing up and emphasizing on the inheritance of Chinese traditional art, and including the Peking opera, Kunqu opera, ancient building, classical Chinese poetry, lyre-playing, chess, calligraphy and painting into the compulsory courses? Especially when China will soon develop the training standards for landscape architecture majors, we should be really careful not to pursue "international level" only. That actually presents a dilemma. I recently read *Intuition of Life: Classical Chinese Music* edited by Mr. Gu-Fang Lin, and I admire Mr. Lin's courage to rise and defend traditional Chinese culture in the music field that we are at obvious disadvantage comparing with the West. Although I still doubt Mr. Lin's special emphasis on cultural integrity: if so, how wide road id left for the communication and understanding between peoples of the world? However the book actually raised a very important question: for those growing up under the Western education system, can the essence of Chinese culture be really understood?

In 2010 we published Mr. Qian Xuesen's three articles on landscape architecture, and actually a lot of academic masters and literary giants are very concerned about and love landscape architecture. Lately Academician Qi Kang donated all his "Landscape and Gardening Studies" lecture transcripts (16 lessons) to our journal. Taking this opportunity, we open up a new column of "Masters Talking about Landscape Architecture" from this issue, and in future we will also take the initiative to invite Chinese and world masters of various professions and trades to participate in and discuss landscape architecture, which will definitely open up our vision, enlighten our wisdom and promote development of the industry.

2011年2期

曾和一位朋友谈起，为了提高设计质量和工作效率，可以开展景观设计的模式化工作。如果把景园设计看作是由科学和技术的确定性决定的工程，这无可非议；如果看作是表达心灵的艺术，则值得思考。将模式化进行到底，只要条件输入得充分，景园设计几乎可以全部交给计算机。艺术（art）这个词本来和手（arm）同源，离开了手，我们无法绘画、弹奏和触摸，无法用实体表达心灵。机器独自如何表达心灵呢？是个疑问。艺术，在思维工业化时代（即信息时代）遭遇了一次否定。否定之否定会否到来呢？又留给我们一个疑问。

后来和另一位朋友谈起计成时，让我再次想起这个问题，按照计成的设计路线是根本不可能出现让机器替代人的感受在现实环境中进行园林设计的想头的。也许，凡尔赛的作者勒·诺特尔听到这个消息会非常兴奋，这难道是东西方思维的差异？或者是东方"落后"的根源？以效率的名义，似乎是无可争辩的，但效率可以侵蚀人的内心，又确实是人们没有看到的。

当周围的人们大多在津津乐道资本主义的优势时，有没有想过这个世界上最有钱的4个人，斯利姆（Carlos Slim）、比尔·盖茨（Bill Gates）、巴菲特（Warren Buffett）和安巴尼（Mukesh Ambani）的财富总和比世界上最贫穷的57个国家的财富还要多（见2011年1月27日《华尔街日报》"可能终结巴菲特时代的四大历史趋势"）？这4个人果真有这么大的本事和贡献吗？我以为除了牛顿、爱因斯坦等对人类有巨大贡献的大科学家，一个靠赚别人钱的人再勤奋，再聪明，也没有这么大的贡献于人类，他们的财富聚集（也包括所谓发达国家的财富聚集）除了先行一步，主要靠的是这个制度。一个能够制造如此巨大且无理的财富差距的制度难道没有本质上的问题吗？

本期以"教育研究"为主题，重点围绕着创造性思维的开启和借鉴国外的经验。由于中小学教育和大学本科机械式评估制度的失误，我们的大学生创造性、自主性、实践性的缺失十分明显，当然理性的（请注意，不是后现代的那种任意的、随想的或狂妄的）批判性思维能力就更差，盲从对象从老师到大师到洋大师，一步步升级，甚至从举止到说话到写文章都要洋泾浜腔调。在这个世界面临巨变，特别是变化主要的动力来自中国之时，中国人不能只是一味地学习，还必须学会批判，包括自我批判，没有批判能力的民族是无法承担如此艰巨的重任的。

《基于"源—汇"景观调控理论的水源地面源污染控制途径——以天津市蓟县于桥水库水源区保护规划为例》一文提出一种值得推广的大地规划方法。"源""汇"本来是大气污染研究中常用的方法，以反映了大气污染物的来源和去向，景观生态学最近开始借用"源""汇"的观念，作为解决"景观格局与生态过程"之间关系的信息化和数理化之难题的一种手段，我觉得很好，我们必须破除对于脱离具体自然规律的图形（格局）的迷信，才能走上科学发展的坦途。

Phase 2, 2011

Once I talked with one of my friends that landscape design can be standardized into modeled work to improve design quality and efficiency. It is beyond reproach if landscape design is regarded as an engineering project based on the certainty of science and technology, but it is worth considering if regarding it as an art for the expression of soul. To the widest extent of modeled work, landscape design could be handled all by computers as long as the input conditions are sufficient. The word "art" has the same origin of the word "arm", and without arms we cannot paint, play, touch, or express our minds in physical forms. How can machines express our minds? It is in doubt. Art, in the era of industrialization in thinking (namely the information age), suffers a negation. Will the negation of negation come? It leaves us a question again.

This question came up to me again later when I talked about Ji Cheng with another friend. According to the design principle of Ji Cheng, the idea of landscape design in real environment with machine instead of people's feelings will never appear. Perhaps the designer of Versailles Le Nôtre would be very excited: is this the difference between the Eastern thinking and the Western thinking? Or the origin of "backwardness" of the East? It seems be indisputable in the name of efficiency, but efficiency can erode people's minds, which is really not seen by people.

When most people around are talking about the advantages of capitalism, has anyone thought about that the sum of the world's wealthiest persons, Carlos Slim, Bill Gates, Warren Buffett and Mukesh Ambani, is even more than the total wealth of the world's 57 poorest countries (from Four reasons Buffett's legacy will die by 2020, *The Wall Street Journal*, January 27, 2011)? Do the four persons really have so many skills and contributions? As I think, except the great scientists with great contributions to human such as Newton and Einstein, those who make money out of others—even though they are clever and hard-working—have not much contribution to human, and their wealth accumulation (including the so-called wealth accumulation of developed countries) mainly relied on the system besides their earlier steps. Does the system that makes such a huge and unreasonable wealth gap have no essential problems?

This month we take "Education Study" as the theme, focusing on the opening of creative thinking and learning of foreign experience. As a result of the failures of the primary and secondary education and the mechanical undergraduate education assessment system, our students are obviously lack of creativity, initiative and practice, and of course rational (please note that it is not any kind of the post-modern, capricious or arrogant ones) critical thinking is even worse. Their blind obedience to teachers gradually rises step by step to masters and foreign masters, and even their manners, talking and writing are all pidgin. As the world is facing dramatic changes—mostly driven by China in particular, Chinese should not just keep learning, they should also learn to criticize, including self-criticism, and the nation without critical ability could not take such an arduous task.

The paper *Water Source Surface Source Pollution Control Approaches Based on Source-sink Landscape Control Theory — A Case Study on Water Source Conservation Planning of Yuqiao Reservoir in Ji County in Tianjin City* proposed a land planning approach that should be promoted. "Source" and "sink" are original a commonly used terms in atmospheric pollution research to reflect the origin and destination of air pollutants, and the concept of "source" and "sink" is recently borrowed by landscape ecology as a solution for the information and digitalizing problem of the relationship between "landscape patterns and ecological processes", which I think is very good. We will step on the easy path of scientific development only we break the superstition on graphics (patterns) that are cut off from concrete natural laws.

2011年3期

由IFLA主导的《全球风景公约》，若再过几百年来回顾，必将被认定是世界风景园林史上最重大的事件，也是我们行业对人类的最重大贡献。所以，本刊将响应IFLA的号召，积极投入"在所有层面上发起了一个自下而上的运动，号召所有会员国和代表采取各种正式和非正式的方法来联系各国相关政府官员及其在联合国教科文组织的代表，希望他们在联合国教科文组织执行理事会上同意将全球风景公约提案提交给联合国教科文组织大会"的活动。当然，我们也会一如既往，对一切舶来的文本采取分析的态度，例如，走向国际风景公约的号召书鼓励人们"把风景看成是既自然又文化的概念，既具体又抽象的存在"的说法，前半句（既自然又文化）反映出西方文化中克服二元化思维方式过程中的进步，后半句（既具体又抽象）则反映出20世纪流行哲学（如现象学、存在主义等）的强大影响。我颇担心，如果风景成为"抽象的存在"，风景将在堕落为那些为自己私利随意玩弄风景（或曰景观）的人的工具的道路上越滑越远。

为配合北京奥运会和上海世博会，两个城市都启动了系统的大规模城市绿地科研计划。本期以"上海世博会生态技术研究"为主题，得到上海绿化和市容管理局的支持，在此深表谢意，其中《生态技术在上海世博园区绿地建设中的综合应用研究》一文，正如本刊审稿专家的评语所言："是对世博生态技术的总计和归纳，对世博技术的后续利用和我国'十二五'都有借鉴和指导作用，对风景园林学科如何对接我国国家战略，解决实际科学问题具有引导作用。"而其他文章，亦可供相关专业工作者参考。

本期的另一专题集中介绍了欧洲绿道。欧盟推进绿道的工作比起我们这个大一统国家还好，值得我们思忖。

同义词、近义词同时存在现象，世界皆然。"绿道"一词，在中国还有绿带、绿廊、绿脉、绿线等说法，在欧美也有greenway、green path、green corridor、greenbelt、trame verte等说法。不过我觉得在科学领域，无实际意义区别的称谓越少越好，如果我们的科学家不去做认真的研究，而是热衷于创造新名词来表现自己，那不如去做一个作家或演员。在中国的风景园林界，大家正在向"绿道"集中，这是好事。我所感兴趣的是，为什么大家很少把"林带"纳入视野。如果说林带太简单，太机械，那为何更简单机械的"树阵"却被许多人青睐呢？

最后，我要特别推荐杰克·埃亨博士的《可持续性与城市：一种景观规划的方法》，这是篇紧跟世界最前沿学术发展的文章，水平很高。也许由于中西思维方法的差异，译文很难读，但我希望对深层学术理论感兴趣的中国读者，认真研读一下这个文本，或者干脆研读原文，以努力去理解其中的深意。

Phase 3, 2011

The IFLA-led *Global Landscape Convention*, if reviewed after hundreds of years, would be certainly regarded as the most important event in the world's landscape architecture history and also our industry's most significant contribution to mankind. Therefore, our journal will answer the call of IFLA and be actively involved in the activity "to launch a movement from the bottom up at all levels and call on the member states and their representatives to contact relevant government officials and their representatives to UNESCO in formal or informal ways in the hope that they agree in the UNESCO Executive Council to submit the proposal of The *Global Landscape Convention* to UNESCO's General Conference". Of course, we will, as always, take an analytic attitude to exotic papers. For example, Toward International Landscape Convention encourages people "to regard scenery as both natural and cultural concept, and both concrete and abstract existence". The first half (both natural and cultural concept) reflects the progress of Western culture in the process of overcoming the dualistic way of thinking, while the last part (both concrete and abstract existence) reflects the powerful impact of the popular philosophy of the 20th century (such as phenomenology, existentialism, etc.). I am quite worried that, if scenery becomes "the abstract existence", it will slip down as the tool of those who play with scenery (or landscape) randomly for their own interests.

To support the Beijing Olympics and Shanghai World Expo, the two cities both launched large-scale urban green space system research projects. This issue takes "Research on Shanghai World Expo Ecological Technology" as the theme and has received the support of Shanghai Municipal Afforestation & City Appearance and Environmental Sanitation Administration, and we are deeply grateful. The paper of *Study on the Comprehensive Application of Ecological Technology in Shanghai World Expo Park Green Space Construction*, as one reviewer of our journal commented, "is the summary and conclusion of the World Expo ecological technology, of reference and guiding value for the follow-up utilization and the 12th five-year plan, and plays the steering function for the landscape architecture discipline to link up with China's national strategy and address actual scientific problems". And other articles could also be reference for related professionals.

European greenway is introduced in focus in another special topic. European Union promotes greenway even better than our big unified country, and we need to think about it.

The concurrence of synonyms and homoionyms exists anywhere around the world. The word "lüdao (greenway)" means lüdai (green belt), lülang (green corridor), lümai (green vein), and lüxian (green line) in Chinese, and also has the expression of greenway, green path, green corridor, greenbelt, and trame verte in Western languages. But I think in the scientific field it is better that the titles or terms with no meaningful difference are fewer. If our scientists do not do serious research and are interested in creating a new term to express themselves instead, they should better be writers or actors. It is a good thing that in Chinese landscape architecture circle we are concentrating on "greenway". What I am interested in is why we rarely bring lindai (forest belt) into our vision. If it is too simple, too mechanical, then why can the even simpler and more mechanical shuzhen (tree array) be the favorite of a lot of people?

Finally, I would highly recommend Dr. Jack Ahern's *Sustainability and Cities: A Landscape Planning Approach*, which is a high level paper keeping in step with the world's most advanced academic trend. The Chinese translation is hard to read, perhaps due to the different Western and Chinese ways of thinking. But I hope Chinese readers who are interested in deep-level academic theories read the paper seriously, or read its original English version if possible, and try hard to understand the profound meaning in it.

2011年4期

正当本期即将合版之际，传来了国务院学位委员会和教育部《学位授予和人才培养学科目录（2011年）》将我们风景园林学科列为一级学科的大喜讯。本刊将以此为中心组织下一期的主题。

本期主题再次触及"文化景观"，通读之后我感到本期主题将是学术性非常强的一期。

正如邬东璠《议文化景观遗产及其景观文化的保护》所指出："虽然文化景观遗产的讨论与申报在西方世界如火如荼，但在本应是文化景观大国的中国却一度没有受到重视，在西方人已经开始认可自然文化有机融合的东方哲学时，中国人自己却不识庐山真面目"，这实在是令中国人汗颜的事。在遗产保护中把自然和文化对立起来，本是西方的传统，cultural landscape的提出让西方人的思维方式有了一个巨大的进步，但反观我国似乎长期存在着对文化遗产的偏爱，有法律、有机构、有经费，而对"文化景观"类型的风景名胜区，关注度则大大不如。中国风景名胜区工作者和本学科学术界最好能读一读。

周小棣等的《从对象到场域：一种文化景观的保护与整合策略》，提出了比"场所"更加深化的"场域"概念，并通过对"场域"的解读，得到文化景观是一个变化发展的过程产物的结论，指出"价值判断"在文化景观实践中的重要性，从而有可能在理论上为中国式的遗产保护（不只是静态保护，更注重活态保护）找到一种解释，颇具开创性。而曹新的《遗产体系与游憩度假体系之辩》，紧密结合当前非常严峻的遗产保护形势，提出了一种实行"价值判断"可操作的建议，可供有关行政管理部门和规划实践部门参考。

城市绿轴、区域绿道，是近来中国绿地建设中非常突出的2种现象；而风能发电的普及也必将对中国大地景观带来巨大的影响。万敏的《我国城市中心绿轴的基本特征与思想探因》，吴隽宇的《广东增城绿道系统使用后情况（POE）研究》和张柔然的《基于景观特征评价的风力农场规划手法——以英国谢菲尔德风力农场规划为例》，分别对其做了研究、介绍或论述，值得大家关注。

我深信，全局性和地域性的关系是风景园林学科的永久性主题之一。对于建筑学出身的同仁来说，植物特别令人头痛，除了华南与华北不一样，北京和天津也不一样；不仅广州和深圳有差别，广州的番禺和从化有差别，甚至一条街的路东和路西也有差别。做大规划的人不可能掌握得那么细，但总得不离谱吧？这里的分界在哪里？张雯婷等的《中国植物极限温度分区的探索性研究》就提供了一种解读，也暗示着一种本学科科研可能的开拓方向。

Phase 4, 2011

On the occasion of the final editing of this issue, great news came up that our landscape architecture discipline is listed as the first-level discipline in *The Discipline Catalog for Degree Granting and Personnel Training (2011)* by the State Council Academic Degrees Committee and the Ministry of Education. This will be the central theme of next issue of our journal.

The theme of this issue involves "cultural landscape" once again, and I feel it is quite strongly academic after reading through the issue.

As Wu Dongfan pointed out in *Discussing on Cultural Landscape Heritage and Its Landscape Culture's Conservation*, "although the discussion and application of cultural landscape heritage is growing vigorously in Western countries, it is not taken seriously for quite a long time in China which should be a great power of cultural landscapes; while the Westerners have begun to recognize the Eastern philosophy that organically integrates nature and culture, the Chinese people themselves do not know the truth," which is really shameful for the Chinese people. It is the Western tradition to set nature against culture in heritage protection, and advancing cultural landscape marks a great step forward of the Westerners' way of thinking, while in China there seems to be a preference for cultural heritage, having laws, institutions and funding, but less attention has been paid to the scenic areas of the "cultural landscape" type. The professional workers of Chinese scenic areas and the academic circle of the discipline should better read it.

Zhou Xiaodi and coauthors' paper of *From Object to Field: A Strategy for Cultural Landscape Protection and Integration* proposes the concept of field, much deeper than place/site. Through the interpretation of field, it is conclude that cultural landscape is the product of the change and development process, and the significance of "value judgment" in cultural landscape practice is pointed out, making it possible to find a theoretical explanation for the Chinese way of heritage protection (not only static protection, but also dynamic protection), which is quite innovative. And Cao Xin's *Discrimination of the Heritage System and the Recreation System* closely combines with the currently serious situation of heritage protection and puts forward a practical proposal of making "value judgment", which could be reference for the relevant administrative departments and planning and practice departments.

Urban green axis and regional greenway are two every prominent phenomena in the green space construction of China, and the popularity of wind power will also greatly impact the landscape of China. Wan Min's *Exploration of the Characteristics and Ideological Reasons of City Center Green Axis*, Wu Juanyu's *Post Occupancy Evaluation (POE) on the Zengcheng Greenway System in Guangdong Province*, and Zhang Rouran's *The Method of Wind Farm Planning Based on Landscape Character Assessment – A Case Study on the Sheffield City Peak National Park Fringes Wind Farm Planning*, have made research, introduction or discussion respectively, which are worth readers' attention.

I am confident that the relationship between wholeness and localization is one of the permanent themes of the landscape architecture discipline. For our colleagues with the background architecture, plants are particularly troubling. Not only South and North China are different, but also are Beijing and Tianjin (both in North China); not only Guangzhou and Shenzhen are different, but also are Panyu and Conghua (both the districts in Guangzhou); and even the two sides of a street might have differences. It is impossible for those who make macro-planning to grasp so finest details, but they should be far away from the normal, right? Where are the boundaries here? Zhang Wenting and coauthors' *Study on Plant Hardiness and Heat Zones in China* provides an interpretation, which also implies a possible development direction of the scientific research of the discipline.

2011 年 5 期

本期以庆祝风景园林成为一级学科为主题，并得到一直关心风景园林学科发展的陈俊愉、吴良镛、齐康、孟兆祯诸位院士和清华、北林的学科带头人的大力支持，我在此代表杂志社深表感谢。

以宙代为基本单位来观察，地球分为太古宙、元古宙和显生宙。显生宙在经历了古生代（Paleozoic）、中生代（Mesozoic）、新生代（Cenozoic）以后，正处在一个根本性的转折时期，杨锐教授在《风景园林学的机遇与挑战》一文中介绍了Thomas Berry的新时代理论，该理论的真正价值，不在于它提出了一个新名词ecozoic，而在于它揭示了这个新时代的基本特点："这个在过去数千年中自我直接管理的行星，现在在很大程度上通过人类的决定来决定它的未来。"

"可持续发展"已是人类对这个新时代的基本共识，但如何做到，还有巨大分歧，基本可分为两派：一派认为必须加强有组织行为；一派主张尽快建设自组织社会。大体上，认为人类已基本掌握真理的拥护前者，认为人类在真理面前还必须十分谦虚地倾向后者。但不管哪派，都同意必须处理好人与自然的关系，所以这是个以"天人关系"为代表的时代。

ecozoic这个新词并不算太好，生态（eco）是地球自从出现生物以后就存在的现象，不等于"天人关系"，就算未来的人类没搞好，毁灭了自己和相关环境，其他生物可能还会存在，eco还要继续下去。我觉得当下这个时代可以叫作"人生代"（manozoic）。

目前，风景园林身处的人居环境这个"大公司"里的多半董事是以代表人的利益为主的大股东，只有风景园林是代表中小股东（这里比喻自然）利益为主的独立董事，当然，合格的董事都必须以公司整体利益为重。但是，正如欧美一些成熟的大公司已经由独立董事占多数一样，长远来看人居环境这个大公司也会慢慢由独董占多数。由此看来，我们从事的是真正"夺天公之造化"的大业，万一搞不好，真恐为天地所不容。

虽然中国被称为"世界园林之母"，虽然80多年前就有陈植、李菊、程世抚等先生分赴日本、欧美学习landscape architecture，并回国后开课，新中国成立后立即有汪菊渊、吴良镛等先生主持设立了本科专业，这些都与当时世界上本学科的发展基本上同步，但不幸在中国城市化开始起飞的20世纪末反而把风景园林这个学科取消了。现在，国务院学位委员会决定风景园林成为一级学科，消息一出，全行业为之欢欣鼓舞，实乃学科之大幸，行业之大幸，国土之大幸。

当此之际，本刊当更为自觉地认识到所负任务之重，更努力地工作，为学科发展做好服务。谨以此表示本刊对风景园林成为一级学科之庆祝。

Phase 5, 2011

With the theme of landscape architecture becoming the first level discipline, this issue has been greatly supported by Academicians Chen Junyu, Wu Liangyong, Qi Kang, and the academic leaders of Tsinghua University and Beijing Forestry University who have always been concerned about the development of the landscape architecture discipline, and I would like to express our gratitude on behalf of the journal.

By observing in the geological time scale of eon and era, the history of the Earth is divided into Archean, Proterozoic, and Phanerozoic eons. The Phanerozoic eon is now in a fundamental changing period after the Paleozoic, Mesozoic and Cenozoic eras. Professor Yang Rui introduced Thomas Berry's new era theory in his paper *Opportunities and Challenges of Chinese Landscape Architecture Discipline*. The true value of the theory lies not in the coming out of the new word ecozoic, but in that it reveals the basic characteristic of the new era: "The planet was in direct self-management in the past thousands of years and now, to a large extent, its future is determined by human decisions."

"Sustainable development" has been human basic consensus on this new era, but there are vast differences on how to achieve it, which could be divided into two main groups: one group thinks that the organized behavior should be strengthened, while the other advocates building the self-organization society as soon as possible. In general, those thinking that human beings have basically grasped the truth support the former, while those thinking that human should still be very humble in front of the truth favor the latter. So this is an era characterized by the "universe-human relationship".

The new word ecozoic is not good enough, for ecology is the phenomenon existing since the emergence of living beings on the Earth (namely since the Paleozoic Era); even if humans do not do a good job and destroy ourselves in the future, other organisms may still exist, and eco will continue. So I think this era can be called manozoic.

Currently, most directors of the board of the "big company" of human habitat environment which landscape architecture is in are heavy stockholders standing for the interest of human beings, and only landscape architecture is the independent director which represents the medium and minor shareholders (here for nature), and of course qualified board directors should all takes into account the interest of the whole company. As some European and American large companies have independent directors as the majority, in the long run the "big company" of human habitat environment will be gradually dominated by the independent directors. In view of this, we are engaged in the real big cause of "competing for the heavenly-created nature", and the world would not tolerate if we fail.

Although China is known as "mother of world gardens", and although there were masters Chen Zhi, Li Ju, and Cheng Shifu who went to Japan, Europe and the United States to learn landscape architecture 80 years ago and opened courses after returning to China, and Wang Juyuan and Wu Liangyong who established the undergraduate program of landscape architecture after the foundation of new China, which kept in step with the world development of the discipline at that time, unfortunately the landscape architecture discipline was cancelled late last century when the urbanization of Chinese cities accelerated. Now the Academic Degrees Committee of the State Council decides to set landscape architecture as the first level discipline, and the whole industry is elated and inspired by the news. It is a great fortune of the discipline, the industry, and the territory.

Upon this occasion, our journal should be more consciously aware of the big mission, and work harder and better serve the development of the discipline. Hereby our journal would like to congratulate landscape architecture becoming the first level discipline.

2011年6期

2011年我们学科喜事连连，先是风景园林成为一级学科，再是争取设立注册风景园林师制度的工作开始启动，再有就是我们刚刚经历了孙筱祥先生90大寿的盛典。孙先生无愧为我国"风景园林和大地规划"（这是孙先生自己选定的对landscape architecture内涵最全面最正确的翻译）学科的泰斗，也是对我国20世纪风景园林规划设计影响最大的导师、学者、大师。回想起来，20世纪50年代到90年代初期，我国各地的公园建设虽然有地域的区别，但基本做法和风格是很类似的，这可能与当时的举国体制和在全国各地主持公园规划设计的人大多数是孙先生的学生有关，但肯定和孙先生的体系非常给力相关。我建议现代园林史可以好好地研究一下这个问题，如果要给这种流派起一个名称，我提议从这里做起，赶上世界潮流，一改我们过去贬低个人，避讳个人贡献的恶习，采用"筱祥风格（xiaoxiang style）"或"孙主义（sunnism）"，因为没有强势的个人地位，现代社会是建立不起来的。而且，这样积累下去，我们的教师、学生，就可以不再一写东西就"言必称希腊"，一开口就都是洋大师、洋主义。

发展学术，是成为一级学科后我们的最重要任务，其中理论体系探索，是最基础的工作，也是最艰苦、最卓绝、最需要深厚功底的工作。本期刘滨谊的《风景园林学科发展坐标系初探》和李树华的《"天地人三才之道"在风景园林建设实践中的指导作用探解——基于"天地人三才之道"的风景园林设计论研究（一）》，是2篇力图建立具有中国自己文化印记的风景园林基本理论体系的大作，2位教授的努力，为广大中青年学者和师生做出了榜样。中国的风景园林只有经过这一步，才能真正确立自己在世界上的地位。

协同学是系统论的发展，提出自组织和他组织的概念，是协同学最大的贡献之一，也是从科学和哲学的最高层次探索人类现代和未来的重要武器。自组织和他组织，已经成为社会学和生态学的两大基本阵营。方志戎和周建华的《人口、耕地与传统农村聚落自组织——以川西平原林盘聚落体系（1644～1911年）为例》是我刊第一次发表关于自组织问题的文章，发表的目的，就是启发探索方向，拓宽研究思路，提高学术水平。

过去我们刊登过余树勋老主编关于老人公园设计的文章，后来也专题讨论过儿童公园的设计。李佳芯和王云才的《基于女性视角下的风景园林空间分析》又从女性的角度探讨了相关设计的问题，说明我们的学科对人的关怀在同步于国家和社会，逐渐进步。同理，刘志强《基于城市安全的园林规划设计编制框架研究》提出"城乡安全绿地系统规划、行政分区安全绿地系统规划、社区安全园林规划设计"的全面城市安全绿地框架，它比单独的"防灾公园"概念更进一步，我觉得应该引起各级行政部门的重视。

刘家麒先生的《建议积极推广〈中国园林绿化树种区域规划〉》写得非常好，也非常重要，值得全国科研和设计界重视。

Phase 6, 2011

This year our discipline enjoyed joyous events again and again, first was landscape architecture becoming the first level discipline, then the endeavors to establish the registered landscape architect system were initiated, and the next is that we have just experienced Mr. Sun Xiaoxiang's 90th birthday celebration. Mr. Sun deserves the title of the leading authority of the discipline of "landscape architecture and land planning" (this is what Mr. Sun chose as the most comprehensive and correct translation of the connotation of "landscape architecture"), and he was also the most influential mentor, scholar, and master of landscape architecture planning and design in China in the last century. From 1950s to early 1990s, the basic approaches and styles of park constructions all over the country were very similar even though there were geographical differences, which might be related with the nationwide system at the time and also that most of the people in charge of the park planning and design throughout the country were Mr. Sun's students, but surely it was linked with Mr. Sun's powerful academic system. I suggest that the modern landscape architecture history could well study this issue. If giving a name to this school, I propose to start from here to catch up with world trends and change our bad habits of demanding perfection and taboos on showing individual contributions, and take the name of "Xiaoxiang Style" or "Sunnism", since modern society would not be established without strong personal position. And if this kind of accumulation continues, our teachers and students could get out of the habits of mentioning "Greece", foreign masters or doctrines once writing or talking.

The academic development is the most important task after landscape architecture becoming the first level discipline, of which the theoretical exploration is the most fundamental, difficult, extraordinary work and also requires most profound foundation. In this issue, Liu Binyi's paper of *Exploration of the Coordinate System for the Disciplinary Development of Landscape Architecture*, and Li Shuhua's paper of *Trying to Explain the Directive Function of the Theory on Tian, Di, and Ren in the Practice of Landscape Architecture Construction(1)*, are two master pieces trying to establish the fundamental theoretical system of landscape architecture featured with China's own cultural imprint. The two professors' efforts have set an example for young scholars and students. Only after this step could China's landscape architecture truly establish its position in the world.

Synergetics is the development of the systems theory, and its biggest contribution is to bring forward the concepts of self-organization and herter-organization, which is also an important tool to explore the modernity and future of human from the highest level of science and philosophy. Self-organization and herter-organization have become two basic camps of sociology and ecology. Fang Zhirong and Zhou Jiahua's paper of *Population, Cultivated Land and the Self-organization of Traditional Rural Settlement – With the Case of the Linpan Settlement System (1644 – 1911) in Chuanxi Plain* is the first one about the self-organization issue ever published by our journal. Our purpose is to inspire exploration directions, broaden research ideas and improve academic levels.

We published our former chief-editor Yu Shuxun's paper on the design of the elderly's park in the past, and later we also set up the special topic to discuss the design of children's park. Li Jiaxin and Wang Yuncai's paper of *Landscape Space Analysis with the Female Perspective* discusses related design problems from the female viewpoint, and it means that our discipline's care about people advances together with the progress of the nation and the society. Similarly, Liu Zhiqiang's *Study on the Formulation Framework of Landscape Planning and Design Based on Urban Safety* proposes the comprehensive urban safety green space framework with "the safety green space system planning of the urban-rural, administrative region, and community", which goes further than the single "disaster-prevention park" concept and I think it should be paid attention to by administrative departments.

Mr. Liu Jiaqi's *Proposal of Actively Promoting China Landscape Greening Tree Species Regional Planning* is very good, and also very important. The research and design communities of the country should attach importance to it.

2011年7期

发动全中国的优秀单位和人才一起办好本刊,是本杂志社的基本宗旨,除了几所著名大学,我们还想邀请优秀企业参与办刊,这样既利于一线人才参与理论开拓,也利于刊物贴近实践。非常感谢中国城市规划院风景园林所率先承接了本期的主题《滨水空间规划设计》,做出了表率。

也许是随着年龄的增长,积累越来越多,审稿时我越来越多地发现重复前人工作的研究,重蹈前人覆辙的作品,不了解情况乱给前人扣帽子的稿件或说辞,好像前人的经验已经全部归零,什么都有待现在的"新"发现。但前人上千年的经验是靠大量浪费、挫折甚至失败积累下来的,弃之不用岂不是对社会的不负责任?

当下世界的主流意识是什么?资本主义?还不够格!至少世界上还有社会主义国家,在西方也有绿党、社民党、无政府主义、个人主义在批判资本主义,原教旨主义的伊斯兰世界更把资本主义视为恶魔。那么,最容易流行的,或说阻力最小的是什么?不外新潮流、新时尚、新思想、新发明、新造型、新风格、新流派、新主义……一言以蔽之,是唯新主义。只要是新,有几个人敢反对?

在唯新主义面前,老东西只有打败仗的份,所以,不顾历史、抛弃历史,逐渐成为时尚。看来,从西方二元对立的思维出发,新老的对立是无法解决的。

回头看看中国,我们有句老话叫"历久弥新"。何谓历久?当然是历史悠久;何谓弥新?弥者漫也,处处透着新鲜。我生活在广州,广州就有点这种味道,但是还是不如港台,虽然港台也没有做到最好。悠久的文化底蕴,并不是说越老越好,而是说尊重前人的经验和积累,不搞唯新主义。

中国人若放弃讲历久弥新,跟着外国人搞唯新主义,愚蠢至极。大量的经验和成就我们有,别人没有,只有傻瓜才不要这份遗产。当然,如果在我们没有的经验方面,如对待德赛二先生,也硬充好汉,至少也是昏瓜。

轻视历史文化,是资本积累无限贪欲的需要,是劳累、压榨和剥削别人的需要。某种意义上,所谓高雅文化是"磨"出来的,是"闲"出来的。人类的发展速度最终会慢下来,生活方式最终会优雅起来,受资本刺激发展起来的文化,或低俗廉价,或奢华虚伪,最终会受到全人类的鄙视和抛弃。

本期有几篇贴近高雅文化的文章,例如吴肇钊的《〈园冶图释〉——掇山篇初稿择页》,户田芳树的《简洁、柔美的水景设计——GreenPia津南》,韩炳越等的《清清云水谣——北川羌族自治县新县城永昌河景观带规划设计》,也许读者能从中体味到某种历久弥新的味道。

《城市公园——城市生物多样性契机——原生乡土植被覆盖城市指导原则》这篇文章中"自从代表着人类文明的机动交通工具问世以来,人类居住范围日益外扩,早已打破城市固有的界限,从城市近郊的乡村扩张至遥远的自然区域"的描写,让我很恐怖:城乡一体化有可能走向负面,也就是消灭自然。

我的基本设想是:人不要占领整个地球,只给自己划出一小块,例如5%,任由自己的利益来主宰，还可以从中分出一块给那些热心搞硬质景观的人。其余的95%，大约一半用于农业生产，另一半还给自然。这样既为自然留出领地，也避免了动物对人的危害（我始终不相信天人合一）。如果让各种生物都进入城市，人也很危险，这恐怕是作者不敢触碰的问题。

Phase 7, 2011

It is one basic aim of the journal to mobilize excellent people and units of all China to run the magazine well and with good quality. In addition to several prestigious universities, we would also like to invite excellent enterprises to participate in running the journal, which is conducive to the involvement of front-line talents in theoretical development and the journal going closer to practice as well. We are very grateful that the Institute of Landscape Architecture, China Academy of Urban Planning and Design is the first to take over the theme of "Planning and Design of Waterfront space", which sets an example.

It is perhaps because of age and experience accumulation that when reviewing papers, I more often find researches that repeat previous work, works that repeat the previous mistakes, and papers or excuses that stick labels indiscriminately on predecessors without understanding of the situation, which seems that the predecessors' experiences have been zeroed and everything needs to be "newly" found now. However, the predecessors' experiences were accumulated after a great quantity of waste, frustrations and failures. Is it irresponsible to the society if they are abandoned?

What is the mainstream ideology of the present world? The capitalism? It is below the standard! At least the world also has socialist countries, the Green Part, Social Democratic Party, anarchism, and individualism in the West are criticizing the capitalism, and the fundamentalist Islamic world even regards the capitalism as the devil. Then what is the most likely popular, or that with least resistance? No more than the new trend, new fashion, new ideas, new inventions, new forms, new styles, new schools, new doctrine. in a word, the "only new" doctrine. As long as it is new, is there any person who dares to oppose?

In front of the "only new" doctrine, old things only have the fate to be defeated. So it has become the fashion to neglect or abandon the history. It seems, from the perspective of the Western dualistic thinking, the confrontation of old and new can not be solved.

Looking back at China, we have an old saying called "being still fresh after long". What is "after long"? Of course it refers to the long history. What is "being still fresh"? It should mean being fresh from all aspects. Guangzhou, where I live, has a little of this flavor, but it is still not as good as Hong Kong and Taiwan, although Hong Kong and Taiwan have not done their best. It is not to say that the older the profound cultural heritage the better it is, and the point is that the predecessors' experiences and accumulation should be respected and the "only new" doctrine should not be followed.

It would be the stupidest for the Chinese to abandon "being still fresh after long" and follow the foreigners to have the "only new" doctrine. The wealth of experiences and achievements is what we have while the others do not have. And of course, it would be also foolish if we pose as a hero at what we do not have much experience, like the science and technology.

Neglecting history and culture is the need of capital to accumulate infinite greed, and the need to exhaust, oppress and exploit others. In a sense, the so-called high culture comes out of "grinding" and "idleness". Human development will eventually slow down, and the lifestyle will eventually be elegant, while the culture stimulated by capital, vulgar and cheap, or luxurious and hypocritical, will eventually be despised and abandoned by all mankind.

This issue has a few articles close to high culture, for example, Wu Zhaozhao's *On the First Draft on "How to Piles up Rockery" of Drawing to Explain Yuan Ye*, Yoshiki Toda's *Simple and Gentle Waterfront Landscape Design – Green Pia Jinnan*, and Han Bingyue and coauthors' *Ballad of Bright Cloud and Clear Water – The Planning and Design of Yongchang River Landscape Belt in the New Town of Beichuan Qiang Autonomous County*, in which readers might appreciate some taste of "being still fresh after long".

In the paper of *Urban Parks – A Chance for Biodiversity in Cities – Guiding Principles for Cities with Former*

Native Forest Cover, the description that "since the invention of motorized vehicles that represent human civilization, the scope of human habitation has been increasingly expanded, and the inherent boundaries of city has been broken, expanding from city suburbs to villages and even to remote natural areas" terrifies me: the urban-rural integration is likely to go negatively, namely eliminating the nature.

My basic idea is: people do not occupy the entire planet, only to draw a small piece for their own, like 5%, to let their own interests to dominate or even separate one part for those enthusiastic people who engage in hard "landscape", while for the remaining 95%, about half for agricultural production and the other half returned to the nature. This will not only allow for the natural territory, but also avoid the harm of animals on people (I still do not believe in the universe-human integration). People would be in danger if allowing a variety of living things all into the city, which is the issue that the author would not dare to touch.

2011年8期

总体上风景园林的科研水平与相关行业（建筑、规划、艺术、生态、植物、地理等）比较起来还不高，当然我们也不必太自卑，比起一些新兴热门行业，如旅游、城市设计、专业公园等来说，我们还是有些本钱的。

如何提高我们行业的科研水平，需要认真研究，我觉得应从基础打起，严格要求。首先，科学研究既然强调从定性向定量、从模糊向精确的转变，科研人员就要培养对数字的严格和敏感。手边有一篇已发表的《世界优秀月季园赏析》，说日本有个花卉纪念公园，有"7 000多个品种和3 000多株植物"，每株植物代表2个品种？！又说大阪城公园的月季园"占地9 000 480m^2，3 200株月季"，每株月季平均占地4亩半？！这类荒诞的数字也经常见于本刊的来稿中，甚至是我参审的某些博士论文中。对此，虽然编辑也有校对的义务，但重要的是我们的作者群首先要培养自己对数字（包括量纲）敏感的素质，也就是看到数字背后的实际意义。

本期发表了《华南地区人居环境高功效植物景观设计与实践》一文，其中"高功效植物"本来可以做成很有创意的新概念，但作者放弃了，对此我虽然可惜，但也同意，因为整个行业还不具备工作基础来建立"高功效植物"的评定标准，没有巨量的基础研究，此事谈何容易。但是，文章将绿量作为评定高功效植物景观设计的主要指标是大可商榷的，这里有两方面的问题：从自然规律看，过密的林带反而对环境的影响作用下降，对此20世纪中叶西方和苏联都有研究，可参考有关造林的书籍；从社会功效看，绿量最高的绿地肯定是人进不去的，难道园林里真的只准人们进入道路广场等硬质地面？这种文章如果出现在植物学、生态学、林业、园艺中，没有事，但如果出现在风景园林中，就显得片面了，因为我们行业的本质是"人居环境"，我们考虑问题需要比纯自然学科复杂得多，当然也难得多。这里有个行业思维习惯问题，同样需要慢慢培养。

《生态与园林的交融——杭州长桥溪水生态修复公园的启示》一文，编委周为教授级高工的复审意见很有见地，为了不影响所谓科技论文的标准体例，我们将其附在文后发表，以期引起学界的关注。湿地公园的设计路线和机制，的确应该引起各地行政主管部门的注意了。

听说明尼苏达大学把所有设计类的学科都并入了风景园林学院，并改名为设计学院，这个消息带给我的感觉是五味陈杂，不知好坏。西方设计学的兴起，引起许多人的欢呼，并将其推展作为己任。当下设计的主流还是源于对几何的崇拜，对几何形体的美感成为设计者引以为豪的天赋，但是它把对自然很有影响和影响不大的设计混在一起，暗藏着一个巨大的危险，即人类将自己干涉天然物体造型的意图，从只限于周边事物泛化到整个地球。况且，这里我们还不谈资本对当代设计的主导作用。

Phase 8, 2011

The overall level of scientific research of landscape architecture is not high comparing with related industries (architecture, planning, art, ecology, plant studies, geography). However, we should not be much self-abased, for we still have some accumulation comparing with the new hot industries like tourism, urban planning, and specialized park.

How to improve the scientific research level of our industry requires careful study, I think the basis should be the starting point, combining with strict requirements. First, since scientific research emphasizes the transformation from the qualitative to the quantitative and from the vague to the precise, researchers must develop and become strict and sensitive to numbers. There is a published paper *Appreciation of the World's Best Rose Gardens*, saying that there are "over 7 000 species and over 3 000 plants" in a Japanese floral memorial park. Can each plant stand for two species? And the paper also said the rose garden in an Osaka park "covers an area of 9 000 480m^2 and has 3 200 rose plants". Does each rose plant occupy an area of nearly 3 000m^2? Such absurd figures are often seen in the contributed articles of our journal, and even in some doctoral dissertations I examined. In this regard, although the editors have proofreading obligations, it is important that our authors should first develop their quality of being sensitive to numbers (including scales and dimensions), that is, to see the real meaning behind the numbers.

In this issue we published the paper of *The Human Living Environment High Efficient Plant Landscape Design in South China*. The "high efficient plant" could have been develop into a creative new concept, but the authors gave up, which is a pity but I agree with that, since the industry has not had the foundation to establish the assessment criteria for "high efficient plant" and it would not be easy without a huge amount of basic research. However, it might be open to question that the paper takes green volume as the main indicator of high efficient plant landscape design, which could be viewed in two aspects: from the perspective of natural law, too dense forest will impact its environmental role, which had been studied in Western countries and USSR in 1950s and could be referred in related books; from the perspective of social function, people cannot get in the green space with the highest green volume—can people really only be allowed to enter the hard surface ground like roads and squares? The paper would be fine if it appears in botany, ecology, forestry, or horticulture, but it becomes one-sided in landscape architecture, since the nature of our industry is "human living environment", and what we consider is more complicated—and of course more difficult—than pure natural sciences. And it is an issue of industrial thinking habit, which needs to develop gradually.

For the paper of *Perfect Combination of Ecological Restoration and Landscape Architecture – Inspiration from Changqiao Stream Eco-Restoration Park*, our editorial board member Professor Zhou Wei's review comments are really insightful and are published after the paper—not to affect the so-called standard pattern of scientific paper, in order to arouse the concerns of the academic circle. The design route and mechanism of wetland park should indeed be paid attention by the administrative departments.

It is heard that the University of Minnesota merged all design disciplines into the School of Landscape Architecture and renamed it into the College of Design. This news brings me mixed feelings, and I do not know it is good or bad. The rise of the Western Design makes many people cheerful and they regard promoting it as their own responsibility. The mainstream of current design is still from the worship of the geometry, and the beauty of geometric forms becomes the proud talent of designers. But it mixed the designs that have big or little impacts on the nature, and a huge hidden danger is that human beings may generalize their intention to interfere with natural forms of objects from the limited surroundings to the whole globe. Moreover, we have not yet talked about the dominant role of capital in contemporary design.

2011年9期

读2011年8月26日的《老年报》，在《齐奥塞斯库强拆毁掉了罗马尼亚首都》标题下有这样一段话："1971年对朝鲜的访问，才是齐奥赛斯库改造布加勒斯特想法出现的一个重要原因。……平壤如同一个宏伟而又规整的兵营一般，充斥着对领袖金日成的崇拜和传颂。盛大的游行和阅兵，穿着鲜艳衣服的可爱少先队员……，一切秩序井然、整齐划一。这使得奇奥塞斯库对布加勒斯特那些庞杂的具有不同时代背景和风格建筑的混搭状况感到厌恶。奇奥塞斯库决心要改造他的国家和首都，要建立一座干净、洁白、光彩夺目的城市。"

我国很多地方官员和老板，还包括多数技术型专家和一些"学者"，可能和奇奥塞斯库的水平差不多，对待地球的态度属于强烈的"他组织"派。这种人如果只插手城市，倒还罢了，因为城市只占陆地面积的1%多点。如果这种人插手到风景园林，那就太恐怖了，因为借口landscape可以把整个地球的land都"管"起来。

当前这个社会，最聪明的人都去搞金融和IT，遗憾的是这2个行业的人在对待地球的态度上大体也和老板及技术专家相似，比奇奥塞斯库好不到哪里去。于是又回到了这几年常常困扰我头脑的问题：如何评价眼前这个地球上的主流社会和主流价值？

敬畏大自然，常存恐惧心，应该是风景园林行业的第一条道德准则。本期朱育帆和姚玉君的《为了那片青杨（上）——青海原子城国家级爱国主义教育示范基地纪念园景观设计解读》虽然只是上篇，但让我感到了风景园林工作者的一颗良心。放眼看去，这种良心对于人类的未来实在是太重要了，这也是我们这个行业的光荣。此外，该文触及风景设计中的若干内核的关键点，代表着我国风景园林设计理论上了一个大台阶，让我更加坚信中国的LA在未来的一二十年里肯定能够在世界上领袖群伦。

本期主题"智慧景区"，密切结合IT的技术发展和风景区管理工作的需要，了解和抓紧这项工作很有必要。将来肯定需要不少人从事这项工作，而且我国的起步基本与世界同步，我国风景和国外的景观特别在和文化的结合上又有很大不同，所以是大有可为的。然而我还是有些隐隐的担忧：智慧景观很可能成为"他组织"行为的一种强大工具。由此又想到：中国人能否在这方面也对世界作出自己的贡献呢？将人的需要和动植物的需要科学又艺术地相结合，可能是风景园林行业最核心的任务，也是最难的结节。如果不碰触这个结节，我们和城市设计没什么本质区别。

其实，王云才《景观生态化设计与生态设计语言的初步探讨》所介绍的国外在生态设计上的蹒跚步履，看起来是生态设计和艺术设计的关系问题，本质还是人的需要和动植物的需要的关系问题。看看世界上真正有良心的潮流，文章对于我们一些只会口头说说"生态优先"之类口号，实际上对生态一窍不通的设计，提出了严重的警告。

Phase 9, 2011

In the *Senior Citizen Post* of August 26, 2011, the article titled *Nicolae Ceaușescu Destroyed the Romania Capital through Dismantlement* said that "the 1971 visit to DPRK is an important reason that Ceaușescu had the idea to transform Bucharest… Pyongyang is like the grand structured barracks, filled with the worship and eulogy of the leader Kim Il Sung. Grand parade and review of troops, lovely Young Pioneers in bright dresses… everything is orderly and uniform. That made Ceaușescu disgusted with the mixture of complex architectures with various histories and styles in Bucharest. He determined to transform his country and the capital, to build a clean, spotlessly white, dazzling city."

Many local officials and bosses of China, as well as the majority of technical experts and some "scholars", is about at the same level with Ceaușescu, and they all take the "herter-organization" attitude toward the earth. It is tolerable if this kind of people intervene in city, since cities only cover around 1% of the land area. But it would be horrible if they intervene in landscape, since they can "control" the whole land of the earth with the excuse of landscape.

In the current society, the smartest people engage in finance and IT, but unfortunately the people in these two sectors generally have the similar attitude with the bosses and technical experts toward the earth, nor better than Ceaușescu. So it comes back to the question that has often plagued my mind in recent years: how the present mainstream society and mainstream values of the earth could be assessed?

To revere the nature and always keep the fear heart should be the first code of ethics in the landscape architecture industry. Although it is only the first part of Zhu Yu-fan and Yao Yu-jun's paper of *For the Poplars Over There(1) – The Landscape Design of the Memorial Park of National Patriotism Education Base in Qinghai Atomic City*, it makes me feel the conscience of landscape professionals. Looking ahead, this kind of conscience is really important for the future of human beings and is also the glory of our industry. In addition, the paper also touches several key points of the kernel of landscape design and represents the landscape design theories of China leaps a big step forward, which makes me more convinced that Chinese landscape architecture will lead the world in the next one or two decades.

The theme of this issue, " smart famous scenic site", closely integrates the development of IT and the needs of scenic site management, and it is very necessary to understand and grasp the work. It will certainly need a lot of people engaged in this work, China and the world are almost at the same starting point, while Chinese and foreign landscapes are especially quite different, so the work is promising. But I still have some concerns: smart landscape is likely to become a powerful tool of "herter-organization" acts. From this I also thought that: can Chinese also make contributions to the world in this aspect? To combine the needs of human beings and the needs of animals and plants scientifically and artfully might be the most essential task of the landscape architecture industry, and also the most difficult nodule. We will have no essential difference from urban design if we do not touch the nodule.

In fact, the staggering gaits of foreign eco-design introduced in Wang Yun-cai's paper of *Landscape Ecological Design and Eco-design Language* appear to be the relationship between eco-design and art design, but are essentially the relationship between the needs of human beings and the needs of animals and plants. Looking at the real world trend of conscience, the paper made a serious warning on the designs that only have lip service of "eco-priority" but actually know nothing about eco-design.

2011年10期

在宿迁举行的第七届江苏省园艺博览会，反映出江南园林故乡的潮流取向，我特别对世界文化遗产"苏州园林"老家的苏州园感兴趣。显然，苏州在这里力图挣脱传统的束缚，创造出一种新的感觉：既在追赶世界的潮流，又能表现地域的文化和传统。不过，正如李敖批评早期的白话文（例如鲁迅）不成熟一样，苏州的这次努力只能算漫漫长路（白话文运动已经将近100年了）上的一次试探：除了入口采用了一点传统符号（景墙的清水磨砖、梅花漏窗、简化了的冰纹户槅，以及被颠覆到地面的小青瓦屋脊式样的台座等）外，进门后已经彻底"现代化"了。看得出，方案是坐在屋子里画平面图搞出来的，以致连近在咫尺的骆马湖都没有借用，主要精力放在平面的分割和地形的微处理上，然后用些不同材料（包括地被和水面）填充地面，偶尔用些植物或硬体分隔或半分隔一下空间。如果没有明显是后来加进去的几幅"大阿福"跳入眼帘，真的以为是在日本或美国的某些小园地。

显然，在传统文化与现代化的结合上，我们比建筑大哥们落后了许多。如果不看重植物的表现，园林的技术含量比起建筑的确较低。中国园林所值得自傲的主要是文化，不是诗人画家，没有一、二十年的积累，根本做不好中国园林。《园冶》的精华在于"景"，在于"文"和"画"，某些人照抄建筑理论发现的所谓"空间往复"仅仅是园林手法的一种。也许，我们面临的任务比建筑界困难得多，因为建筑的内部空间和功能已经现代化了，而且建筑的内外可以有某种分别表现的自由，正如同样信仰的人可以有不同外貌。而园林的内部空间就是其外部显现，古今中外都挤在这里就出现了深刻的冲突。更困难的是，古代园林所服务的文化内容和生活方式，例如诗词、歌赋、昆曲、燕集、琴、棋、书、画等，已越来越脱离大众。那么，我们就只有彻底西化一条路了吗？

我当然不这么看，主要原因是西方文化不能拯救人类，反而在加速人类的灭亡。前几天教师节，我的一个学生说：老师，我觉得人类正在挖大坑，有些人还在拼命地挖。这个比喻太好了！听到这个比喻，人类真的应该反省一下：我们怎么会走到这一地步？而每个地球人也应当想想，你是在拼命挖？还是在被动挖？还是在想法填这个坑？

中国园林的希望，就在于未来整个人类对现代社会秩序和生活方式的颠覆之上。到那时，现代看来非常强大的若干流行思想会被视为笑谈；而一些现在被嘲笑为迂腐的深刻思想，会再度惠及人类。当然，社会的这样一种"回归"是在更高的一个层次上，探索这个更高，就是我们的任务。

谨以此献给本月召开的中国风景园林学会在学科成为一级学科后的第一次年会。

Phase 10, 2011

The 7th Horticultural Exposition of Jiangsu Province held in Suqian reflected the trend of the hometown of Southern gardens. I am particularly interested in Suzhou Garden from the hometown of the World Cultural Heritage "Suzhou gardens". Obviously, Suzhou tried to create a new sense by breaking the shackles of traditions here: not only catching up the world trend, but also showing the regional culture and traditions. However, as Li Ao criticized that the vernacular Chinese writings (e.g. those of Lu Xun) in the early period was immature, this effort of Suzhou could only be considered as an attempt in the long way (the vernacular Chinese movement has been on for almost 100 years): the inside was totally "modern", except some traditional signs at the entrance (such as scenery wall of water levigation bricks, plum decorative windows, simplified ice form household frameworks, and ground pedestals in Chinese tile ridge styles). It could be seen that the plan was done by sitting in the room and drawing plane figure, with the result that the scenery of the very close Luoma Lake was not borrowed, and the main efforts were focused on the flat segmentation and the micro-processing of terrain, followed by filling the ground with different materials (including groundcovers and water surface) and occasionally separating or half-separating space with some plants or hardware. If without several pieces of "Big Afu" — obviously added later — in sight, it could really have been thought as some small gardens in Japan or the United States.

Clearly we, landscape architecture, are left far behind by architecture in the combination of traditional culture and modernization. If not emphasizing the performance of plants, the technical content of gardens is indeed lower than architecture. It is culture that Chinese gardens are proud of, and it would be impossible to do gardens well if not being poets or painters or without accumulation over one or two decades. The essence of *Yuan Ye* is "scenery", and is "culture" and "art", while the so-called "space shuttle" from those copying the architecture theories is just one of the gardening skills. Perhaps our task is much more difficult than the architecture circle, since the inner space and function of architecture have been modernized, and the inside and the outside of architecture have the freedom to have respective expressions, just like people with the same faith can have different appearances. However, the inner space of a garden is exactly its external display, and the profound conflict comes when the ancient, modern, Chinese, and foreign elements crowd in one place. And it is even more difficult that the cultural content and living styles ancient gardens served, like poetry, Kunqu opera, dinner and gathering, lyre-playing, chess, calligraphy and painting, have become increasingly divorced from the general public. So, is total Westernization the only way we have?

I certainly do not think so, mainly because Western culture can not save mankind, but to accelerate the extinction of mankind. In the Teacher's Day a few days ago, one of my students said: "Sir, I think humans are digging a big pit, and some people dig even harder". Great analogy! Humans should really reflect upon hearing the analogy: how can we come to this point? And every person should also think that, are you digging actively, or digging passively? Or are you trying to fill this hole?

The hope of Chinese landscape architecture lies in the subversion of the social order and lifestyle of the entire mankind in the future. By then, the modern popular — and seemingly powerful — thoughts would be regarded as jokes, while the deep thoughts derided as being antiquated now would again benefit the mankind. Of course, such kind of "return" of the society is in a higher level, and it is our mission to explore this higher level.

I would like to dedicate this to the first annual meeting of Chinese Society of Landscape Architecture after landscape architecture being listed as the first-level discipline.

2011 年 11 期

我国第六次人口普查报告：2010年11月1日零时中国大陆"居住在城镇的人口为665 575 306人，占49.68%；居住在乡村的人口为674 149 546人，占50.32%。同2000年第五次全国人口普查相比，……城镇人口比重上升13.46个百分点"。也就是说，中国的城市化已经走完了一半路程，而且若依然依照每10年城镇人口比重上升13%的速度来发展，不到20年中国城市化就完成了。

这是个令我们行业忧虑的数字。第一，行业当下的"繁荣"所剩时间已经无多；第二，在走完一半路途时，中国风景园林的现代化，或现代风景园林的中国化，还只处于起步阶段，这将意味着中国城市化阶段的风景园林建设成果，大部分都是一些青涩的果实，我们将无颜面对我们的后人。

其实，这城市化后半段对我们行业的资金投入，也将面临问题。《中国房地产报》总编室主任薛晖在其《房地产融资困局》一文中写道："2006年，中国住房按揭贷款已到2.2万亿元，也就是说，开发商与地方政府（土地供应者）当年就分享了百姓今后数十年高达2.2万亿元的财富。这就是金融杠杆的伟力。"从最后一句话来看，薛先生基本对金融资本还是持赞美态度的。虽说对于"寅吃卯粮"的态度，自然经济和资本经济从来不同，但谁都对"寅吃卯粮"的限度没有很好地研究，故而，向后代举债大军还在不断壮大，其中有：各级政府的举债性投资，各种公司的股份和债券，广大个人的举债性消费（按揭仅仅是其中的一部分）等。放眼世界，各国政府无一不用增发钞票来掩盖这个事实，难道我们真想把子子孙孙的"剩余价值"全都用光才算了事？一旦社会实现举债和还贷的平衡，所谓金融杠杆的虚拟光环也将消散，还是得回到"挣多少花多少"的年代，顶多预支一点发展所带来的余额。须知，那时肯定是慢发展社会，其速度还不如现代所谓发达国家，除非我们也去剥削贫穷国家。

显然，如果我们想通过出口来继续行业的繁荣，就必须尽快完成中国风景园林的现代化，拿出既能符合现代社会需求，又具理论前瞻和艺术优势的自己品牌（主义、流派）。希望还是有的：我们的主力院校不再仅仅以紧跟洋人和在洋人那里拿奖为荣，而把自己站在世界前沿作为目标；我们的主力设计院的制度设计，除了关注产值，也要关注人才和产品的素质，同时关注理论的发展；我们的主力企业，特别是上市公司，凭借世界风景园林史上难得的雄厚财力，创造品牌，冲出国门，以源于中国的风景园林在全球的发扬光大为己任，为民族、也为世界的可持续演进作出贡献。

当然，刊物和媒体也要为自己做好定位。虽然为大家找到一面可以共同高举的大旗并不容易，作为中国风景园林学会的学刊，本刊必须为做好自己的本职工作有所筹划，有所作为。更困难的是，要举起和金融资本及其社会文化斗争的旗帜，似乎有唐·吉诃德的举止之嫌。3年前当我在荷兰站在巨大风车下面听着那扇叶呼呼啸叫，才真正体会到古道西风瘦马上的唐·吉诃德的渺小、勇敢和可笑。每思至此，屈原式的悲情便涌上心头：路漫漫其修远兮，吾将上下而求索！

Phase 11, 2011

China's sixth census reports: at 0:00 on November 1, 2010 in China, "the population living in cities and towns is 665,575,306, accounting for 49.68%; population living in rural areas is 674,149,546, accounting for 50.32%. Comparing with the fifth census in 2000… the proportion of urban population increased by 13.46%". In other words, China's urbanization has run half the distance, and if the proportion of urban population still increases by 13% every ten years, urbanization in China will complete in less than two decades.

This is a worrying number for our industry. First, the industry's current "boom" has not much time left; second, upon the half road of urbanization, the modernization of Chinese landscape architecture, or Chinese modern landscape architecture, is still only in its infancy, which means the landscape architecture construction achievements in the urbanization period of China are mostly unripe fruits and we will be discredited in the face of our future generations.

In fact, our industry's capital investment in the latter half of urbanization will also face problems. Xue Hui, Director of the Editorial Department of *China Real Estate Business*, wrote in his article *Real Estate Financing Dilemman* that: "In 2006, China's housing mortgage loans has added up to 2.2 trillion yuan, that is, developers and local governments (land supplier) shared the 2.2 trillion yuan that year, the wealth of people over the next few decades. This is the mighty force of the financial leverage." It seems, from the last sentence, Mr. Xue basically takes the praising attitude toward financial capital. Although the natural economy and capital economy have different attitude toward "eating May's grain in April", no one has well studied the limit or scale. Therefore, the group borrowing from future generations is growing, including: debt investment of local governments of various levels, shares and bonds of various companies, the majority of personal consumption loans (mortgages are only a part), and so on. Looking at the world, governments are all inputting additional money to cover this fact. Do we really want to use up the "residual value" of our grandchildren? Once the society achieves the balance of debt and repayment, the virtual glory of the so-called financial leverage will disappear, and we will get back to the days of "making both ends meet", at most advancing some remaining sum of development. We should know that it will certainly be a slow development society and its speed is not good as modern so-called developed countries, unless we also exploit poor countries.

Obviously, if we want to continue the prosperity of our industry through export, the modernization of Chinese landscape architecture should be completed as soon as possible, and our brand (doctrines, schools) that meets the needs of modern society and has theoretical vision and artistic advantage should come out. We still have the hope: our major institutions are no longer just following foreigners, and they take it as their target to stand in the leading edge of the world; their system design should focus not only value, but also quality of talents and products, and theoretical development as well; our major companies, especially listed companies, with the world's unprecedented strong financial resources, build up brands and go overseas, take the responsibility to carry forward Chinese landscape architecture around the world, and contribute to the sustainable evolution of the nation and the world.

Of course, journals and media should also have their own good positioning. Although it is not easy to find a big flag that we can hold high together, as the academic journal of Chinese Society of Landscape Architecture, we try our best to get prepared and make a difference. And it is even more difficult to raise the flag of fighting against financial capital and its social culture, which is suspected of the Don Quixote-type behavior. Three years ago when I stood under a huge windmill hearing blades whirring in the Netherlands, I really felt the tininess, bravery and funniness of Don Quixote on his lean horse down the worn path. Every time I think about this, the Qu Yuan-style pathos will rise in my heart: the road ahead is long and rugged, but I am determined to explore it myself.

2012年

2007年　2008年　2009年　2010年　2011年　　2013年　2014年　2015年　2016年

2012年1期

这几天，各大媒体的主编们都忙着写"回顾""寄语"一类的东西，除了极少数国家的一些傻乎乎的自由派，大部分表现出一片幽叹和迷茫。其实这个世界的发展轨迹还是很清晰的：金融资本和强权都不应该是人类的主宰，靠剥削来过奢侈或懒惰的生活也是人类的罪恶。人类真的想"永续"生存，只能走改造当前社会的路。而这个主动权，现在的确掌握在中国人手里。能否把这件事做好，绝不是那些只懂"解构"的人能操作的。

前两天翻阅一本"沙发刊物"，厉害，光是赠阅就每期几万本，于是对我们就提出一个问题：奢侈是不是文化？我想，不好一概而论。大原则，耗费大量心智的作品不一定定义为奢侈，而耗费大量资源的奢侈肯定是罪恶。这样，对于过去历史上只有极少数人享受的奢侈，我们也找到了一个宽松以待的口实，因为对地球生态影响不大。

最近一件事对我刺激很大。作为改革开放的地标广州白天鹅宾馆要进行景观改造了，于是聘请了若干世界上著名的设计团队，被选中的团队自然人才济济，各个精英。当问及他们如何理解岭南园林时，回答是：我们计划用1个月的时间来研究岭南园林。如此肤浅的回答，典型地代表了当前世界主义精英们在哲学、历史、艺术、地理、社会学等方面的水平。让他们来观察世界，当然看不清楚。如果任他们扫荡全世界的文化，不知多少人类的积累毁灭在他们手中，不啻是一次世界范围的"文化大革命"。

本期好文较多，有中国风景园林学会2011年会上的发言和5篇优秀论文，有孟兆祯先生在"纪念钱学森诞辰100周年暨风景园林与山水城市学术研讨会"上的主旨报告等。但是我觉得最值得向大家推荐的，是（美）伊丽莎白.K.迈耶的《可持续之美，外观的性能——宣言之一》。由于原文较长，我们将分3期发表。此文的主题极好：美与生态的关系。这是西方大多数所谓现代景观学者一直在打马虎眼的地方，也是其理论体系的致命弱点；它让我们看到，所谓的现代景观理论体系千疮百孔，内部也有很大分歧，真正有水平的"先进"者很少，这对某些人动辄就大肆鼓吹的"美国先进"论，是沉重打击；显然，作者对landscape的理解，绝不是汉字的"景观"，文章尖锐地指出，景观界习惯地不准确地或者胡乱使用词语，是多么地糟糕；作者以西方少有的明智看到，新自由主义的实质，是为资本利益服务，并由此指出：被许多人赞美的麦克哈格的千层饼法，竟然成为开发商榨干地利的工具。对于属于总结的第三部分（将发表在2012年第3期），整个体系还是不错的。遗憾的是，她对中国园林毫无了解，许多问题中国人早就解决了，她还要苦苦探索。

为此，我推荐另一篇很有深度的文章：阴帅可和杜雁的《以境启心　因境成景——〈园冶〉的基础设计思维》。

Phase 1, 2012

These days the chief editors of major media are busy writing something like "reviews" and "messages for the new year", but most of them are expressing their sighs and confusions except some silly liberals of very few countries. In fact, the development path of the world is still very clear: the financial capital and power should not be the dominator of human, and it is also human evil to live a luxurious and lazy life through exploitation. Transforming the current society is the only way if humans really want a "sustainable" living. And this initiative now indeed rests in the hands of the Chinese people. Whether this thing can be done well could not be operated by those who only know "deconstruction".

When I read a "sofa magazine" a couple of days ago, it is terrible to know that tens of thousands of copies of every issue are just complimentary copies, and then there is a question for us: Is luxury culture or not? I think better not to lump together. The general principle is that the work completed with a lot of ideas and efforts is not necessarily defined as a luxury, but the luxury costing a lot of resources is certainly evil. In this way, we also find a loose pretext for the luxury enjoyed by only a handful of people in the past history to enjoy the luxury, because it affected little on the earth's ecology.

Recently there is one thing irritating me a lot. The White Swan Hotel, a symbolic landmark of the reform and opening up in Guangzhou will have a landscape transformation. And a number of world renowned design teams are invited, and the chosen team is naturally a galaxy of talents, all elites. When asked about how they understand the Lingnan garden, they answered that: we plan to spend a month's time to study the Lingnan garden. Such a superficial answer typically shows the current cosmopolitanism elites' level in philosophy, history, art, geography, sociology, etc. There is no doubt that they cannot make it clear when observing the world. If giving full reins to them to wipe out the cultures of the whole world, I do not know how much the human accumulation would be destroyed in their hands, and that would be a world-wide "Cultural Revolution".

There are many outstanding papers in this issue, including the keynote speeches on the 2011 Annual Meeting and five winning papers of the Outstanding Paper Award, Mr. Meng Zhaozhen's keynote report on the Symposium of Commemorating the 100th Anniversary of Qian Xuesen's Birthday and on Landscape Architecture and Shan-shui City, etc. And Elizabeth K. Meyer's paper of *Sustaining Beauty. The Performance of Appearance: A Manifesto* in Three Parts is the most recommended one. As the original paper is long, we will publish in three issues. The subject of the paper is excellent: the relationship between beauty and ecology. This is what most of the so-called Western landscape scholars act dumb at, and also the fatal weak point of their theoretical system. It allows us to see that the so-called theoretical system of modern landscape is scarred and battered and internal divergences are also large, and the real high-level "advanced" ones are rare, which is a heavy blow for those prone to vigorously advocate "American advancement". Obviously, the author's understanding of landscape is never the Chinese word "jingguan", and the article pointed acutely that it is too bad that the landscape industry has the habit to use words inaccurately or randomly. With the wisdom, rare in the West, the author saw that the essence of neo-liberalism is to serve the capital interests, and thus pointed out: the most praised "Layer – cake model" of Ian McHarg unexpectedly becomes the developers' tool to drain land profits. As to the summarizing third part (to be published in March), the entire system is good. Unfortunately, she has no understanding of Chinese gardens, and she is still trying to explore many problems that Chinese people have already solved long ago.

For this reason, I recommend another in-depth paper: Yin Shuaike and Du Yan's paper of *Enlightening Heart from Ambience, and Shaping Landscape by Ambience – On Basic Designing Thoughts of Yuan Ye*.

2012年2期

艺术总是在极简和极繁之间游动。极简主义肯定不是艺术的终点，当下极简主义的流行，是因为人们都去忙着赚钱，剩下一点时间搞搞艺术，当然以简单为主。格式塔主义似乎在探求形式审美的终极，对此我不大相信，认为其结论顶多是形式审美原理的一个部分。如果人类不再忙于征服别人和大自然，专注"心灵"一点，"极繁"就会在某种程度再度兴旺起来。极繁的代表应该是汉字，据说总数达10万多个，康熙字典收录了47035个，而这些字又有从甲骨文到草书的若干种变化，乃至个人风格的变体。在西方，苏美尔的楔形文字总数不过500来个，埃及人的象形文字也不过800来个，到了现代，仅仅26个拉丁字母中还有几个和阿拉伯数字混淆不清，看来西方人在线条造型方面想象力是有限的。

最让我反感的一种现象是一提到西方就毫不思索地在前面加上"先进"二字。当下西方的总体先进的确是事实，除了在精神方面对自由和民主的探求已经成为人类的宝贵资源（这是因为它们很符合资本的需要，至于平等与博爱则基本是虚伪或空谈，因为和资本利益经常冲突），在自然科学方面也明显领先。但总的看来，西方的先进都表现在"为了自己"和"征服自然"两个方面，而在"族类融合""人类延续"乃至"人与天调"方面的思考与实践，比起我们老祖宗来逊色明显。

我有个想法：讲求"线而空"的艺术，能最大程度地保护自然，比之讲求"实体"的艺术，更符合未来"人与天调"的时代要求。为此，我请有兴趣的读者研读一下本期《达其性情，形其哀乐——解读中国传统园林与书法之线条情感审美通感》一文，看看我们有多少人能从对西方体系的一厢情愿的膜拜中解脱出来。当然，复古不是我们的意图，我们还是主张综合中西文化的优势，去开拓人类的未来。

在我的记忆中，以孙筱祥、孟兆祯等先生为代表的老师们对学生设计能力的第一要求，是看到设计图就能凭想象"进入"这个空间，"看到"这个地方将是什么样子，包括视野、地形、线性、坡度、植被、建筑、色调等，并感觉到空间的尺度和特性，以及在游走过程中的变化。我不知道现在还有多少学校是这样教学生的，但可以肯定有大量学生是没有这个功力的。《园林的融合与创新——以菏泽曹州牡丹园改造工程为例》介绍的老园，应该属于缺乏空间想象力的设计。对这种园子的改造，比新建一个要难得多。

广东人有句话："有钱大家赚"，它比起同样产自商业社会而饕餮无度的"我的是我的，你的也是我的"来，先进许多，属于近来在世界上勃兴的"双赢（win-win）"思想。我很高兴地向大家推荐《人与自然的双赢——苏黎世锡尔河滨河地区景观更新规划研究》一文，人类不能把自然全变成属于"我的"。

Phase 2, 2012

The art is always swimming between the minimalism and the extreme complication. The minimalism is certainly not the end of the art, and the current popularity of minimalism is because that people are busy making money and the time left for art is little and the simple things are given priority. The Gestalt doctrine appears in the search for the ultimate form of aesthetics, which I do not quite believe, while I think that its conclusions are at best a part of the formality aesthetic principle. If human beings are no longer busy conquering others or the nature, focusing more on "the spirit", "the extreme complication" will revive to some extent. The representative of extreme complication should be the Chinese characters, said to be in a total of over one hundred thousand, 47,035 ones included in *Kangxi Dictionary*, and the characters also have several variations from the oracle inscriptions to the cursive handwriting and variations from personal style changes. In the West, the number of Sumerian cuneiform is only about 500, the Egyptian hieroglyphics around 800, and to the modern, in the only 26 Latin alphabets, several are even confusing with the Arabic numerals. It appears that Westerners' imagination in the line shape is limited.

A phenomenon that I resent most is describing with the word "advanced" without hesitation when mentioning the West. It is indeed a fact that the West is generally advanced. In addition to that the spiritual pursuit of freedom and democracy has become the valuable human resources (this is because they are in line with the needs of the capital, while equality and fraternity are basically hypocritical or empty talk since they often conflict with capital interests), they are also clearly in the leading place in natural sciences. But in general the advancement of the West is all shown as "for themselves" and "conquering the nature", and they are obviously left far behind of our ancestors in the thinking and practices of "race integration", "human continuation", and "harmony of man with the universe".

I have an idea: the art with emphasis on the "linear and hollow" can maximize the protection of the nature and is more in line with the requirements of the times of "harmony of man with nature" comparing with the art with emphasis on the "entities". To this end, I ask the readers to read through the paper of *Expression from Emotion and Form out of Emotion – Interpretation of the General Aesthetic Emotion of Lines in Traditional Chinese Gardens and Calligraphy*, to see how many people can get rid of the wishful worship of the Western system. Of course, the vintage is not our intention; we still advocate integrating the advantages of Chinese and Western cultures to explore the future of mankind.

In my memory, the principal requirement for students' design ability by teachers represented by masters Sun Xiaoxiang and Meng Zhaozhen is to "enter" the space and "see" the place with imagination when looking at the design plan, including vision, topography, line, slope, vegetation, building, color tone, etc., and feel the scale and characteristics of the space, as well as changes in the visiting process. I do not know how many institutions are teaching students in this way, but certainly a large number of students do not have this skill. The old garden introduced in the paper of *Integrated Design Concept in Landscape Restoration – A Case Study of Caozhou Tree Peony Garden Reconstruction Project in Heze City* should be classified as the design lacking of spatial imagination. The transformation of this kind of garden is more difficult than building a new one.

The Cantonese saying "the money we all can earn" is more advanced than the gluttonous "mine is mine and yours are also mine" produced from the business community, and belongs to the "win-win" thinking recently flourished in the world. I am very pleased to recommend the paper of *The "Win-Win" Between Human and Nature: A Study on Waterfront Renewal of the Sihl Area in Zurich*. The human cannot simply make the whole nature "mine".

2012年3期

刘滨谊教授推荐并组织了本期的主题，还领衔翻译了埃卡特·兰格等人的《视觉景观研究——回顾与展望》一文，这是我企盼多年的回顾文章，译文质量也不错。文章里面谈到数理化的西方景观体系遇到两大难点：1）"迹象表明没有一个自成一体的评估框架。一边是客观主义者或物理学派（基于专家的），另一边是主观主义者或心理学派（基于个人的）"。2）"方法学上很难处理，它涉及所见的景观是来自一个给定的景点，还是看这个实际景点本身（"从四周看此景点"与"从此景点向四周看"的问题）"。依我看，第一个困难是不可克服的，第二个难题本质是个增加计算维度的问题，理论上是可解的，但实操上还是个麻烦事。

景的特性是业界的核心问题之一，也是设计中最重要的问题，其中视觉的权重无疑是最大的。文中可见，国际上对景的特性问题研究在认识论上有了较好的进展，特别是能认识到过去的景观理论存在很大的缺陷，忽视了对动态、多因素（如听觉、嗅觉）、相互关系等的研究。但在具体技术上近年来的进展不大，仅就视觉特性来讲可能还需要等待IT技术的大飞跃。

在讨论主题的名称时，我们遇到了麻烦。Visual是"观"，landscape是"景"，所以visual landscape就是"景、观二字之重合部分"，指景的视觉特性，可是现在landscape自己就把"景观"二字占完了，弄得汉字"视觉景观"在概念上自我缠绕，犯了大忌。最后我们还是抛弃了直译，采用了意译的"景观的视觉性"。关于汉字的"景"和"风景"，正好王其亨和杨锐两位教授各自的团队分别发来了深究其含义的文章，在此一起发表，期望引起业界的深思。

西方为了逻辑的清晰和准确，为了避免模糊和多歧，推出了"非此即彼"的概念模式及其一套思维方法，从而构造了一个基本完美的逻辑体系，这是西方体系能够征服大多数知识分子的基本原因。

但这在现实中实在是很有限的情况，比如非热即冷，可偏偏碰到不冷不热，或者何为热，何为冷的难题。那么尊重现实的中国人，即使非常重视"正名"，也只要求"名实相符"即可，并不要求一定"非此即彼"。有人提出"同心圆"是中国的概念模式，即从概念的核心向外发散，逐渐淡化，对此我大体同意，但不将其绝对化，因为中国人也在符合实际时使用"非此即彼"类型的概念。这样，中国的概念体系就可以存在大量的交集空间。比如风景园林，在现在的西方体系中就很难确定其是工科？理科？农科？文科？可在中国人心中，风景园林指向哪些东西本来不是问题。反过来，真正按西方体系的要求，景观就不符合概念的严谨性，至少，它是主观还是客观？"学界"就从未想清楚过。本期发表了《可持续之美，外观的性能》的最后一部分，整个文章所纠结的核心问题，亦在于此。

Phase 3, 2012

Professor Liu Binyi recommended and organized the thematic papers of this month, and also led to translate Eckart Lange and his coauthors' paper *Visual Landscape Research. Overview and Outlook*, which is the review article that I have looked forward for many years and the quality of translation is also good. The article talked about the two major difficulties the scientific Western landscape system encountered: 1, "In assessing the visual quality of the landscape there is no sign of a single unified assessment framework. There is on the one hand the objectivist or physical paradigm (expert based approach) and on the other the subjectivist or psychological paradigm (lay person based approach)"; 2, "That is methodologically difficult to deal with relates to seeing something from a given viewpoint vs. seeing the actual viewpoint (the problem of 'view-of' and 'view-from')." In my opinion, the first difficulty is insurmountable, while the nature of the second problem is about to increase computing dimension, which is theoretically solvable but a trouble in practical operation.

Landscape character is one of the core concerns of the industry and also the most important part of design, in which the weight of vision is undoubtedly the biggest. It could be seen in the paper that there has been good progress in epistemology for the study of landscape characters around the world, in particular to recognize that there were a lot of deficiencies in the past landscape theories, and research on dynamic, multi-factors (e.g. hearing, smell) and their relations was neglected. However, there was little progress in specific technical development, and the big leap of IT technologies is awaited visual characters and other aspects.

We are in trouble in discussing the name of the topic. "Visual" means "guan", "landscape" means "jing", so "visual landscape" is the "coincidence part of jing and guan" and refers to the visual character of jing. But "landscape" itself was expressed as the phrase "jing guan" in Chinese, making the Chinese characters "shi jue jing guan" self-winding in conception, which committed a taboo. Finally, we abandoned the literal translation, using a paraphrase of the "visual landscape". As to the Chinese character "jing" and "feng jing", we have papers on their connotations from Professors Wang Qiheng and Yang Rui and their teams respectively. They are published together in this issue, in the expectation of arousing the thinking the industry.

For logical clarity and precision, and in order to avoid vague and multi-manifold, the West has introduced the concept of "either this or that" and a set of thinking ways, and thus a basically perfect logic system has been constructed, which is the essential reason why the Western system can capture the minds of the majority of intellectuals.

But this is really very limited in actual circumstances, such as not hot mean cold. How about the problem of being neither cold nor hot, or what is hot or cold? The reality- respecting Chinese just requires the "matching of name and reality", not necessarily for "either this or that", even though they attach great importance to the "rectification of name". It is claimed that the "concentric circles" is a Chinese conceptual model, namely diverging outward from the core of the concept and gradually fading out. Generally I agree to that, but it is not absolute, because the Chinese also use the "either this or that" kind of concept in proper situations. In this way, the Chinese concept system has a lot of intersection space. Landscape, for example, it is difficult to determine whether it is engineering, science, agricultural, or liberal art, in the Western system. But in the hearts of the Chinese people, it is never a problem about what landscape refers to. In turn, with the real requirements of the Western system, landscape is not in line with the rigor of concept. At least, it is subjective or objective? The "academic circle" has never thought about it clearly. In this issue the last part of *Sustaining Beauty. The Performance of Appearance* is published, and the confusing problem of the whole paper is also about this.

2012 年 4 期

上期我们讨论到西方和中国在概念方法上的区别，以及西方逻辑体系的严谨性。细心的人会想到，中国有没有逻辑体系？从《易经》《道德经》等文献可见，中国人很重视"道"，即发展规律，现实的逻辑。中国的逻辑体系是由"道—关—网"体系组成，即"线路—节点—网络"体系，其中："道"是由许多"名"（概念）连接出来的，对于众多的"名"按性质进行分类，最基本的分类就是阴和阳，八卦则是一个分类系统；"关"又称为"门户"，线系统通过"关"连接为网系统，避免了仅仅考虑单线发展的误区。特别重要的是，无论是道、关、网，都要分别主次：对道和关要抓住主要矛盾，讲究"抓牛鼻子"；对网要区分纲和目，讲究"纲举目张"。而这个体系的正确与否，这个体系如何避免混乱，不是由自身的严谨性决定，而是由是否合乎现实决定，即由实践考验来决定。这样的体系，成功避免了西方体系由于人类自大带来的许多烦恼，但也容易存在许多漏洞，会被一些骗子利用。

本期主题"小尺度设计"很有意思。我的理解，这里的"小尺度"其实就是"人的尺度"。正如王向荣教授指出："10公顷的尺度，最容易出效果的设计尺度，但大设计院却没有兴趣做，究其原因还在于经济上的考虑。"我们可以想想，如果没有人的尺度上的精心处理，只有粗糙或单调的大景观安排，这是个人民的环境，还是强权的环境？最近我在从化温泉镇买了一套公寓，准备彻底退休后"诗意地栖居"。这里青山环绕、林木苍翠、空气一级、水质一级，是广州环境质量最好的区域，可惜小区的开发商颇具乡气，于是有些别墅业主就想动手改造，无奈坚决执行开发商命令的保安们一概不允，为的是维持面貌统一的大景观。由此我想到，无论东西方，为什么一些老村、小镇很耐看，很有人情味？因为那里一方面有某种一致的文化氛围笼罩，一方面是每户居民自己动脑筋的群体智慧结晶。而至于统一大景观的场所有哪些能够令我长久徘徊，至今我还想不起来。

谈到文化，本期《云光落茗杯：晚明文人尚茶之风对园林的影响》讲述的是真正的文化，也是真正的学术，当然，这不是用SCI标准所能评价的学术。当然，文化一直是演变的，《两宋园林中方池现象研究》就对广为流传的中国园林不喜欢方塘石洫说提出质疑。其实，当今仅存的古代园林中，方池除文中所说的岭南外，还大量存在于西北西南地区，如成都武侯祠、草堂、罨画池，甚至新繁东湖、新都桂湖，也属于不大规整的方形水池。故对其研究也许能大大改变人们对中国传统的认知。

文化的继承和发展，已是老话题，现在的大问题是具体如何做。《传承·演变——江苏徐州铜山区娇山湖公园规划设计解读》和《重建老城区居民滨河景观体系——记北京东南二环护城河休闲公园设计的思考解题过程》在这方面做出的努力，值得向大家推荐。

Phase 4, 2012

Last month we talked about the differences of the Western and Chinese conceptual approaches, as well as the rigor of the Western logic system. The careful ones could have thought that, is there a Chinese logic system? It could be seen from The Book of *Changes*, *The Tao Te Ching* and related literature, that the Chinese attach great importance to "tao"(road), namely the law of development, the practical logic. The Chinese logic system is formed with the "road-gateway-net" system, namely the "line-node-network" system, in which: "road" comes from the linkage of many "names" (concepts); the various "names" are classified according to their properties, and the most basic classification is yin and yang, while the Eight Diagrams is another kind of classification; "gateway" is also called "door", and lines are linked into nets through "gateways", which avoids the misunderstanding that only considers one-way development. Of particular importance is that, regardless of the road, gateway, or net, the primary and the secondary should be distinguished respectively: for road and gateway, the primary aspects should be seized, which pays attention to "grasping the key link"; for net, the outline and the items should be distinguished, and "a lucid exposition of an outline" is emphasized. Whether the system is correct or not, and how the system avoid confusion, are decided not by the rigor of the system, but by whether it is realistic, namely by the practical test. Such a system successfully avoids the many troubles of the Western system from human arrogance, but there are many loopholes that could be exploited by cheaters.

The current theme of "small scale landscape design" is very interesting. I understand the "small scale" here is actually a "human scale". As Professor Wang Xiangrong pointed: "The 10 ha. scale is the most effective design scale, but large design institutes have no interest to do that, and the reason is for economic considerations." We can think that, if there is no careful handling on the human scale, only rough or monotonous landscape arrangements instead, is this the people's environment, or the environment of the power? Recently I bought an apartment at the hot spring town of Conghua and planned to have a "poetic dwelling" after retirement. The area has green hills, green trees, perfect air, and perfect water, and with best environmental quality in Guangzhou. Unfortunately, the community developer is rather rustic, so some villa owners wanted hands-on transformation, but the security guards resolutely implementing the developer's command did not allow that, in order to maintain the face of a unified landscape. This brings me to think that, regardless of the East and the West, why is some of the old villages and towns very engaging and with human touch? It is because there is a consistent cultural atmosphere and also because it is the group wisdom of all household brains. So far I am unable to call to mind any place that only with a unified landscape could make me linger for a long time.

As to culture, the paper of *Sunrays Passing the Clouds and Dropping on the Teacup: Influences of Scholars' Tea Fondness on Gardens in Late Ming Period* talks about the real culture, and also is real academic, which could not be evaluated with the SCI standards. Of course the culture is always in evolution, and the paper of *Study on Square Pools in Song Dynasty Gardens* queries that Chinese gardens are known to dislike square pools. In fact, in the existing ancient gardens, in addition to Lingnan area mentioned in the paper, there are also many square pools in the Northwest and Southwest of China, such as the Temple of Marquis, the Cottage, and Yanhua Pond in Chengdu, and even Xinfan East Lake and Xindu Gui Lake are also irregular square pools. Therefore the research may be able to significantly change people's awareness of Chinese traditions.

Cultural inheritance and development is an old topic, and now the big question is how to do specifically. The papers of Inheritance and Evolvement—*The Landscape Design of Jiaoshan Lake Park in Tongshan District of Xuzhou, Jiangsu Province* and *Reconstruct the Riverside Landscape System for Old District Residents — The Train of Thought & Solution to the Southeast 2nd Ring Road Moat Leisure Park Design* have made efforts on this, and are worth to recommend to our readers.

2012年5期

最近，国务院常务会议决定将PM2.5纳入空气监测标准，这必将对中国的城市建设带来巨大和深刻的影响。PM2.5是测定大气中直径≤2.5um的悬浮颗粒物含量的指标，这些悬浮颗粒的最大特点是能长久飘浮在空中，其净化的途径只有靠降水洗刷和植物叶面的黏附。从城市规划来讲，也就是除了安排好化工企业，只有扩大绿地和减少汽车的拥塞2条途径。著名大气环境专家、北京大学环境科学与工程学院院长张远航估计：即使按照世卫的标准（与美国标准比稍微松一些），加入PM2.5后，中国空气质量达标的城市将从现在的80%下降到20%。显然，我国现有的城市绿地和交通用地指标下根本不可能解决PM2.5问题，但要让各级官员认识到这个问题，没有十几年的教训是做不到的。那时，我国的城市化已经差不多快完成了，不过"也好"，再大拆大建一次，正好又是一批GDP，中国的城市化过程以及我们行业的繁荣，又被延长了20年。善哉？孽哉？

根据《莫斯科绿地系统规划建设经验研究》一文的介绍，莫斯科1975年规划公共绿地（主要是公园）面积近200km^2，约26.1m^2/人；1999年进一步扩大了绿化用地，总面积增加到350km^2，2011年提出新的城市发展规划，仅自然综合体（不包括公园等）面积规划达到250km^2，人均40m^2。上述数字，肯定不包括我们所谓的附属绿地。所以莫斯科人均绿地面积当在70m^{2}以上，已经超过了广州现在的实际人均城市用地面积。有什么好的解决办法吗？是否可以展开一场大讨论？因为这是百年大计的事情。

在学科地位基本确立之后，我们现在的最大问题是法规，可以说，这是千年大事。一些学者已经开始研讨相关课题，本期《北京风景名胜区调研》、《我国园林绿化行政管理体系调查与分析》和《美国风景园林师执业注册法述评》3篇文章就是证明。

尽量将现代科学方法应用于我们学科，大家从本期目录中就可以感到这股大潮。作为学会的学刊和行业唯一的（双）核心刊物，为大潮摇旗呐喊推波助澜是我们的天然使命，我们非常欢迎这方面的来稿，特别是高水平的来稿。何谓高水平？是指真正理解并正确使用所采用的理论思想和研究方法。现在有股很不好的风潮，没有思想和理解力，乱套用一些"方法"或"模型"就制造"结论"，而这类"研究结果"一旦被刊登又被多次引用，贻害无穷。我刊一定努力避免在这股歪风中成为助纣为虐者。

从学术上深入研究中国园林，达到世界高度，甚至领袖群伦，始终是本刊的核心任务之重者。《从画到园林——苏州古典园林画面转换研究》的作者思维敏锐，抓住了中国造园手法的本质问题，有望从中发展出一整套学说。而《传统园林空间中的留白思想解析》则直触中国传统思维的一个难点，因为要建立合乎现代标准的理论体系，首先要定义园林的"留白"，有兴趣的读者可以试试，就知道这里的难度，也可以发些小文章来讨论。

Phase 5, 2012

Recently, the executive meeting of China State Council has decided to include the PM2.5 into air monitoring standards, which would bring tremendous and profound impact on China urban construction. PM2.5 in the indicator of the content of suspended particulate matter with the diameter ≤ 2.5um, and the biggest feature of the matter is its long floating in the air, and the purification way only relies on rainfall washing and plant leaf surface adhesion. In terms of urban planning, in addition to proper arrangements for analogous chemical companies, expanding green space and reducing vehicle congestion are the only two ways. The well-known atmospheric environmental expert and Dean of the Institute of Environmental Science and Engineering of Peking University Prof. Zhang Yuanhang estimates: according to WHO standards (a little looser than U.S. standards), and adding the PM2.5, the percentage of Chinese cities up to the air quality standards will decrease from 80% to 20%. Obviously, under the existing urban green space and transportation land index, it is impossible for China to solve the PM2.5 problem, but officials at all levels should be made recognize the problem, and it could not succeed within a decade of lessons. At that time, China's urbanization will have been almost completed, but "it may as well", another major demolition and construction will bring another batch of GDP, and China's urbanization process and the prosperity of our industry will also be extended for more than 20 years. Good? Or bad?

According to the introduction of the paper *Research on the Experience of Moscow Green Space System Planning Construction*, the planning public space (mainly parks) of Moscow in 1975 was almost 200km^2, about 26.1m^2 per capita; the green land was further expanded in 1999, with a total area increased to 350km^2; the new urban development plan was raised in 2011, and just the natural complex (not including parks) planning area reaches 250km^2, 40m^2 per capita. The above figures certainly do not include what we call the subsidiary green space. Therefore, Moscow's per capita green area should have been over 70m^2, which has exceeded the real per capita urban land area in Guangzhou. Are there any good solutions? Shall we start a wide-range discussion? Because it is a fundamental task crucial to generations to come.

After the disciplinary status is basically established, our greatest problem is the laws and regulations, which we can say is a millenary event. Some scholars have begun to research on related topics, which is proved by the following three papers: *The Investigation on Beijing Famous Scenic Areas*, *The Survey and Analysis of China's Landscape Administrative Agencies*, and *Research on Landscape Architects Licensure Laws of United States of America*.

Modern scientific methods should be applied at best to our discipline, and you can feel this tide from the table of contents of the current issue. As the journal of the Society and the only double core journal of our industry, it is our natural mission to support and push the tide, and we welcome papers about this aspect, especially high-level papers. What is high level? It refers to the real understanding and the proper use of theoretical ideas and research methods. Now there is a bad agitation that "conclusions" are made by blind application of "methods" and "models", without any thinking or understanding, and such "research findings" once published and repeatedly quoted would cause endless troubles. Our journal always avoids holding a candle to the devil in this unhealthy trend.

To study Chinese landscape architecture in depth academically, and reach the world level and even hold a leadership position is always the important part of the core tasks of our journal. The author of the paper From *Picture to Garden – The Study of Transformation between Traditional Landscape Painting and Suzhou Classical Garden* has a sharp thinking and captures the essence of Chinese gardening practices, and is expected to develop a set of doctrines from it. *The Analysis of the "Blank Space" in Traditional Garden Space* hits the difficult point of Chinese traditional thinking, since the landscape "blank" should be defined first to establish a theoretical system in line with modern standards. The interested readers can have a try and will know the difficulties, and can also contribute some short articles to discuss.

2012年6期

终于拥有了一套还算可以的音响,坐在5.1系统的中央,评品着乐师们对声场的精心处理,我似乎回归到1983年在几乎无人的九寨沟环境中构思规划的心境。偶然起身走出声场去添点茶,突然发现声音变成平面的了,场没有了,境消失了,这种感觉怎么那么熟悉?噢,原来很像身临苏州园林后翻看苏州园林画册的感觉。由此我想到,我们的风景园林艺术设计体系一直建立在二维画面基础上,甚至是基于小孔成像原理的透视图基础上,是否太脱离实际了?是否从一开始就走歪了?现在不是讲"回归原点"吗,如果回归"造境"的原点,我们应该怎么办?

本期的主题"绿道",我的看法还是一种造境,在大地上创造一个集自然和人文美好事物和舒适环境为一体的生活和生态通道网络。广东在开始打造珠三角绿道网络时,出现过将建设绿道视为建设自行车道的倾向,后来注意了生态问题,并贯彻到全省绿道规划中,但由于所谓大部制的行政体制分割,林业系统又单独提出一套建设"生态林带"的规划,这个把人类生活和生物生态2套生境系统分割处理的作法,是否合理,值得其他地方思考。

2011年中国成立了生物多样性保护国家委员会,这是我国首个从国家战略上协调环境维护和可持续发展的机构,是个大好事。在最近举行的第一次会议上,国务院副总理、生物多样性保护国家委员会主席李克强强调:保护生物多样性是保护环境、实现可持续发展的重要任务和基本内容,要按照科学发展的要求,弘扬生态文明理念,加大生物多样性保护力度,推进绿色发展,建设人与自然和谐相处的美好家园。这个理念和我们专业的目标十分符合。我也特别拥护会议"对重要生态功能区、生态环境敏感区和脆弱区等,应划定生态红线,禁止与保护无关的开发活动"的决策,因为我特别担心有些规划工作者妄图"规划整个大地"的宏伟理想是否会加速人类的灭亡。让我有些不解的是,为何不直接成立国家生态委员会,是不是以为生物多样性保护可以替代国家的生态问题?毕竟"重要任务和基本内容"并不等于全部内容。

上面谈了不少"宏大"问题。说到本期文章,为了满足基层工作同事们的要求,我们推荐几篇很"具体"的文章:《当前园林植物景观营造常见问题》谈到了地方上的一些倾向;《北京周边地区绿化树木死亡原因分析及防治对策》不仅是作者在北京周边地区多年实践经验的总结,对各地正在北京轰轰烈烈进行的第八届园博园建设也很有助益。

5月,罗哲文先生告别我们,25年前,我在四川省建委工作时曾陪伴罗先生回故乡宜宾视察,因此得以向罗先生请教许多问题,受益匪浅。正当本期即将截稿之际,又惊闻本刊创刊的重要推动者恩师陈俊愉院士撒手人寰,不胜悲怆。他们一生都在为人类的美好未来而辛勤工作,而眼下的地球,却被日益猖獗的资本和专制所笼罩。此时到天国冷眼观察世界的演变,未必不是对灵魂的安慰。谨以此作为与二位先生的告别语。

Phase 6, 2012

Finally I have a set of fairly good stereo system. Sitting in the center of the 5.1 sound track system, and appraising musicians' careful handling of sound field, I seem to have returned to the state of mind of conceiving planning in the almost no one environment of Jiuzhaigou in 1983. When occasionally getting up out of the sound field to add some tea, I suddenly found the sound become flat, and there were no sound field and the environment disappeared. How does this feel very familiar? Oh yes, it feels very like looking at the album of Suzhou gardens after having been in the gardens personally. Thus, I think, our landscape art design system has been built on the basis of two-dimensional image, or even based on the perspective view from the pinhole imaging principle. Is it too divorced from reality? Or have we been on a wrong way from the very beginning? With the popular talking of "back to the origin", if returning to the origin of "building environment", how should we do?

The theme of the current issue is "greenway", which in my opinion is a kind of building environment, building the living and ecological greenway network integrating natural and humanity good things with the comfortable environment on the earth. When Guangdong Province started to build the Pearl River Delta greenway network, there was a tendency to regard the greenway construction as building bike lanes. Later the ecological problems were noticed and included into the province-wide greenway planning. But because of the segmentation of the administrative departments, the forestry system put forward an independent set of "ecological forest belt" planning, which dealt with the two sets of living environment systems of human living and bio-ecology separately. It is worth the thinking of other places whether it is reasonable.

The Biodiversity Conservation National Committee was established last year in China, and it is great that this is China's first organization to coordinate environment protection and sustainable development from the national strategic level. In the first meeting recently held, Vice Premier and Chairman of the Committee Li Keqiang stressed: the conservation of biological diversity is the important task and basic content of protecting the environment and realizing the sustainable development; we should be in accordance with the requirements of scientific development, promote the concept of ecological civilization, strengthen the biodiversity protection efforts, promote green development and build the beautiful homes of the people in harmony with nature. This idea is very much in line with our professional goals. I particularly support the conference's decision "to define the ecological read line for important ecological functional areas, eco-environment sensitive areas and vulnerable areas, and prohibit all development activities that have nothing to do with conservation", since I am much worried that some planners' grand ideal of vainly attempting to "plan the whole earth" will accelerate the destruction of mankind. I am a little puzzled that, why not establish a national ecology committee directly? Or is it thought that the national ecological problems could be replaced by the protection of biological diversity? After all, the "important task and basic content" does not mean all the contents.

Above are discussions of many "grand" matters. Regarding the articles of this month, in order to meet the requirements of colleagues from the practical levels, we recommend a few very "specific" papers: *Frequently Asked Questions in Current Plant Landscape Construction* talks about some local tendencies; *The Analysis of Causes of landscape Tree Death in Beijing Surrounding Areas and Prevention Measures* is not only the summary of the author's years of practice and experience in Beijing surrounding areas, but also very helpful for the current spectacular constructions of various regions in Beijing for the eighth garden expo.

Mr. Luo Zhewen bid farewell to us in May. 25 years ago when I worked in Sichuan Provincial Construction Committee, I accompanied Mr. Luo to visit his hometown Yibin, and got the chance to seek advice on many questions from him and learned a lot from him. On the occasion of drawing to the deadline of the month's issue, we were shocked and sorrowful that the mentor

Academician Chen Junyu, who contributed greatly to the launch of our journal, also died. They had been working hard throughout their lives for a better future of mankind, but now the earth is shrouded by the increasingly rampant capital and autocracy. It might be a comfort the soul to keep the coo-hearted and objective observation on the evolution of the world from the heaven. We sincerely take this as the farewell words to the two professors.

2012年7期

装修虽是家事，其中有些亦可启发思考。最近买了些橱柜和卫柜，没过多久就发现许多门关不上了，原来是门和门框尺寸严丝合缝，但使用后，材料稍一膨胀就造成这样的后果。想来，现在大厂都是数控机床，问题出在设计图的出图人远离实际。那我们行业有没有这类问题呢？或许还没有凸显，因为大多数建筑木作还主要靠手工，靠师傅的经验，而将来呢？

关于"艺术的永恒题材"，有一个著名的判言是"爱情"，除此之外还有没有？我以为有，那就是"善恶"。风景园林艺术的好处之一是尽量表达"善"，善有善报，所以从事此行业的人，大都得到"长寿报"。然而其中有没有"恶"？也有！例如仅仅为满足眼球的观感去修剪植物、改造地球，又如打着"生态"的旗号而自己对生态一窍不通等。更有设计师，为了自己出名，出尽怪招，不顾人民的实际需要，不顾财富的巨大浪费，北京的"三鸟"就是这方面的典型。当然，只是设计师干不成这种"大事"，还需要为虎作伥者的助阵，看一看谁在围绕着这些东西起舞，就知道为"伥"者有些什么类型了。尽管这类怪物原则上都是"舶来品"，但并不代表人类文化发展的主流，而人性、实际、亲切、友好、关爱的艺术，才是文化涌动潮流的永恒主体。我感到，《停留空间设计与公共生活的开展——扬·盖尔城市公共空间设计理论探析（2）》一文所反映的设计思潮，才代表人类的良知。

所以从行业的长远发展看，行业必须自律，这里还有大量深入和艰苦的工作要作。例如，我们期望风景园林教育的基本规范尽早出笼。从技术层面讲，本期《风景园林制图标准体系构建及应用探析》是所有设计工作者应该读一读的文章，图纸的缺陷和混乱，是设计者自己思维不清晰、不全面或不正确的反映。从管理方面讲，改变我国目前"在注册建造师和工程咨询、工程监理领域依然没有从市政公用行业中分离出来。另一方面，国家行政管理部门对于园林绿化行业属性甚至建设工程类型的认识是不明确的，导致了国家和地方在立法上对园林绿化表述不清，在风景园林行政管理过程产生了一定的混乱"（见《园林绿化行业属性探讨》）的状况，也是一项艰巨的任务。

"遗产"，似乎不应是我们这个拥有久远历史民族的弱项，而现实却常常相反，这个现实又反过来迫使我们在"救急"和"治本"中间不得不先选择救急。本期主题"矿业遗产"由北京大学主持，是抓得比较及时和准确的，具有强烈的前瞻性。阙维民教授为此写了《中国矿业遗产的研究意义与保护展望——兼〈中国园林〉矿业遗产组稿导言》，写得非常之好，我觉得可以成为此后类似文章的范本，希望处于行业高层和尖端的人士都能读一读。

Phase 7, 2012

Although the home decoration is a household thing, some aspects of it can also be thought-provoking. Recently I bought some kitchen and toilet cabinets, and it did not take long to find many doors cannot be closed. The problem is that the doors and frames are almost the same size and slight expansion of materials will cause such consequences. It is supposed that the big manufacturers are now using CNC machine tools and the problem lies in the designers of drawings losing contact reality. Then are there such problems in our industry? They might not have been prominent, since most of the construction woodcrafts are still mainly handmade, relying on the experiences of the qualified workers, but how about in the future?

As to "the eternal theme of art", a famous answer is "love", and are there any others else? I think it should be "good and evil". One of the benefits of the landscape art is to express the "good" to the greatest extent, and good will be rewarded with good, so most of the people engaged in this industry enjoy longevities. However, is there is any "evil"? The answer is yes, such as trimming plants and transforming the earth just for the eyes' perception, or knowing nothing about ecology while holding up the "eco" banners. It is even worse that some designers want to win fame and use many weird tricks, regardless of the actual needs of the people and regardless of the huge waste of wealth. "Beijing National Stadium, National Centre for the Performing Arts, CCTV Building" in Beijing are such bad examples. Of course, only the designers could not accomplish such kind of "big events", and there are also someone to assist the evildoers. We can know such kinds of assisting ones by looking around who are jumping for joy on these "events". Despite that these monsters are principally "exotic", they does not represent the mainstream of development of human culture, while the art of human nature, practicality, warmth, friendliness, love and care are the eternal subject of the surging cultural trend. I think that the design trend reflected in the paper of *Design of Staying Space & Emergence of Public Life – The Study on Jan Gehl's Theory for Public Space Design (Part 2)* represents the conscience of mankind.

So for the long-term development of the industry, the industry must be self-discipline, and there is a lot of deep and hard work to do. For example, we expect the basic norms of landscape architecture education comes out as soon as possible. Technically speaking, the paper of *Research on the Construction and Application of Drawing Standard System in Landscape Architecture* is worth all designers' reading. The defects and chaos of drawings reflect the designers' unclear, incomplete or incorrect thinking. Talking from the management aspect, it is also a daunting task to change the situation that in China "the registered architect system, engineering consulting and project supervision are still not separated from the municipal utility industry. On the other hand, the national administrations do not have a clear understanding of the industry properties and construction engineering types of landscape greening, leading to the unclear expression of landscape greening in national and local regulations and the chaos in the administrative management of landscape architecture." (As in *A Study about Industry Properties of Landscape Greening*)

"Heritage", it seems not the weaknesses of our nation that enjoys a long history, but the reality is often the contrary, and this reality, in turn, forces us to have to first select the emergency between "emergency" and "cure". The current theme of "mining heritage", organized by Peking University, is timely and accurate and of strong forward looking. Professor QUE Weimin wrote the paper of *Research Significance and Conservation Perspective of Mining Heritage in China – Preface of Editing the Papers on Mining Heritage for Chinese Landscape Architecture* for this. It is a very good paper, which could become the model one for future similar papers, and I hope those at the high and strategic level of the industry could read it.

2012年8期

本期封面采用梅花,一是为了配合主题——纪念挚爱梅花的陈俊愉先生,二是在隐隐可预的不善气候逐渐显现之际,期望国人以梅花精神激励自己,以此为立于世界之林之本。

读着一篇篇记述陈先生感人事迹的文章,我深深为他炙热的工作热情,对学生的深切关怀,对科研的高度严谨所感动,更为陈先生的崇高精神力量所折服。特别是陈先生94岁高龄时当着许多老同事、老学生的面,公开承认自己过去一些事做得不对,表现出他的灵魂已经得到彻底净化,我相信陈先生进入天国时的唯一遗憾是还有一些工作未全部完成。

作为一个遗产大国,2012年本刊5期和7期分别以"文化景观"和"矿业遗产"为主题,但由于问题太多太杂,好像意犹未尽,所以本期又组合了"世界遗产与可持续发展"、"西湖世界文化景观"和"矿业遗产"3个专题,总之都围绕着"遗产"进行探讨,展开了比较深入全面的研究,期望能对改善当前我国还很不完善的遗产工作有所贡献。在这里,我要对为组稿付出辛勤劳动的同济大学韩锋教授,北京大学阙维民教授和清华大学杨锐教授致以深深的谢意。特别是杨锐教授的《中国的世界自然与混合遗产保护管理之回顾和展望》一文,把纷乱的问题整理得相当清楚,读此文可以对我国整个遗产工作有个全面清晰的理解,并启发思考。对这方面感兴趣的读者读了文章之后,还可以将想法和建议写给本刊。

一晃4年过去了,一场新的奥运开幕大戏在伦敦上演。伦敦和北京的开幕式各有鲜明特色,相较之下我更喜欢伦敦的一些。北京讲了很多"新与旧",力图告诉人们很多"结论性"的东西;而伦敦着力于"善与恶",给人以更大自由发挥空间:对农业社会"进步"为工业社会如何评价?"善与恶"的鲜明区分是否仅仅是儿童的梦幻?而互联网之父出场的结尾反倒触发我的思考:这个新时代将带来的是另一场"进步",还是"危机"?当然,2次开幕式也有相通之处,北京传递给人们的信息是"和谐",而伦敦想传递给人们的是"爱"。

让伦敦可以骄傲的是她可以拿出近几十年许多影响世界的文化成果展现在奥运开幕式上,可我们呢?别的不说,仅就本来可以让中国人骄傲的风景园林来说,这几十年我们拿出了什么可以影响世界的作品或理论?怎么办?大致有3条路可走:其一,继续向老祖宗求教;其二,追逐全球化的潮流;其三,开辟一条新路。我想这第三条路若要达到新的辉煌,不应仅仅是在前2条路之间进行调和与折中,我们完全可以回到风景园林最初的原点——创造宜人环境,重新审视古今中外的一切,从中发现新的坦途。当然,这是非常艰巨和困难的任务。

Phase 8, 2012

We use plum blossom on the cover of our journal this month, to fit in with the theme—commemorating Mr. Chen Junyu who loved plum blossom, and also to expect that on the occasion of indistinct bad situation gradually emerging the Chinese people could be motivated with the plum blossom to become the leading part of the world.

Reading through the articles accounting the touching stories of Mr. Chen, I am deeply moved by his passion for work, deep concern for students, and high strictness on scientific research, and also impressed by his lofty spiritual power. Especially he admitted some of his past doings were not quite right at the age of 94 at the presence of many old colleagues and old students, showing that his spirit has been thoroughly purified. I believe that the only regret when Mr. Chen enters heaven is that some work is not completed yet.

As a big country of heritages, our journal organized the themes of "cultural heritage" and "mining heritage" in May and July editions respectively. But there are too many and too complex topics to discuss and the two editions of the journal seem not enough, this month we organize the three themes of "World Heritage and sustainable development", "The West Lake as the world cultural landscape heritage", and "mining heritage", which are all discussion about "heritage", with more deeper and comprehensive researches, expected to contribute to the improvement of the incomplete heritage undertakings of China. I hereby express my sincere gratitude to Professor Han Feng of Tongji University, Professor Que Weimin of Peking University and Professor Yang Rui of Tsinghua University for their industrious contributions in organizing the papers. Especially Professor Yang Rui's paper of *Retrospect and Prospect: Protection and Management of World Natural and Mixed Heritage in China* makes the chaotic problems very clear. Reading through it could help have a comprehensive and clear understanding of the heritage undertakings of China and trigger thinking. Readers interested in this topic are welcome to write their ideas and suggestions to our journal after reading the paper.

Fast went four years, and a new opening of the Olympic Games was staged in London. London and Beijing's opening ceremonies have their own distinct characteristics, and in comparison, I prefer the London one. The Beijing one talked a lot about "new and old", trying to tell people a lot of "conclusive" things, while the London one focused on "good and evil", giving people greater freedom and space to play: how to evaluate the agricultural society "advancing" into the industrial society? Is the clear-cut distinction of "good and evil" just children's fantasy? The ending with the appearance of the father of the Internet triggers me thinking: will the new era bring another "progress", or "crisis"? Of course, the two opening ceremonies have it in common that "harmony" was the information Beijing passed to people while London would like express "love" to people.

What London can be proud of is that she can come up with many world-influence cultural achievements in recent decades and show in the Olympic opening ceremony, but can we do? If nothing else, only landscape architecture that the Chinese people could have been proud of, have we got any world-influence works or theory in recent decades? What to do? Generally, there are three choices: first, continue to ask for advice to the ancestors; second, chase the tide of globalization; third, open up a new path. I think if we want to reach a new glory on the third way, it is not enough to just reconcile and compromise between the previous two ways, and we can absolutely return to the initial origin of landscape architecture—create a pleasant environment, re-examine all ancient and modern things and discover a new road out of them. Of course, it is a very arduous and difficult task.

2012年9期

作为科协主管下的学会学刊，本刊理当将科学精神作为核心指导思想，当然，基于本学科的优势，打破科学与艺术，人与自然之间的隔阂，也是我们义不容辞的义务。目前各个学科都以能够在国家自然科学基金项目中拿到项目为荣，而我们学科刚刚被确立为一级学科，由于曾长期被置于不利的地位，在这个基金中得到立项的明显偏少，这是很多学者极为关心，也是非常头痛的事情。《2002—2011年风景园林学科国家自然科学基金项目立项分析》一文及时抓住了这个大事，应该对推进本学科的科研大有助益，也希望其他学者围绕这个问题来稿发表高见。

从本期目录中就可以感到，近来本刊来稿中科研的分量逐渐加重，符合自然科学研究规范文章的比例逐渐增加，这对我们维系"核心期刊"的地位非常重要。但同时，在一线从事实操的同事们提出他们感到越来越难懂，疏远感日增，过去我刊和大家的亲密关系有所变化。作为主编，我也为此发愁。当前我们采取的方法是增加页码，以维持原来比较实际的文章的数量，同时又能增加科研的比例。此外，也希望作者在写作时多从读者方面考虑，让文章尽量容易读懂，叙述简洁清楚。能不能找到更好的途径，切盼大家来信提供良方。

科学为什么能得到大家的信任？因为严格，严于律己。不严格要求的"科研"，是假科研，而假科研是会害人的，发表这样的文章，不是扶助后进，更不是百家争鸣。特别是在当下，一旦发表，就会有人引用，遗患无穷。科研腐败，好人都痛恨，甚至咬牙切齿，为何？就是因为它撼动了一个民族比道德还基础的精神——求真！

求真，就是追求真理。彻底的求真，包括在真正理解和应用一些科学原理和方法的基础上，对其应用范围和深度也提出存疑。我最近在审定一篇文章时表示，至今没有任何主义可以解决一切问题。实际上我最想说的是：我们毫无必要对任何洋人和古人的说法彻底顶礼膜拜。其实，也没有任何一个科学结论可以解决一切问题，包括爱因斯坦的质能公式，更何况作为初级阶段的牛顿定律。在另一篇文章的审批意见中我表示：对系统论，也可以存疑。系统论的黑箱方法，的确大大提高了我们的思考和研究速度，但也颠覆了传统科学对"机制"的追求。正是由于放弃了对"源头"的探索，忽视真实的反应过程（机制）的研究，单纯依赖统计来了解情况和寻找规律，依赖"反馈"来调控，才造成了问题的积累，投入的加大，浪费的"合理"，乃至"涌现"的大量涌现。

在我的理想中，风景园林完全可以成为科学和人性的高度结合点，应该是最有哲学高度，最具科学精神，最合人类良心的学科。当下那些只求显摆自己，玩弄所谓"创意"，藐视人民大众，通过浪费大量资源来攫取个人私利的流行偶像，不过是人类历史长河中的一片垃圾，一簇泡沫。

Phase 9, 2012

As the journal of Chinese Society of Landscape Architecture under the supervision of the Association for Science and Technology, our journal should naturally take the scientific spirit as the core guiding ideology. Of course, it is also our bounden duty to break the barriers between science and art and between man and nature based on the disciplinary advantages. At present various disciplines are proud to be able to get the projects funded by the National Natural Science Foundation. Since the landscape architecture discipline has just been established as the first level discipline, and the discipline has been placed in an unfavorable position, the funded projects in the discipline are obviously less, which is really a headache greatly concerned by many scholars. The paper *Analysis of Accepted and Funded National Natural Science Foundation Projects of China in LA Field from 2002 to 2011* timely captures this major event and it should be helpful for promoting scientific research in this discipline, and we hope other scholars could join the discussion about this question and contribute their invaluable comments.

You can feel from the table of contents of the current Issue that the scientific research papers have been weighing greater and the proportion of papers in line with the natural science research norms has been gradually increasing, which is very essential for us to maintain the position of "core journal". However, our colleagues in the practical frontiers feel more and more difficult to understand the papers, the feeling of estrangement increases continuously, and the close relationship with the journal in the past has changed. As the chief editor, I am also worrying about this. Our current approach is to increase the pages, in order to maintain the original practical papers while at the same time to increase the proportion of scientific research papers. In addition, we also hope the authors could take the readers into account when writing to make the papers easy to read as much as possible and express in a concise and clear way. Please do write us if you could give us better suggestions and solutions for this.

Why can science get people's trust? For its strictness and self-discipline. "Research" not strictly required is fake research, which will be harmful, and these kind of papers are not to help the juniors, nor to let more than a hundred schools of thoughts contend either. Especially in the moment, once published, the papers will be cited, causing consequences of maximum seriousness. Corruption in scientific research is hated by many people, why? It is because it shook up the spirit that is even more fundamental than morality—seeking truth!

Truth-seeking is the pursuit of truth. There are also doubts about the application scope and depth about thorough truth-seeking, including on the basis of fully comprehending and applying certain scientific principles and methods. I recently mentioned when examining a paper that so far no doctrine can solve all the problems. I actually want to say: we have no need to pay homage to what any foreigners or ancients said. In fact, there is no any scientific conclusion that can solve all problems, including Einstein's mass-energy equation, let alone the Newton's law at the initial stage. In the examining comments of another paper I indicated: systems theory can also be in doubt. The black box method of systems theory is indeed greatly improve our speed of thought and study, but also subverts the traditional scientific pursuit of the "mechanism". It is due to giving up the exploration of the "origin", ignoring the study of real reaction process (mechanism), relying solely on statistics to understand the situation and look for law, and depending on "feedback" to regulate, that leads to the accumulation of problems, the enlarged investment, the "reasonable" waste, and the massively occurrences of "emergence".

My ideal landscape architecture can become a highly integrated point of science and humanity, and is the discipline of the highest philosophical level, the most scientific spirit, and most human conscience. The current popular idols that just show off themselves, play with the so-called "creativity", show contempt for the masses of the people, and grab personal interests by wasting a lot of resources are just garbage in bubbles in the human history.

2012年10期

我希望大家都来关注吴人韦先生等人的《走向联合国的"国际风景公约"》一文，同时注意2个时间点：一年以前（2011年）和一年以后（2013年）。

一年前，国际非政府组织"环境法比较国际中心"（简称CIDCE）法国利摩日会议和国际风景园林师联合会（IFLA），分别将关于"国际风景公约"的提案正式提交给联合国可持续发展大会（"里约峰会"）秘书处，这标志着南北发展差别和东西方文化差异的殊途同归，可望成为人类社会迈入风景文明的重要里程碑。这2份文件作为国际社会的重大合作成果，正式启动了"国际风景公约（ILC）"的引擎。

一年后，联合国教科文组织第37届大会将在巴黎召开，届时"国际风景公约"有望成为大会的正式议题。正如文中指出：这将促进人类的共同财富——"风景"作为人类环境品质的可持续发展，还标志着"风景"作为人类对环境的一种认识和行为准则，突破了过去的局限性，扩大为"城市、乡村和郊野范围内的、优秀的、日常的，乃至退化了的风景"，上升为物质形态、人类感受和社会文化之总和。

文中"标志着……东西方文化差异的殊途同归"一句特别引起我的兴趣，我把它看得比"南北发展的差异"还重要。我越来越觉得，人类为了不因内斗而加速自己的毁灭，必须走上世界大同，特别是"东西方文化差异殊途同归"的道路；而为了不因破坏环境而加速自己的毁灭，必须尊天重史，敬人虚己，走上"多元文化"的道路。初看起来，这似乎是个"二律背反"的命题，但经过仔细分析，其实并没有达到如此高度，关键是人类是不是按照当前的"全球化"（即主流受金融资本控制的）轨道来走上世界大同的道路。

从"国际风景公约（提案）"的基本内容来看，这是一个力图摆脱金融资本负面影响的文件，反映的是人类的良知。正因如此，在当前这个世界，它必然在联合国众多文件中处于一个相对比较劣势的位置。这在我国也是一样，听说有人主张我国用"旅游法"来替代"风景名胜法"，当我听到这个信息时，并没有特别惊异，因为这个世界里用华丽的辞藻掩饰金钱私利的现象到处皆是。

我以为世界大同还是走中国文化的道路较好。《人文园林》（总第4期）上施奠东先生在"六一泉"一文讲道：欧阳修自号六一居士，客有问曰，六一，何谓也？居士曰，吾家藏书一万卷，集录三代以来金石遗文一千卷，有琴一张，有棋一局，而常置酒一壶。客曰，是为五一，奈何？居士曰，以吾一翁，老于此五物之间，是岂不为六一乎！

欧阳修的思维方式，代表了达到中国文化高峰的核心思维方式——"物我融一"。以此种思维对待世界的人，绝不会以榨干别人和地球的财富和资源作为自己的生活理想，这正是东西方文化的根本差异。

我现在最关心的问题之一便是，中国的教育何时能摆脱西方体系的羁绊？结论之一是：待中国真正成为世界强国之后。

Phase 10, 2012

I want everyone to pay attention to Mr. Wu Renwei and co-authors' paper *The International Landscape Convention toward the UN*, and to two time points as well: a year ago(2011), and a year later(2013).

A year ago, the international NGO Centre International de Droit Comparé de l'Environement (CIDCE) French Limoges Conference and International Federation of Landscape Architecture (IFLA) respectively submitted the proposals on "International Landscape Convention" officially to the Secretariat of United Nations Conference on Sustainable Development (Rio+20 Summit), which marked the differences between South- North development and East-West cultures achieved the same, and was expected to be an important milestone of mankind into the landscape civilization. These two documents as the significant results of cooperation of the international community officially launched "International Landscape Convention (ILC)" engine.

A year later, the 37th UNESCO General Conference will be held in Paris, when the International Landscape Convention is expected to become the official agenda of the conference. As pointed out in the paper: it will promote the sustainable development of the common wealth of mankind "landscape" as the quality of human environment. It also marks that "landscape" as human's understanding and code of conduct of the environment breaks through the limitations of the past, expands to "city, village and countryside, excellent, day- to-day, and even degraded landscape", and rises into the sum of material forms, human feelings and social cultures.

The sentence in the paper "…marked…the differences between East-West cultures achieved the same" particularly catches my interest, and I put it more important than "the differences between South-North development". I feel more and more that mankind must embark on world harmony to avoid accelerating their destruction due to internal fighting, especially the path of "East-West cultural differences achieve the same"; to avoid accelerating their destruction due to damaging environment, mankind should respect the nature and history, respect others and humble themselves, and embark on the road of "multiculturalism". At first glance, this seems to be a "antinomy" proposition, but after careful analysis, in fact it does not reach such a high level, and the key is if mankind goes on the road of world harmony via the track of current "globalization" (mainly under the control of financial capital).

Seen from the basic contents of "International Landscape Convention (proposal)", it is a file trying to shake off the negative impact of financial capital, which reflects the conscience of mankind. For this reason, in the present world, it is inevitable in a relatively disadvantage position among many documents of the United Nations. It is the same situation in China as well, and I heard someone advocate of using the "Tourism Law" to replace "Scenic Site Law". I was not particularly surprised at this, since the phenomena of using flowery rhetoric to disguise monetary interest are everywhere in this world.

I think it would be better to take the Chinese cultural road toward harmony. It is mentioned in Mr. Shi Diandong's paper "Liuyi Spring", published in *Humanities Landscape* (General Issue 4) that: Mr. Ouyang Xiu titled himself as Liuyi (six ones) Jushi (hermit). When asked what Liuyi (six ones) means, he said: "I have one wan (ten thousand) volumes of books, one thousand volumes of inscriptions since the Xia, Shang, and Zhou Dynasties, one qin (musical instrument similar to zither), one set of go, and often one pot of wine." Then it was questioned those were five ones, why? He said, "I am an old man, aging along with those five ones, were those not six ones?"

Ouyang Xiu's way of thinking represents the core thinking way of the peak of Chinese culture — "the integration of mankind and the material world". Those who treat the world with this kind of thinking will never take draining the wealth and resources of others and the earth as their ideals in life, and it is the core of the fundamental differences between the Eastern and Western cultures.

One of the issues I am most concerned about now is when China's education will get rid of the fetters of the Western system? One of the conclusions is: when China truly becomes a world power.

2012年11期

迄今为止，人类认知真理的途径，除了宗教式的上帝启示之外，就只有2种：严谨的逻辑或实践的检验。我们也讲悟性，但悟出来的东西最后还是得交给实践或逻辑来检验。

现代的教育，已经很久不教逻辑了，好像数学可以解决逻辑思维问题，其实不然。自担任主编以来，我发现大多数理工科学生的文章，形式逻辑一塌糊涂。形式逻辑当然不能替代辩证逻辑，但不讲形式逻辑就如同不懂算数的人想学习高等数学，如何可能？

西方思维的传统主导是理性主义。所谓"理性"，产生于争辩，就是讲道理，而讲道理必须逻辑严密，不能自相矛盾。于是，概念、内涵、外延、分类、推理、结构、范式等思维方法陆续出现，这就造成西方思维越来越严谨，思想越来越沉重，方法越来越细密，眼界越来越狭小的倾向。到了现代，有些人发现这一套方法继续下去已经越来越难，便在解放思想的大旗下，叛离理性，甚至叛离了对真理的追求，陆续出现了后现代、解构主义、感性主义、个人主义等思潮。

中国人则敬天悯人，把自己视为天地中的一分子，中国思维走的是注重整体，注重大局、注重关系、注重绵密的变化，特别是变化的条件和关节的路子，对逻辑的严谨并不苛求。中国文化之所以能够延续数千年，没有出现可能导致民族灭亡的大问题，更重要的是由于以"实事求是"为纲的基本思想方法，正是由此，"实践是检验真理的唯一标准"的思想在中国很容易得到广泛认同。从长远看，中国人不必像西方人那样费力，中国人更容易靠近真理。

所以，我们做研究如果不愿意经受严格枯燥的逻辑训练，就应该走认真了解事实的路子。总不能2条路都不走，只凭自己海阔天空地随想，那不是科学，而是艺术。虽然艺术也是接近真理的一条路子，但其产品是不是真理，最后还是得交给实践或逻辑来检验。

本期的主题"中国风景名胜区制度建立30周年"是以记录这30年的历史事实为主要目的的。历史当然不可能百分百的准确，但是，由于作者们都是事件的亲历者，甚至是相关工作的主要领导者，根据我的经验（我也多少经历过其中的过程），这一组文章的准确性是极高的，可以作为后世了解这段历史的基本材料。其中《国家遗产·集体记忆·文化认同》一文的主旨与其他文章略有不同，主要是介绍联合国教科文组织（UNESCO）提出的"世界遗产"和"普世价值（OUV）"概念的真正含义。我觉得针对此概念，我国有2种倾向：一是用世界遗产的标准审视地域文化，实际上破坏地域文化；一是力图把普世价值并不突出的地域文化申报为世界遗产，不明白世界遗产的意义。我很欣喜地看到文中有"UNESCO所提出的有关OUV的10条并不成熟的标准"这样的话，在国内还很少有人行文达到如此高度！

此外，《景观都市主义的涌现》一文让我们全面地了解了当下风行世界的景观都市主义的历史和基本思想，我在此予以特别推荐。而《美国城市开放空间规划方法的研究进展探析》和《凝视风景——眺望前方与关注当下》二文，互为补充，有助于我们理解国外的主要动态，并与《景观都市主义的涌现》有异曲同工之效。

Phase 11, 2012

So far, in addition to the religious revelation of God, there are only two ways of human cognition of truth: the rigorous logic, or the test of practice. We also talk about the perception, but the things out of perception finally get to the practice or logic to test.

Modern education has not teach logic tor a Iona time, as if mathematics can solve the problem of logical thinking. In fact, after becoming the chief editor, I have found that most of the science and engineering students' papers are a mess in formal logic. Formal logic of course is not a substitute for dialectical logic, but no formal loqic is just like someone not knowing counting wants to get into the higher mathematics, how can it be possible?

Rationalism leads the tradition of Western thinking. So-called "rationalism" comes out of argument, namely reasoning, and reasoning must be logical and not self-contradictory, so the ways of thinking of concept, connotation, denotation, classification, reasoning, structure, and paradigm emerged, which resulted in the tendencies of the Western thinking becoming more and more rigorous, increasingly heavier thinking, increasingly detailed methods, and increasingly narrower vision. In -the modern times, some people found that this set has become increasingly difficult to continue, and they betrayed the rationalism, and even the pursuit of truth, under the banner of the emancipation of the mind, and the post-modernism, deconstructivism, sensual doctrine, and individualism emerged one after another.

However, the Chinese people respect the nature and regard themselves as part of the heaven and earth, and the Chinese thinking pays attention to the overall, focuses on the overall situation, focuses on the relationship, focuses on the detailed changes, especially the conditions of changes and ways of joints, not demanding in the rigorous logic. That Chinese culture are able to continue for thousands of years and there is no big problem that may cause the nation to perish more lies in taking the basic idea of "seeking truth from facts" as the key links, and it is thus that the idea of "practice is the sole criterion for testing truth" is easily widely recognized. On a long view, the Chinese people can get close to the truth more easily without the strenuousness of the Western people.

So, if we do not want to undergo the strict and boring logical training, we should take the approach of a good understanding of the fact when doing research. It is impossible not to take either ways but just random thoughts, and then it is not science, art instead. Although the art is also a way to the truth, whether its product is the truth or not has to be tested via practice or logic.

This issue's theme "the 30" Anniversary of the Scenics Areas System of China" aims to record the 30 years' history. History of course cannot be one hundred percent accurate. Since the authors witnessed the events, or even were the main leaders of the work, according to my experience (I more or less have experienced the process), this series of papers have extremely high accuracy and could be the basic materials for future generations to understand this history. Among them, the paper of National Heritage, Collective Memory and Cultural Identity has a slightly different theme, mainly introducing the real meaning of the concepts of "world heritage" and "outstanding universal value (OUV)" proposed by UNESCO. I think there are two tendencies about the concepts in China: one is to examine the regional culture with the world heritage criteria, but in fact undermine the regional culture; one is to try to declare the regional culture with no obvious OUV as world heritage, not knowing the meaning of world heritage. I am very pleased to read the words in the paper like "10 immature standards of OUV proposed by UNESCO", and domestic papers have seldom reachedsuch a high level.

In addition, The Emergence of Landscape Urbanismlets let us have a comprehensive understanding of the history and basic idea of the world's current popular landscape urbanism, and I particularly recommend it. The papers of Analysis of the Process of Urban Open Space Planning Approaches in USA and Viewing Landscapes— The Look-out's and the Look-at's complement each other for us to understanding foreign trends, and have a similar effect with the above-mentioned paper The Emergence of Landscape Urbanismlets .

2012年12期

十八大报告将生态文明建设提升到战略层面，与经济建设、政治建设、文化建设、社会建设并列，构成中国特色社会主义事业"五位一体"的总体布局。党章中对此作了这样的阐述："树立尊重自然、顺应自然、保护自然的生态文明理念，坚持节约资源和保护环境的基本国策，坚持节约优先、保护优先、自然恢复为主的方针，坚持生产发展、生活富裕、生态良好的文明发展道路。"这是中华民族永续发展的大好事，也是我们行业的福音。

建设和维护生态良好、环境美丽、生活舒适的室外人居环境，是风景园林行业的宗旨。建设这种环境是走尊重自然、顺应自然和保护自然的路线，还是一味追求人造路线，在学科内部还是有分歧的。2008年我在美国看到一本科普杂志，主题是鼓吹未来的"绿色工厂"：一座数十层的钢架玻璃大楼，正方形底边大约30m长，每层都是农场，还若有其事地计算了这个大楼的年产量。看了我就发笑：除了最上面的一层，下面（除了边缘一小圈）哪有足够的阳光？没阳光哪里有光合作用？用光伏发电带动电灯照明吗？为何不在安装光电电池的地方直接种东西？反正老天（太阳）只给地面那么一点能量，你建100层也增加不了，除非建3000多层，高到积雨云之上，躲开了阴云的遮蔽，但这么做可能危害更大，因为影响了大气环流。

近来看到一些学生的参赛作品，类似的"人定胜天"式畅想还真不少。学生是无罪的，我怕的是得到老师们的鼓励。生态和景观不一样，是自然规律主持，不懂基本的自然规律，想靠任意狂想来解决生态问题，让我想起来1958年"大跃进"时的流行口号"人有多大胆，地有多大产！"这种疯狂思想的恶果，是经历了这段历史的人有目共睹的。

2012年是我国明末的造园家和理论家计成诞辰430周年，而大约380年前，计成的伟大著作《园冶》问世。为了纪念风景园林史上的这个重要事件，中国风景园林学会和武汉大学共同主办了一场国际研讨会。会议办得很好，学术性很强，也很有收获。会议将出版论文全集，本刊也将分2期把会议上最精彩的文章发表出来，以尽早与读者分享。

380年前，那是个什么时代？正是明末崇祯年代，一方面政治腐败，一方面经济活跃。近读吴梅村的《张南垣传》，十分惊异，张南垣也是这个时代的人，每年竟然有数十家来求他造园，这种状态甚至可能延续到康熙初年。在《园冶》里我们也可以感到，当时的园林爱好者可谓成群结队。如果不是中国人窝里斗，能有些改革思想，一直发展下来的中国现在会是什么样子？有人说历史不能假设，不能假设还有什么思想的空间？不能假设研究历史对将来还有什么用处？所以，我热切盼望我们的读者，不论在自然科学还是在人文学术领域，不论对中对西，对古对今，不要迷信什么名言语录，一切都通过脑子想一想，再作判断，这样就能发掘出人类历史上真正的思想家和好人，找到我们应该尊重的人和事。如果问我，我的经验是：拒绝一切蛊惑。

Phase 12, 2012

The report on the 18th National Congress of CPC Eighteen elevated the ecological civilization construction up to the strategic level, side by side with the economic construction, political construction, cultural construction, and social construction, forming the "five in one" overall layout of the socialist cause with Chinese characteristics. The CPC constitution elaborates as this: "… to foster the ecological civilization idea of respect for the nature, harmony with the nature, and protecting the nature, to adhere to the basic state policy of conserving resources and protecting the environment, to adhere to the principles of giving priority to thrift, conservation, and natural recovery, and to adhere to the civilized development road with production and development, an affluent life and sound ecology." This is a very good thing for the sustainable development of the Chinese nation, and also good news for our landscape architecture industry.

The aim of the landscape architecture industry is to build and maintain the ecologically good, beautiful, and comfortable outdoor habitat environment. Building such an environment, to take the route of respect for the nature, harmony with the nature and protecting the nature, or to take the route of the blind pursuit of artificial effects, there are disagreements inside the discipline itself. The theme of a popular science magazine I read in the United States in 2008 was advocating the future "green factory": a steel and glass building with dozens of floors, a square bottom with each edge about 30m long, all floors are farms, and the annual production of this building is also calculated as if it is being. It really gave me a laugh: apart from the top layer, how can there be enough sunlight at the following layers (except the narrow edge around)? How can there be photosynthesis without sunlight? Is the electric lighting driven by the photovoltaic power generation? Why not directly grow plants at the place to install photovoltaic cells? Anyway, God (the sun) only gives the ground such amount of energy, and you cannot increase it even if you build 100 floors, unless you build over 3000 floors, higher above the cumulonimbus and out of the dark clouds shadowing, but doing so may be more harmful because of the influence of atmospheric circulation.

Recently I have seen some students' competition works, and there are really a lot of "man-can-conquer-nature" imaginations. The students are innocent, and what I am afraid of is if they are encouraged by their teachers. Different from landscape architecture, ecology is managed by the natural laws. Not understanding the basic laws of nature and seeking any fantasy to solve ecological problems reminds me of the consequences of the crazy thinking of "more guts you have, more harvests you get", the popular slogan in the "Great Leap Forward" movement in 1958, which is obvious to people who experienced it.

This year is the 430th birthday anniversary of Ji Cheng, the famous gardener and theories in the late Ming Dynasty, while about 380 years ago Ji Cheng's masterpiece *Yuan Ye* (Craft of Gardens) came out. To commemorate this important event in the history of landscape architecture, CHSLA and Wuhan University co-sponsored an international seminar. It was a good meeting, greatly academic and rewarding. The complete collection of papers of the seminar will be published, and our journal will publish some most wonderful articles of them in two issues to share with readers.

380 years ago, what was the times? It was under Emperor Chongzhen's reign in the late Ming Dynasty, when there were political corruptions and also vigorous economy. When reading Wu Meicun's *Zhang Nanyuan's Biography recently*, I am very surprised that Zhang Nanyuan was also at this period, and every year dozens of families asked him to design gardens, which even continued into the early years of the reign of Emperor Kangxi in the Qing Dynasty. We can feel in *Yuan Ye* that garden enthusiasts could be described as in droves. If Chinese were not in internal fighting and were somehow reform-minded, how well could China develop up to now? It is said that history cannot be assumed, but could there be any space of thinking if without assumptions? What is the use of history study for the future if without assumptions? Therefore, I fervently hope

that our readers, whether in the natural sciences or in the humanities academic fields, whether to China or to the West, to the ancient or the present, do not have any superstitions upon famous quotations, and think over all things before making a judgment, so that we would be able to discover the real thinkers and good men in the history of mankind, and find the people and things that we should respect. If you ask me, my experience is: reject all demagogues.

2013年

2007年　2008年　2009年　2010年　2011年　2012年　　2014年　2015年　2016年

2013年1期

2012年11月18日，住房和城乡建设部以建城〔2012〕166号印发了《关于促进城市园林绿化事业健康发展的指导意见》。文件开宗明义说明，这是"为全面贯彻落实党的十八大精神，进一步深入落实科学发展观，大力推进生态文明建设，加强城市园林绿化规划设计、建设和管理，促进城市园林绿化事业健康、可持续发展"。并明确了城市园林绿化的性质和功能是"为城市居民提供公共服务的社会公益事业和民生工程，承担着生态环保、休闲游憩、景观营造、文化传承、科普教育、防灾避险等多种功能"。我认为，文件讲得很对，很好，而且，对当前城市经营、业绩工程、面子项目、违背科学、领导意志、铺张浪费、单纯景观、玩弄理念、忽悠群众等现象中的不良倾向，开出了药方，其中绝大部分内容，对于风景园林行业中的非事业领域也有指导意义。

现在，人们在谈论"顶层设计"。如何把风景园林的顶层设计做好，既是个体制问题，也是个科学问题，虽然有难度，但比起整个改革开放来说，还是容易得多，关键在于是否真正尊重科学。对此，本刊理当主动担起重任，今后更应注重理论体系的科学性、完整性建设，强化理论与实践工作的结合，主动关注现实中的问题，认真挖掘深层次的根本性问题，为全行业的科学化、法制化与思想、艺术的完美结合，做出贡献。

2012年12月1日中国航天员中心宣布：我国首次受控生态生保集成实验取得圆满成功。2名参试乘员在密闭实验舱内进行了为期30天的科学实验，利用种植蔬菜的光合作用，一方面为乘员提供所需氧气，同时净化他们呼出的CO_2，完全实现大气的"自给自足"。其中我最注意的是种菜面积为$36m^2$，折合每人$18m^2$，这和过去我们常用的温带地区一个人呼吸需要$10m^2$左右的林地或$25m^2$的草地的数据是基本相符的。但考虑到城市还需要平衡5~10倍的各种有机燃料的耗氧，所以城市里这点绿地是远远不够的，但也不能指望城市氧气自给自足。这个指标，应该是城市建设顶层设计的最基本数据。10年前我曾写道：假设我国在城市化程度达50%时，由于形势所迫，决定城市的人均绿地（主要用于建设组团隔离带和新鲜空气通道）和道路指标（为了发展民族汽车工业和推动国民经济增长的需要）各增加$10m^2$，设这部分土地的2/3需由拆迁解决，拆迁地区的容积率为3。又设此时我国的总人口为14亿，则总的拆迁量为280亿m^2（见《城市开敞空间规划的生态机理研究》，本刊2001年第4、5期》）。现在，仅这些建筑的基本造价就在30万亿元以上，大约等于全体中国人一年所创造的GDP。所以建议我国政府对此予以高度重视，组织力量进行重点研究。

Phase 1, 2013

On Nov.18 of 2012, Chinese Ministry of Housing and Urban-Rural Construction issued the document ([2012] No.166) "*Guidance on Promoting the Healthy Development of Urban Landscaping* ". The document comes straight to the point that it means to "fully implement the spirit of CPC's 18th National Congress, further implement the concept of scientific development, vigorously promote the construction of ecological civilization, strengthen the planning and design, construction and management of urban landscaping, and promote the healthy, sustainable development of urban landscaping", and makes clear that the nature and function of urban landscaping is "the social public welfare and livelihood project to provide public services for city residents, and shoulders the multiple functions of eco-environmental protection, recreation, landscape construction, cultural heritage, science education, and disaster prevention". I think that the document is good and right, and writes out a prescription for the current city operation, vanity projects, image projects, violation of scientific laws, leaders' dominance, extravagance and waste, pure landscape, playing with concepts, cheating the masses, and other adverse tendency, and the majority of the document's content is of guidance value for the business sectors of the landscape architecture industry.

Currently people are talking about the top-level design. How to make a good top-level design is a system problem, and also a scientific problem. Although difficult, it is much easier compared to the reform and opening up as a whole, and the key is whether science is really respected or not. As to this, our journal should take the initiative to take the task, and should focus more on the scientific and integral construction of the theoretical system, strengthen the combination of theory and practice, be concerned about the reality problems actively, seriously dig deep-seated fundamental problems, and contribute to the perfect integration of science, legislation, thoughts and art.

The Astronaut Center of China announced on Dec.1of 2012 that China's first controlled integrated ecological life protection experiment achieved a great success. Two participants conducted a 30-day scientific experiment in an enclosed cabin, using the photosynthesis of growing vegetables to provide necessary oxygen for the crew and purify their exhaled carbon dioxide and fully realized the "self-sufficiency" of air. I note that the growing vegetables area is $36m^2$, equivalent to $18m^2$ per person, which is basically consistent with the past data of $10m^2$ of forest or $25m^2$ grassland for one person's breathing requirement in the temperate regions. Taking into account the city also needs to balance the 5~10 times oxygen consumption by a variety organic fuels, this volume of green space is far not enough, and it should also not be expected to be self-sufficient in urban oxygen. This indicator should be the most basic data of the top level design of urban construction. Ten years ago, I wrote: supposing when the degree of urbanization in China reaches 50%, the city's per capita green space (mainly for grouped isolation belts and fresh air channels) and road indicators (to meet the needs of developing national automotive industry and promoting the growth of national economy) have to be increased by $10m^2$ respectively due to the pressing situation, and 2/3 of this part of land are solved by housing demolition and relocation, then the floor area ratio in the demolition area is 3. And also supposing China's total population is 1.4 billion, then the total amount of demolition is 28 billion m^2(*Study of the Ecological Mechanism of Urban Open Space Planning, Chinese Landscape Architecture*, Issue 4-5, 2001). Now just the basic cost of these buildings are more than 30 trillion yuan, approximately the one year GDP of the entire Chinese people. So we recommend that Chinese government should attach great importance to this and put great efforts into the study of it.

2013 年 2 期

我有个预感，不知对不对。我们学科在一派大好形势中，泥沙俱下，一股将会最终毁掉我们学科的浊流正在从暗处走向公开，即完全不顾自然规律和社会财富分配，靠大肆耗费民脂民膏和狂妄地改造自然来制造景观。山东的一个项目，计划在一片平坦滩涂上人工堆出一个100多米高的大山。广东则反其道而行之，硬要在一座分水岭挖出一个"比杭州西湖还大"的人工湖。除了有官员作后盾，这些工程的设计者只能用"无知者无畏"视之。

2012年12月起，网上突然爆发了园林广告热潮，通过百度，这些广告可以渗入许多网页，其中有一个设计教学软件的广告最为显眼，吹嘘30天就可以教会你景观设计。气愤之余，回头冷静地想一想，如果我们学科的规划只是些斑块廊道，我们的设计只是些平面构成，我们的材料只是些砂石水泥和几十种植物，然后再学会点"创意"和"营销"手腕，就可以到处炒作，30天当个景观设计师也不是不可能。而这种设计师是在建设美丽家园，还是在浪费和毁害地球和社会资源？鬼才知道。这说明，我们必须加快注册风景园林师制度的建设，赶快制定专业的基本要求，加快推进教学水准评估的进程。

2012年欧洲景观协会主席来访，我问他，他们协会是否想规划整个欧洲大地，他先说"是的"，但似乎马上意识到我问题的深意，补充说："我们和建筑师规划的不同是我们主张多学科联合规划。"老实说，让外行"专家"做规划当然很危险，即使单靠少数"大家"来做特大尺度的规划，也会很危险，因为没有一个人能够拥有所需的自然、工程、艺术、经济和社会的全部知识和天分。在人类整个知识水平还相当有限的前提下，较好的方法，一是多学科的联合；二是尽量缩小人类干预自然的尺度。

以这种观念审视本期主题"棕地修复"，我猛然感悟到这是我们学科20世纪最重要的贡献，是比"景观都市主义"重要得多的学科进展。为了尽量缩小人类伸向自然的手臂的长度，充分利用已被人们废弃的土地是人类的必经之路。景观都市主义是一柄双刃剑，既可以用来把自然请进城市，也可以用来把城市规划设计的一套伸向自然。拿中国现行的70年建筑寿命计算，如果不实行棕地修复，1 000年后，城市和各种建设用地的占地可能达到地面的1/3！这是好事还是恐怖的前景，取决于人类那时是否真的成为上帝，即掌握了宇宙的一切奥秘。而以"一切都是为我所用"为基本理念的自私人类果真能够负担起上帝的职责吗？这是最直截了当的哲学根本问题。所以，若干年后，棕地修复可能是风景园林行业最主要的工作。由于我们还处于新兴国家的地位，在国内，人们还不太注意棕地修复，及早唤起人们对未来发展的注意，正是学术刊物的主要标志。

Phase 2, 2013

I do not know if I have the premonition right: while our discipline is in a good situation and the bad mix with the good, a burst of turbidity current that will eventually destroy our discipline is going public out of the dark, which is producing landscape by wantonly consuming the hard-won possessions of the people and arrogantly transforming the nature in total disregard of the natural laws or the distribution of social wealth. One Project of Shandong Province plans to pile up an artificial 100-meter-high mountain on the flat beach, while Guangdong Province goes in an exactly opposite way by digging on a watershed to create an artificial lake "larger than the West Lake in Hangzhou". In addition to the officials' backing, the designers of these projects can only be viewed as "the dauntless ignorant".

Since December 2012, landscape architecture advertising suddenly boomed online, and the advertising could penetrate into many pages through baidu.com. A design education software advertising was the most prominent one, boasting to enable you to design landscape in 30 days. But thinking after anger, if the planning of our discipline is just some patches and corridors, our design is just plain, our materials are just gravel, cement and dozens of species of plants, together with some tricks of "creativeness" and "marketing", then the hype can be everywhere and being a landscape architect after 30 days is not impossible. However, are this kind of designers building the beautiful home or wasting and ruining the earth and social resources? God knows. This shows that we must speed up the establishment of the registered landscape architects system, quickly develop basic professional requirements, and accelerate the process of teaching standards assessment.

When the European Landscape Association President visited us in the year of 2012, I asked him whether their association wanted to plan for the whole European land, he first said "Yes", but it seemed that he immediately realized the profound meaning in my question, and he added: "the difference between us and architecture planning is that we advocate multi-disciplinary joint planning". Honestly, planning by layman "experts" is very dangerous, and it is also very dangerous to make mega-scale planning only by a small number of "masters", since no one can have all the required knowledge and talent of nature, engineering, arts, economy and society. Under the premise that the level human knowledge is quite limited, the better approach is to joint multi-disciplines or to minimize scale of human intervention in the nature.

Examining the current issue's theme "brownfield transformation" with this concept, I suddenly came to realize that this was our discipline's most important contribution in the 20th century, the scientific progress that is much more important than "landscape urbanism". To minimize the human intervention in the nature, making full use of the abandoned land is the only way of mankind. Landscape urbanism is just a double-edged sword, which can be used to invite the nature into city or stretch the set of urban planning and design into the nature. If calculating with China's current 70-year building life, without brownfield transformation, the area of cities and constructions will occupy one third of the total land of the earth in 1000 years. Whether this is a good thing or a dreadful prospect, depends on whether the human would really become God and master all the mysteries of the universe. Can the selfish humankind with "all for my use" as the basic concept really shoulder the responsibilities of the God? This is the most straightforward fundamental question of philosophy. So, a few years later, brownfield transformation would probably be the most important work of the landscape architecture industry. In China that is in the status of an emerging country, less attention is paid to brownfield transformation, and it is the main symbol of the academic journals to evoke early attention to future developments.

2013年3期

如果说"棕地修复"是风景园林学科对远景人类的重要贡献（参见2013年2期的"主编心语"），本期主题"绿色基础设施"概念则是20世纪风景园林学科对当代人类的重要贡献。

中国的"风景"和"园林"概念，都源于对人与自然同体关系的理解，即使改造自然，它对自然的"手术"从来不狠，强调的是尊重"本性"，依顺"自然"（这里，本性和自然是同义词），也就是顺从于"道"。这种理念不妄图掌握终极真理，不想当上帝，也不必需现代科学的体系支撑。实际上它认为"道"也是变化的，当然这种变化的尺度极大，时间极长，所谓"天不变，道亦不变"，正是以反叙述的方式阐述的相对论。我想，这就是李约瑟命题"为什么现代科学，伽利略时代这个新的实验的哲学，与早期的皇家学会，只发生于欧洲文明而不产生于中国"的解答。

而LA的概念，是将人的头脑中对真理、秩序、美观等的理解加之于land的技术，是基于对人的能力的信仰，这种能力是上帝赋予人类的特权。正是基于这种信念，"天赋人权"的理念才能得到立足，自由、民主的体制才能得以建立和传播。然而，正像"绝对的权利产生绝对的腐败"一样，谁来监督人类的行为呢？自然吗？在这里自然仅仅是被利用和控制奴隶而已，那么只有上帝来监督人类。在这个最终节点上，人类没有选择，没有自由。这就是西方古典思维体系的整体结构。

我无意断定两种思维体系的优劣。我只是想指出绿色基础设施的概念较好地融合了中西两种思维：基础设施是人工作品，而绿色则是自然的赋予，绿色基础设施既不能全靠人工（至少人类还不掌握叶绿素和光合作用的奥秘），也不能全靠自然（森林并非适宜的人居环境）。它从"抓主要矛盾"的角度抓住了风景园林行业在现代城市的最主要作用，既避开了"城市森林"之类概念的荒诞，也绕开了汉字"景观"之类过分强调视觉刺激的概念的麻烦。当然这个概念并不能完全替代我们学科，特别是无法从社会和文化角度来替代。

本期主题是年初由同济大学倡议的，并在很短时间内就完成了栏目组稿，在此表示衷心感谢。刘滨谊教授等撰写的《城市绿色基础设施的研究与实践》，是近年来对绿色基础设施发展和状态的最全面回顾，可供有志于此项研究的同仁作为最基本的参考资料，当然也是想全面了解我们行业的读者很好读物。栏目中有一篇由伦敦奥林匹克公园植物景观的设计者詹姆斯·希契莫夫（James Hitchmough）撰写的《城市绿色基础设施中大规模草本植物群落种植设计与管理的生态途径》，它批评了过分强调本地性和自然生态型的植被设计的流行理论，很有趣。文章的翻译比较困难，译者需要对生态学和城市规划设计都有较深造诣，所以我们发表了原文。

在"名家名作"栏目中尹豪撰写的《吹起自然化种植的号角——威廉姆·罗宾逊及其野生花园》一文，反映出罗宾逊的观点与希契莫夫是不一样的。我们将其同期发表，用意之一是让大家感受到国外的学术争论气氛，还是那句老话，希望推进我国的学术讨论。我总感到这不是难事，如果用这个例子来观察中国，一大批人都可以成为"思想家"，遗憾的是中国没有做到，原因何在，值得所有中国人思考。

Phase 3, 2013

If "brownfield transformation" is the important contribution of our discipline to the future of human beings (see the chief-editor's words of last month), the current theme on the "green infrastructure" concept is our discipline's important contribution in the 20th century to the contemporary human beings.

The Chinese concepts of "landscape" and "garden" are derived from the understanding of the same body relationship between human and nature, and even If transforming the nature, its "surgical operation" on the nature is never ruthless, stressing the respect for the "inherent quality" and following the "nature" (here, the inherent quality and the nature are the synonymous), namely being obedient to the "way". This idea is not a vain attempt to grasp the ultimate truth, does not want to be the God, and does not need the systematic support of modern sciences necessarily. In fact, it considers the "way" as changing, and of course, the scale of the change is big and the time is very long, so-called "day unchanged, neither the way", said the same, which is exactly the theory of relativity expounded in the anti-described manner. I think this is the answer to Dr. Joseph Needham's question "why modern sciences—the new experimental philosophy of the Galileo era and the Royal Society of the early days only emerged in the European civilization rather than in China".

The LA concept is the technology to put the understanding of truth, order, and beauty in human mind on to land and is the belief in human's ability, and this kind of ability is a God-given human privilege. It is based on this belief that the concept of "natural rights" became possible and the system of freedom and democracy could be established and spread. However, just as "absolute power corrupts absolutely", who is going to monitor the behavior of the human? Nature? Here the nature is just utilized and controlled and slaved, and then only the God oversees human. In this final node, human beings have no choice, no freedom. This is the overall structure of the Western classical thinking system.

I do not intend to conclude the pros and cons of the two thinking systems. I just want to point out that the concept of green infrastructure better blends the Chinese and Western thinking ways: infrastructure is artificial works, while the green is given by the nature, and green infrastructure could neither be fully man-made (at least human beings have not grasped the mystery of chlorophyll and photosynthesis), nor fully given by the nature (forest is not a desirable living environment). From the perspective of "grasping the principal contradiction", it catches the main function of landscape architecture industry in the modern city, avoiding the absurd concepts like "urban forest" and the trouble of concepts like the Chinese term "jing guan" (scenery) emphasizing too much on visual attractions. Of course, this concept could not completely replace our discipline, which is especially irreplaceable socially and culturally.

The theme of this issue was initiated by Tongji University early this year, and related papers were collected in a very short period, and here I would like to express my gratitude to the university. *Studies and Practices on Urban Green Infrastructure* by Professor Liu Bin-yi and coauthors is the most comprehensive review of green infrastructure development and status in recent years, and could be used as the most fundamental reference for those interest in this study, and of course, it is also a good reading material for readers who want to have a comprehensive understanding of our industry. Also in this column, *Applying an Ecological Approach to Extensive, Designed Herbaceous Vegetation in Urban Green Infrastructure*, by the London Olympic Park plant landscape designer Professor James Hitchmough, criticizes the popular plant design theory that puts too much emphasis on local and natural eco-vegetation and is very interesting. The translation of the article requires great attainments in ecology and urban planning and design, so we provided the original English version as well.

In the column of "Masterpieces by Celebrities", *Call on Naturalized Planting—William Robinson & Wild Garden*, by Yin Hao, reflects that Robinson's viewpoint is different from Hitchmough's. We publish them on the same issue, and one intention is to make everybody feel the foreign academic debate atmosphere, and as the saying goes, hoping to promote the academic discussion in China. I always feel that this is not difficult. Using this example to observe China, a large number of people can become "thinkers", but unfortunately, it is not so in China. The reason is worth all the Chinese people's thinking.

2013年4期

很多时候，我感觉以偏执为民族特性的日本人搞的西文汉译，会造成我们对欧美文化的偏见和误解。我常举例的有：把architecture译为缺失了设计心思的"建筑"（虽然"房子"是主流），把landscape译为没有了土地的"景观"（虽然"视觉"是主流），把aesthetic译为忘记关注丑的"美学"（虽然"关注美"是主流），但仅仅关注主流绝不符合辩证和全面的要求。这就如同把"龙"译为dragon，只剩下凶恶的一面，所以洋人无法理解中文"小龙儿""小龙女"的美好。

最近我在思考revolution翻译为"革命"是否准确。Revolution的本意是roll back，垂直方向讲是翻面，水平方向讲是转向。后来词义逐渐扩张，其含义的"左倾"，是革命、造反；其含义的"右倾"，是反复讨论、深思熟虑。将其翻译为革命，则只剩下了"左倾"含义。而中国大众对革命的理解，更加偏激，直接与"取别人性命"发生联想，这是历届大革命中常见的现象，于是roll back成为"折腾"。

"文化大革命"初期，北京林学院（现北京林业大学）出过一本刊物《园林革命》，运用"不是属于无产阶级就是属于剥削阶级"的"阶级分析法"把园林专业说成"封、资、修"大杂烩，除了一点生产功能和生态作用，其他一无是处。但当真地革了园林的命，园林专业被取消，北林被下放到云南农村，那些革命派头头们大多设法逃避了。现在想起来，这场闹剧当真十分可笑。所以，一些事可以revolution，但不等于非得革命。革命的最大危险在于缺乏深思熟虑和得当的安排，可有时细节就是要决定事物性质，例如在关键时刻，一点点误差就决定事情朝哪个方向倾倒，所谓蝴蝶效应，一旦倒错了，再想扳回来可就难了。我觉得，历史学家要是把这个问题搞明白搞透，能够有利于人类摆脱无谓的折腾，那才是功莫大焉。

本期有关植物的内容比重较大，主题是"风景园林植物景观设计"，此外还有"风景园林植物"和"植物园"专题，显然这是从风景园林的生态意义逻辑推演的必然结果。本期我最看重的2篇文章"从麦克哈格到斯坦尼兹——基于景观生态学的风景园林规划理论与方法的嬗变"和"中国绿道与美国Greenway的比较研究"，还是与生态密切相关。

生态主要是个科学问题，也是道德问题，这里最不需要忽悠，可偏偏许多论调扎堆大忽悠。我最反感的是对天文、气象、地质、水文、土壤、生物等自然规律所知甚少，对其间的关系和演变趋势毫不了解，以为画出几个斑块几条廊道，把几张"千层饼"一叠加，就解决了生态问题的倾向。最近听说国内正在掀起一股"碳汇林"热，有人趁机提出要用生长快、病虫少、不易着火的阔叶林替代针叶林，其结果又是一场对自然的大浩劫。其实，不管什么林，只要最终转化不成煤炭，腐烂后其含有的碳还是返回大自然再循环，哪里能形成什么碳汇。由此我又想到野草，生长得快腐烂得也快，长远来看，适宜种树的地方还是种树较好。最近有人提醒我：你们在大力推广城市湿地的时候，有没有注意如何防止血吸虫病再度爆发？提出过什么措施没有？我无言以答，而由此联想到对于登革热和禽流感等，我们同样没有研究过对策，便急于决策。至于用种草来吸收土壤中的有毒元素，在没有研究出来如何处理这些草的方法之前，不过是一场秀。秀这种东西和科学态度是死对头。秀在中国特别盛行，乃是因为它十分适应那些急于出名和获得商业利益的人的需要，也非常适应那些急于做出业绩的官员的需要。

Phase 4, 2013

Many times I feel that the translation of Western terms into Chinese characters by the paranoid Japanese leads to our prejudices and misunderstandings of the European and American culture. The examples I often raise include: "architecture" is translated as "jianzhu" (building) that lacks the meaning of design ideas (even though "house" is main body), "landscape" as "jingguan" (scenery) that lacks the meaning of land (even though "visual image" is the main body), and "aesthetic" as "meixue" (the study of beauty) that forgets to pay attention to the ugly (even though "concern about beauty" is the mainstream), but focusing only on the mainstream does not mean the dialectical and comprehensive requirements. It is like "long" (Chinese word of dragon) translated as "dragon", only the evil side left, so foreigners cannot understand how beautiful the Chinese terms "xiaolong'er" (little dragon boy) and "xiaolongnü" (little dragon girl) are.

Recently I am thinking whether "revolution" translated as "geming" (change of life) is accurate or not. The original meaning of "revolution" is "to roll back", namely turning upside down vertically or turning direction horizontally. Later its meaning gradually expanded, and the left-leaning meaning is to rebel while the right-leaning meaning is to discuss repeatedly and well thought. Translating into "geming" (change of life) keeps only the left-leaning meaning. The understanding of "geming" (change of life) of the general public of China is more extreme, directly associated with "taking the lives of others", which was often seen in all previous revolutions, so "to roll back" it "to toss about".

In the early stages of the "Cultural Revolution", Beijing Forestry University published a publication named Landscape Architecture Revolution, which adopted the "class analysis" that "that does not belong to the proletariat belongs to the exploiting class" and regarded the landscape architecture discipline as the "feudal, capital and revisionist" hodgepodge without any merits apart from some production and ecological function. But when there was really revolution in landscape architecture, the landscape architecture discipline was cancelled and the university was moved down to the rural area of Yunnan Province, most of the heads of those revolutionaries managed to escape. Now thinking of it, this farce was really ridiculous. So, some things can have revolution, but not to take or change their lives. The greatest danger of "taking or changing lives" is the lack of careful consideration and appropriate arrangements, but sometimes details determine the nature of things, and a little error at the crucial moment, for example, decides which direction the things turn to. This is the so-called butterfly effect, hard to get back on track after once perversion. I think if historians thoroughly understand this issue, unnecessary tossing could be thrown off, which would a remarkable achievement.

A large proportion of the contents of this month is about plants, and the theme is "Landscape Architecture Plant Design", plus special topics of "Landscape Plants" and "Botanical Garden", which are the inevitable result of logical deduction from the ecological significance of landscape architecture. The two papers that I value most, *From McHarg to Steinitz: Transmutation of Theories and Methods in the Field of Landscape Architecture Based on Landscape Ecology*, and *The Comparative Study on the Chinese Greenway and the American Greenway* are also closely related to ecology.

Ecology is mainly a scientific issue, and also a moral issue, where needs no fudge, but a lot of arguments gather here to get people hoodwinked. I most dislike the trend of those who know little about the natural laws of astronomy, meteorology, geology, hydrology, soils, and biology, do not understand their relationships and evolution trends, and think of solving the ecological problems by drawing some patches and a few corridors and overlapping several layers. I recently heard that the "carbon sinks forest" is hot in China, and some people take the opportunity substitute the coniferous forest with the fast-growing, less-pest and hard-to-ignite broadleaf forest, and the result would a catastrophe for the nature. In fact, no matter what kind of forest it is, as long as the final conversion is not coal, the carbon it contains would return to the nature for recycling after being rotten, and there would be no carbon sink. Thus I think the weeds grow fast and rot fast as well, and in the long term it is better to plant trees in the places suitable for trees. Some people reminded me: when you vigorously promote urban wetland, have you paid attention to preventing the schistosomiasis from breaking out again? What would be the measures? I have no answer to this, and thereby I associate that for dengue fever and avian flu, we was anxious to make decision before we studies the countermeasures. As for planting grass to absorb toxic elements in the soil, it is just a show before the method of dealing with the grass id found out. Show is the sworn enemy of the scientific attitude. Show is especially prevalent in China, since it meets the needs of those who are anxious to gain fame and commercial interests also the needs of the officials who are eager to obtain results.

作为一级学科，我们必须主动申请参与国家的顶层设计，包括国土战略、城市战略、生态规划和相关的国家重点科研项目。我能够想到的，大概有如下几个方面。

1）研究和设计城市绿地指标。2012年冬，东部大面积城市出现的严重灰霾，除机动车尾气排放等原因外，也突显了我国城市绿地严重不足的问题。历史上著名的大气污染事件，如伦敦的烟雾事件和洛杉矶的光化学污染等，人家都没有等待，马上就着手城市和地区顶层设计，难道我们还要再等待事件的重复出现来验证中国城市绿地的稀缺吗？

2）从城市绿地的科学结构入手，研究改造我国现行城市规划的理论和模型，并进而提出远景城市的理念。这是个涉及未来2/3以上国民幸福指数的大问题。

3）以科学的城市模型为基础，参与研究我国城市建设用地指标。我国严苛的城市用地指标，主要是出自"人多地少"的粗放观察，但国家从来没有对此种思维定式的后果进行系统的研究。可从以下方面深入探讨：（1）严苛的城市用地指标造成的危害和损失，包括城市污染、交通堵塞、健康危害、医疗费用、房价高涨、贫富差距、改造成本、资源损耗等方面；（2）城市用地指标与国土规划的平衡；（3）粮食战略；（4）人口与占有建设用地的关系与政策等。

4）参与全国生态系统建构的研究和设计。

据新华网报道，习近平主席不久前指出："要善于运用底线思维的方法，凡事从坏处准备，努力争取最好的结果，做到有备无患、遇事不慌，牢牢把握主动权。"相当长一段时间来，我们的科研工作放任机械式思维在科学主义的大帽子下到处横行，恶果累累：研究题目越小越孤立越好，结论越模型化（绝大多数都是仅适用于特定条件下机械的线性模型）越好，格式越八股越好，连表格都非得是最简单机械的三线表。其实这种思维还表现在行政管理、评价评定、教育思想、战略制定、国际关系等各个方面。习主席这段话，仿佛一股清风拂面，让我感到辩证法思想又回来了。如果我们能用辩证法审视科学，在科学的基础上进行辩证思维，坚持下去，中国的事情一定能办好。这个思想，也完全适用于顶层设计。我们参与顶层设计，不是说必须由我们说了算才好，而是主动参与到国家战略的综合平衡中，争取对国家长远来说最好的结果。

此外，学科自身也要进行顶层设计。现在我们拥有本科以上专业的大学将近200所，其中的1/10左右拥有博士点，这是多么巨大的一股科研力量！但是我们的研究大多是分散的、重复的、低效的、低水平的。如何整合这股力量？研究的核心和目标？研究的计划？怎样开启这股人马的智慧和激发其创造能量？行业的公益性和商业性之关系？全球视野与国情的关系？都属于顶层设计的内容。甚至由谁来进行组织这项工作，本身就是顶层设计。从国情出发，我觉得首先由政府机构发动比较好，这样有利于尽快解决最顶层的难题，次一层的事情，可以交给学会接手。

当然，对顶层设计本身也要进行辩证考察。仅仅从上到下地推论，这种思维本身就是一种机械逻辑，非辩证的。所以学科还需要大批人员从事基层的工作，并从实际出发，对上层乃至顶层作出反馈。这种作法本身就是中国传统的"实事求是"思想和辩证思维方法的统一。

Phase 5, 2013

As the first-level discipline, our landscape architecture discipline must take the initiative to apply for participation in the country's top-level design, including territory strategy, urban strategy, ecological planning and related national key scientific research projects. What I can think about are mainly the following aspects.

1)Research and design of urban green space indicators. The serious fog and haze over the large areas of the eastern China cities last winter has clearly demonstrated the serious shortage of urban green space in Chinese cities. In the historic air pollution events, such as the London smog and photochemical pollution in Los Angeles, the people there immediately proceed the top-level design of cities and regions without waiting. Do we still have to wait for the event repeated to verify urban green space scarcity?

2)Start from the study of the scientific structure of urban green space to research the transformation of China's current urban planning theories and models, and put forward the concept of the future city. This is a big problem related to more than 2/3 of the gross national happiness index in the future.

3)Take the scientific urban model as the basis, and participate in the study of urban construction land quota. China's rigid urban land use quota mainly comes out of the extensive observation of "the limited land for a big population", but the state has never had a systematic study of the consequences of such a mindset. In-depth discussion could be about: (1) the damage and loss caused by the rigid urban land use quota, including urban pollution, traffic congestion, health hazards, health care costs, rising house prices, the gap between the rich and the poor, the cost of renovation, resource depletion, etc.; (2) the balance of urban land use quota and territorial planning; (3) food strategy; and (4) the relations and policy of population and per capita land for construction.

4)Involve in the research and design of the national ecological systems construction.

According to Xinhua.com, President Xi Jinping recently pointed out: "…to be good at the use of the bottom-line thinking, get prepared for the worst situation, and strive for the best results, so as to be prepared, not panic, and firmly grasp the initiative…" For quite a long period of time, our research work has let the mechanical thinking unchecked under the name of scientism, which has caused numerous serious consequences: the research topics are as smaller and more isolated as possible, the conclusions are as more modeled (the vast majority of which are mechanical linear models only available in certain conditions) as possible, the formats are as more stereotyped writing as possible, and even the tables have to be the most simple mechanical three-line tables. In fact, this kind of thinking is also reflected in the administration, evaluation assessment, educational philosophy, strategy development, international relations and other aspects. The words of President Xi, as if a cool breeze blows, make me feel that the dialectical thought is back. If we use the dialectics to examine sciences and have the dialectical thinking on the scientific basis and stick to it, the things in China will surely succeed. This idea is also fully applicable to the top-level design. Our involvement in the top-level design does not mean that we have the final say, and instead we should take the initiative to participate in the overall balance of the national strategy and strive for the best long-term results of the country.

In addition, the discipline itself must have the top-level design. Now we have nearly 200 universities with the landscape architecture programs for undergraduate study or above, about 1/10 of them with doctoral programs, and this is really the great scientific research strength. But most of our research is fragmented, duplicated, inefficient, and low-leveled. How can we integrate this force? What are the core objective and goal of the study? What is the research plan? How can we open the wisdom of these people and stimulate their creative energy? What is the relations between the public interest and commercial interest of the industry? What is the relations between the global perspective and the national conditions? These are all part of the top-level design. Even about who is the one to organize the related work itself is also the top- level design. From the national conditions, I think that it is better to be launched by government agencies, which will help solve the top-level problems as soon as possible and the sub-layer things could be handled by the landscape architecture society.

Of course, the top-level design itself should be examined dialectically. The way of thinking of the inference only from top to bottom is a mechanical logic, not dialectical. So our discipline also need a large number of people engaged in the work of basic levels and giving feedback to the upper an even the top-level from the practical situations. This method itself is the unity of the traditional Chinese thinking of "seeking truth from facts" and the dialectical thinking approach.

2013年6期

现代LA是由硬质景观（建筑学、规划学）和软质景观（园艺学、林学）两股潮流汇合而成的。而掌控理论主导潮流的，最初主要是建筑学派，现在又加入了生态学派。

我读景观学的文章，总归摆脱不了主客体之间二元分离的感觉："我"是从旁观察对象（景观），以我的知识积累和我对对象的了解，对景观施行策划、规划、设计和决策。在这里，我不是大自然之子，我在代行上帝的职权，我来安排一切。在我国，持有这种心态的还有一些官员和大老板。当然，我们可以改进这种方法，如把"我"改成"我们"，把"风景园林师"改成"多学科专家群"……可是这一套体系无论如何改造，也摆脱不了人类的自以为是。

人们若想真正皈依东方文化，回到自然之子的状态，我能看到的前景只有一种：把自己的生活深深扎根于某一个具体环境中，让自己时刻感觉着上帝的脉搏，自然的律动。但是，只有"土著"才能做到这点。从这种视角观察，风景园林师不应该再自作主张，而应考虑让当地人自主地进行规划设计，风景园林师可以在规划之初和他们"同吃同住同劳动"一段时间，并在规划过程中利用自己的专业知识协助当地人做好他们的土地规划、设计和护理。也就是说，大部分风景园林设计师的工作，最终应该向"学生、教师和设计服务员的综合体"转移。

这当然只是一种粗略的设想，实际问题则复杂得多。比如城市的"土著"就是市民，而市民对自然的理解和关心是非常薄弱的，市民很洋，不土，故而他们缺乏自然之子的品质。如果说，人类正从自然之子异化，变成地球的异数，那么市民就是这种异数的典型。好消息是许多市民想改变这种状态，走出一条"反异化"的道路，力图多亲近亲近自然，风景园林师的主要职责之一就是为这种人做好服务。坏消息是许多建筑师、规划师、风景园林师、资本或权利的掌控者则在努力向着这种异数极致迈进。

难道地球这种令人悲哀的前景就没有办法改变了吗？还是有的，这就是在顶层设计阶段就把市民的能力范围尽量压缩。比如控制城市和建设用地的总面积不超过地球陆地面积的3%~5%，在这里让科学和工业充分发展，让人们做自己的上帝，过足征服自然的瘾，特别是其中还可以分出3%（即陆地总面积的1‰左右）留给那些以自我表现为己任的艺术家，由着他们自由发挥。城市以外，应该是上帝和上帝之子——"土著"们的天下，其中，如果把不到一半留给上帝之子们，以满足人类对第一产业的需要，还能留下一半以上还给上帝，那就很好了。风景园林师可以从"有权决定哪些东西该归属上帝，哪些地方可以接近上帝"之中，感到自己的无上荣耀。

风景园林师的权利从何而来？从受教育而来，从接受科学、艺术和道德教育而来。在最近的一系列学术活动中我们深深地感受到，中国的风景园林教育正在面临巨大的变迁。本期的主题"风景园林专业教育"是刘晖教授主持的，共发表了7篇文章，其中有从全局出发高瞻远瞩的宏大叙述，也有细致而微的深入思考，还有迄今为止最准确叙述了学科在美国和法国发展轨迹的译文。在此我们向西安建筑科技大学表示感谢。

Phase 6, 2013

Modern landscape architecture is formed by the confluence of the two trends of the hard landscape (architecture and urban planning) and the soft landscape (horticulture and forestry). The one that controls the theory and dominates the trend dominates has been the architecture school, and now and the ecology school joins.

When reading landscape architecture articles, I always could not get rid of the feeling of the dualistic separation of subject and object: "I" observe the object (landscape), and use "my" accumulated knowledge and understanding of the object for the planning and design of the object and related decision-making. Here, I am not a child of nature, I take the authority of the God, and I will arrange everything. In China, some government officials and big bosses have such mindset. Of course, we can improve this method, such as changing "I" into "we", changing "landscape architects" into "interdisciplinary experts", etc. However, any transformation of this set of system could not break away from the human-opinionated pattern.

If people really want to be converted to the oriental culture and return to the state of a natural son, there is only one prospect I can see: rooting their own lives deeply into a specific environment to keep the feeling of the God's pulse and the natural rhythm all the time. However, only the "indigenous" can achieve that. Observing from this perspective, landscape architects should never give a free hand, and should consider allowing the local people to carry out planning and design independently. Landscape architects can stay together with them for a while at the beginning of planning, and help the local people to do the land planning, design and maintenance in the planning process with their professional knowledge. In other words, most of landscape architects' work should ultimately be transferred to "the complex body of student, teacher and design service".

This is of course only a rough idea and the actual problems are more complicated. For example, the citizens are the "indigenous" of the city, and their understanding and concerns of nature are very weak and they are very modern instead of indigenous, so they lack the quality of nature's son. If we say that human beings are alienated from the nature's son into the Earth's anomaly, then the citizens are the typical case of this anomaly. The good news is that many people want to change this condition, take the "anti-alienation" way, and try to get close and more intimate to nature, and one of the main responsibilities of landscape architects is to serve and help such people. The bad news is that many architects, planners, landscape architects, capital or power controllers are making efforts toward the extreme of such anomaly.

Is there no way to change this sad prospect of the Earth? The answer is positive, which is to compress the citizens' power range in the top-level design stage. For example, the total area of urban and construction land is controlled to be within 3%-5% of the land area of the Earth, where the full development of science and industry is enabled and people can act as their own God to satisfy themselves from the addiction of conquering nature. Especially 3% of that (around 1‰ of the total area of the Earth) could be left for those artists who take self-expression as their responsibilities, give them the freedom to play. Outside the city, it should be the world of the God and the sons of God the "indigenous", and if a small half of it could be left for the sons of the God to meet human needs of the primary industry, and the big half it could be returned to the God, that would be very good. Landscape architects could feel the supreme glory from their rights to decide what should be returned to the God and where could be closer to the God.

Where do landscape architects' rights come from? From education, the scientific, artistic and moral education. From the recent series of academic activities, we deeply feel that China's landscape architecture education is facing enormous changes. The theme of this month "Landscape Architecture Education" is chaired by Professor Liu Hui, and seven papers are published, including grand narrations from the overall perspective, delicate deep thinking, as well as the translations of by far the most accurate account of the disciplinary development tracks in the United States and France. We hereby express our gratitude to Professor Liu and Xi'an University of Architecture and Technology University.

2013 年 7 期

我们学科的根在于land。所以我多次强调，日本人将landscape翻译成景观，是丢掉了land，而吴良镛先生将其翻译为"地景"，非常准确，将概念指向那些现在或将来漂亮美好的土地。

中文"园林"二字，按我的理解，是指实体，不是指造园活动。而用一类实体作为研究对象，是最扎实稳当的学科，此其一；其二，与landscape专指land的视觉特性不同，园林还包括了land上所拥有的voice、smell、touch、wind、temperature、humidity、air pressure等一切涉及人类感知的元素，以及这些元素本身的活动、生灭、表情、规律……

"园林"的主要弱点，在于其篱笆性，小圈子性。所以再将风景（含城景、水景、声景……）扩进来，是个大好事。然而，"风景园林"似乎还有弱点，对此我还没想清楚。一个例子，是它和城市空间的关系，总不能讲城市的室外空间都是风景园林的对象吧。难道非要再搞一个"城市设计"学科来填空吗？也许吧。无论如何，城市设计对地球生态和人居环境的影响，显然都没有风景园林大，而且过于突出设计，肯定会忽视更为重要和讲求科学的地球管护（stewardship），所以很难成为第一流的学科。

Landscape起源于美感和艺术，它进入科学领域则源于地理学。通过地理学的研究，人们才能知道一块land过去如何，现在为何是这个样子，并依据科学（包括自然与社会）研究的成果推测其将来的面貌。而landscape architects擅长的是塑造一块land的面貌，有些landscape architects注意这个塑造过程中的科学道理，这是我们应该提倡的，而有些人则只顾表达自己的想象。我的看法是，后者不是绝对不可以，但一定要严格控制其能够影响的地域，例如总量最好不超过地球陆地表面面积的1‰。

由此看来，geographers（地理学家）和architects（建筑师）对待LA的态度是有别的。反过来，LA对二者则应该持平等和尊重的态度，当然，界中有人向地理学倾斜些，有人向建筑界倾斜些，都是可以的。可是一定要看清楚，风景园林学不是地理学与建筑学的机械相加，而应该是二者（或许还有规划、园艺、艺术等）融于一体。

本期的主题"传统城镇工业遗产"是由北京大学阙维民教授主持的，篇篇文清理顺，逻辑清晰，格式严谨，为我们做出榜样，为此我要向阙教授表示衷心感谢。这个课题对于大多数风景园林师而言，是相当生疏的，但正如我在2013年2月的"主编心语"中说的，如果不实行棕地修复，1000年后，城市和各种建设用地的占地可能达到地面的1/3！所以如何对待遗产，必将成为风景园林学的重要内容。让我们从现在起就认真地向地理学家学习一些相关的基本知识和研究方法。

Phase 7, 2013

Land is the root of our discipline. So I have repeatedly emphasized that the Japanese lost "land" when translating "landscape" into "jing guan" (scenery and view), while it is very accurate that Mr. Wu Liangyong translated "landscape" into "di jing" (land scenery) and pointed the concept to present or future beautiful good land.

The Chinese word "yuanlin" (garden), as I understand, refers to the entity instead of gardening activities, and using a group of entities as the research object makes it the most solid and reliable discipline. And secondly, different to "landscape" that specifically refers to land, "yuanlin" includes all elements related to human perception, like voice, smell, touch, wind, temperature, humidity and air pressure, as well as the activities, birth and death, facial expressions and laws of the elements themselves.

The main weakness of "yuanlin" is its fences and coterie, so it is great to expand it to include sceneries (urbanscape, waterscape, soundscape…). However, "fengjingyuanlin" (landscape architecture) also seems to have weakness, which I have not thought clearly. An example is its relationship with urban space, since we cannot say that the city's all outdoor spaces are objects of "fengjingyuanlin". Should we then engage in an "urban design" subject to fill in the blank? Maybe. In any case, the impact of urban design on the Earth's ecology and human habitat environment is obviously not bigger than "landscape architecture", and too much emphasis on design will definitely neglect the more important and scientific stewardship of the Earth, so it is difficult to become a first-class discipline.

Landscape originated in aesthetics and art, and it enters the field of science because of geography; since people can only know the past, present and future of land through the study of geography and through the achievements of scientific (natural and social) research. Landscape architects are adept at shaping the face of a land. Some landscape architects note the scientific truth of this process of shaping, which is what we should promote, while others simply express their imaginations. In my view the latter is not absolutely impossible, but we must strictly control the area it can affect, such as no more than one-thousandth of the Earth's land surface area.

Therefore, the attitudes of geographers and architects towards landscape architecture (LA) are different. Conversely, LA should hold the balanced and respect attitude to both of them. Of course, it is ok that some professionals of our discipline and industry are more inclined to geography while some others are more inclined to architecture. But we should be clear that landscape architecture study is not the mechanical sum of geography and architecture studies, while it should be the blending of the two (and perhaps planning, gardening, art, etc.).

The topic of this month "Traditional Urban Industrial Heritage" is hosted by Professor QUE Weimin of Peking University. The papers are logically clear and smooth and in rigorous format, which set an example for us, and I would like to express my sincere gratitude to Professor Que. This topic is quite unfamiliar to most landscape architects. But as I said in the February issue that urban and construction land will occupy one-third of the Earth surface after 1000 years if we do not have brownfield restoration, how to deal with heritage will definitely become an important part of landscape architecture. So let us carefully learn some related basic knowledge and research methods from geographers.

2013 年 8 期

我对当前相当热闹的生态规划抱有很多疑问，这类规划大多数顶多只能称为生态改善规划，无法定量，无法找到最优。

人类对自然规律的掌握，要远远好过对社会规律。当前的科技水平，地域的自然生态研究已经达到总体上可以基本量化的水平。有了这个基本前提，我认为，生态规划应该是评价其他规划的基础，生态规划图应该是进行各项规划以及地域综合规划的底图。这种生态规划的目标是：（1）从对生物讲，地域达到生物产量最大，物种指数最高；（2）从对人讲，空气、水、食品的质量达到最好，数量最多。这张规划图，就可以作为其他规划的底图，以此为依据就可以对其他"建设规划"对生态造成的破坏进行量化计算，最后经过统筹平衡，争取在综合规划中将人类对地域生态的影响降到最低。

这种生态规划思路，是对现行规划方法的重大突破。

生态规划工作，不是对一切问题进行统筹，生态规划不是把一切都考虑进来的规划，这个地球上还没有人有能力做出这种完美的规划。当下通行的总体—分项—总图的工作步骤，以及所谓多学科联合规划，实际上是一种协商制度，远远够不到科学的高度。

至于所谓能够兼顾各个方面的"千层饼"叠图法，从长远看，会被后人讥笑成小孩子的游戏。当然，改造千层饼是非常艰巨的任务，这涉及学科间融合的进展，以及巨量计算的实现和普及。其最后结果，可能是对西方传统分块思维模式的彻底反思。

本期的主题"减量设计"，构思产生于2012年10月召开的中国风景园林学会年会上东南大学成玉宁教授的提议。但对这个涉及本学科深层次和本质的问题，一旦动笔，就会觉得千头万绪，难度极大，所以成教授的组稿工作非常辛苦，特此致谢。

减量设计是对自然影响的减量，不是设计的工作量减量，更多情况下是要求设计工作增量，即通过大量的科学研究和认真的仔细设计，减少无谓的工程量，减少对自然的干扰和影响。我对《基于数字水文地形模型的景观水系优化设计——德国埃尔廷根–宾茨旺根段多瑙河河流修复》一文非常欣赏，它最典型地说明了上述问题，此外，将减量设计与现代科技发展紧密结合，也是本文给我们的重要提示。

据我所知，20世纪五六十年代的"新岭南风格"园林，以山庄旅舍、华南植物园等为代表，就曾经到达过这样的层次。例如：为了减少土方，建筑非常注意原地形的微小变化，同时也丰富了空间趣味；为了抵御波浪对湖岸的冲击，仔细地考虑了湖岸形状、湖面宽度和风向的影响，再选择湖岸坡度和种植类型等。走减量设计的路线，是对金钱诱惑的抵制，其基本态度是对客观规律、对现场环境、对前人努力的尊重，它是科学精神和中国谦虚美德相结合的必然。为了鼓励这种精神，还需要我国的科技政策、评价方法、设计报酬等方面进行配套改革。

Phase 8, 2013

I have a lot of questions on the current boisterous ecological planning, and most of them can only be called ecological improvement planning and cannot be quantified, and the optimal one cannot be identified.

Human mastery of the laws of nature is much better than on the laws of society, and at the current level of technology, geographic natural ecological studies have reached the overall basically quantifying level. With this basic premise, I think, ecological planning should be the basis for evaluation of other planning, and ecological planning figure should be the base map for all the planning and local integrated planning: the goals of the ecological planning are: 1, in terms of biology, the region reaches maximum biomass, and the species index is the highest; 2, from human beings, the qualities of air, water and food reach the best and largest quantity. Then this plan can serve as the base map for other planning, and on this basis, the damage to the ecology by other "construction planning" can be quantitatively calculated and finally integrated and balanced to minimize the human impact on local ecology in the comprehensive planning.

This ecological planning idea is a major breakthrough in the current planning methods.

Ecological planning is not coordinating all issues, nor putting everything into account, and no one on this earth has the ability to make this the perfect plan. The current prevailing "whole- subentry- general drawing" work steps, and the so-called multi-disciplinary joint planning are actually a kind of consultation system, far from the science level.

As to the so-called "layered pastry" overlay method that can take into account all aspects, in the long run, will be later ridiculed as children's play. Of course, the transformation of "layered pastry" is a very difficult task, which involves the interdisciplinary integration progress, as well as the realization of massive computing and popularity. The end result may be the complete reflection on the Western traditional block mode of thinking.

The theme of this issue is "minimization design", and the idea comes from the proposal by Professor Cheng Yuning of Southeast University in the CHSLA annual meeting held in October 2012. However, this topic involves the deep level and nature of our discipline, and once you begin, you will find a multitude of things and extreme difficulties, so we are very grateful to Professor Cheng for organizing the papers.

Minimization design is the reduction of impact on nature, not the reduction of design workload, and in more cases the design workload is increased, namely reducing unnecessary workload and minimizing the interference and impact on nature through a lot of careful scientific research and design. I appreciate the paper *Optimizing Landscape Water System Design Based on Hydraulic Digital Terrain Model – The Danube River Restoration in Ertingen-Binzwangen, Germany* very much. It is the most typical illustration of the above problems, and in addition, it also gives us the important tip to closely combine minimization design with modern science and technology development.

As far as I know, the "New Lingnan Style" gardens in the 1950s and 1960s, represented by Mountain Villa and South China Botanical Garden, have reached such a level. For example: to reduce earthwork, the architecture paid great attention to small changes in the original terrain and also enriched the space interest; in order to resist the impact of waves on the shore, the influence of lakeshore shape, lake width and wind direction were carefully considered before selecting lakeshore slopes and planting types; and so on. The way of minimization design is to resist the lure of money, and its basic attitude is to respect the objective laws, on-site environment, and the efforts of predecessors, and it is inevitable in the combination of the scientific spirit and Chinese modesty virtues. To encourage this spirit, China should have supporting reforms in science and technology policies, evaluation methods, and design remuneration.

2013年9期

芒福德（Mumford）把1900年至今划为新技术时代，并在传统的经济三要素中添加了第四要素"创新"，有人用"从手制造到脑创造"来概括这个时代的特征。我觉得，芒福德不过是一家之言，至于把古人视为"动手不动脑"更是编造。最近奥巴马提出的"大数据时代"，却的确让美国人抓住了先机。大数据将会彻底改变整个人类的思维、判断、教育、生活方式和社会结构等，给人类带来只有蒸汽机发明可以比拟的根本性变革。

历史上任何新时代的到来，总会有人兴奋，也有人因难以预测后果而感到恐惧悲哀。至于这个时代是让人类进步还是衰退，只有历史可以作出结论。我觉得，大数据时代的关键在于人类和自然以及现实之间被大数据隔离，由于逃避了事实检验这一关，人性中"恶"的潘多拉盒子可能被随意打开。过去，荒诞的思维因为不断与事实碰撞，影响有限，而在大数据时代人人都可以随便取得数据并引以为据，言之凿凿，网上的言论已初现此象。所谓精英们也在巨量信息面前束手被擒，无力抓住要点，无力形成自洽的系统，东拼西凑，任意拷贝，常常前言不搭后语，自相矛盾，这类现象大量表现在近些年的论文、论坛、讲座之中，而别人也没有时间去管这些"闲事"，即使管了，过了不多久就被其他信息冲淡了、忘却了，谬论照旧四处流窜。另一方面，乐观正直的人们又在勇于直面种种现实，努力设法一步步地克服新时代的难题。邬峻教授《拉法耶实验——突破网络割据的21世纪创新工厂》一文，就是这种努力的体现，也是本刊涉及这个问题的起步。文中提及的学术界小圈子化现象，本质是"精英"们在巨量信息面前束手被擒的一种反映。

本期的主题"世界自然遗产与国家公园"，由清华大学杨锐教授主持，比较全面地介绍了我国和世界的自然遗产保护工作。其中庄优波博士主笔的《我国世界自然遗产地保护管理规划实践概述》一文，高屋建瓴地总结了迄今为止我国自然遗产保护进程，有利于对今后的工作进行顶层设计，值得向从事此类工作的同仁推荐。联系到上述大数据时代的问题，我为风景园林事业总是密切关系着自然，没有被彻底人为化和digital化而窃喜。由此又联想到风景园林有可能在人类思想史上建立一番丰功伟业，我环视了一下，在各种知识中，风景园林可能是在人与自然的平衡中处于最佳位置的学科。

最后，我向从事城市绿地系统规划的同仁们推荐浙江省义乌市城市规划设计研究院院长杨立新高级工程师的文章《编制城市绿地系统规划时应采用软微风玫瑰图》，他通过实践，对2001年我联名李敏在本刊发表的《城市开敞空间规划的生态机理研究》提出了重要修正，建议用软微风玫瑰图替代常年风玫瑰图，很好，很对，在此我表示深深的谢意。

Phase 9, 2013

Mumford has designated the period from 1900 up to now as the new technology era, and added the fourth element "innovation" into the three traditional economic elements, and someone summarizes the characteristics of this era with "manufacturing with hands to creating with brains". I think it is just one statement of Mumford, and regarding the ancients as those "with hands instead of brains" is fabricated. However, the "big data era" recently proposed by Obama does allow the Americans to seize the opportunities. The big data will fundamentally change the entire human thinking, judgment, education, lifestyle and social structure, etc., bringing to mankind the fundamental change that can only be compared with the invention of the steam engine.

Upon the coming of any new era in history, someone got excited, while others feel frightened and sad due to the unpredictable consequences. As to whether the era makes humankind progress or decline, only history can conclude. I think the key to the era of big data lies in the separation of human with nature and reality by the big data, and because of the escape from the point of reality check, the Pandora's Box of "evil" in human nature may be freely opened. In the past, the impact of absurd thinking was limited since it constantly collided with the facts, while in the era of big data, everyone can easily get the data and cite that data and sounds affirmative, which has been seen on the internet. The so-called elites are also captured by the huge amount of information, unable to grasp the point or form a self-consistent system, and just scrape together and copy from somewhere, often confused and contradictory, and this phenomenon is manifested in a large number of recent papers, forums, and lectures. Others do not have time to care about these "big deal", and even cared, they will soon be diluted by other information and forgotten, while the fallacies flee hither and thither as usual. On the other hand, the optimistic and honest people have the courage to face realities, working to overcome the problems of the new era step by step. Professor Wu Jun's paper *Lafayette's Experiment: Innovation Factory across "Cyberbalkanization" in the 21st Century* is a manifestation of this effort, and also the initial step of our journal on this issue. The academic coterie phenomenon mentioned in this paper is also a reflection that the so-called elites are captured by the huge amount of information.

The topic of this month "World Heritage and National Park" is hosted by Professor Yang Rui of Tsinghua University and a more comprehensive introduction to natural heritage protection of China and the world is made. The paper *Overview of Management Planning Practice for Chinese World Natural Heritage Sites* by Dr. Zhuang Youbo summed up the natural heritage protection progress of China from a foresight perspective, which is beneficial for the future work of top-level design and is recommend to our colleagues in this field. Concerning the problems in the era of big data mentioned above, I am quite happy that our landscape architecture cause is always related to the nature and have not been entirely man-made or digitalized. Thus thinking that the landscape architecture cause might perform miraculous feats in the history of human thought, I look around and find that landscape architecture is probably the discipline in the best position of the balance between human and the nature than other subjects.

Finally, I would like to recommend to our colleagues in urban green space system planning the paper *The Soft-breeze Rose Diagram as the Main Base in Urban Green Space System Planning* by senior engineer Yang Lixin, Director of Zhejiang Yiwu City Urban Planning Institute, who makes an important amendment through practices to the paper *Research on the Ecological Mechanism of Urban Open Space Planning* by Li Min and me, published in our journal in 2001. It is very good and right, and I would like to express my sincere gratitude .

2013年10期

第九届中国（北京）国际园林博览会，无疑是历届中规格最高的一届，其影响也肯定是最大的。本期以"第九届中国（北京）国际园林博览会"为主题，邀请作者介绍了北京、江苏、上海、杭州、武汉、济南等地的优秀作品，以供读者鉴赏，或作为参观、学习时的参考。此前，我们介绍过"岭南园"，以后，也将会继续发表这方面的文章。

艺术与科学的关系，历来是争论不休的问题。我的看法是：艺术的深处是人性，艺术是人性的表现。科学必须依靠思维，而思维，特别是深度思维，也是人性的一个方面，是唯一让人能够确信真理（应该是相对真理）的这一方面。所以，艺术包含了思维，但不等于高于思维，如果用接近真理作为高低的标尺的话。虽然有时艺术想象更加靠近真理，但这种真理还未经过证明，不能视为已经确立。这样看，把一切艺术都用阶级分析来对待，是一种大闹天宫式的体系荒谬，试图用对"特殊性"的机械理解实现对"共性"的否定。本期夏成钢先生的《达意、传神、纳新——第九届园博会北京园设计思路》，讲述了他在第九届园林博览会的"北京园"的设计和施工中，有关继承和发展皇家园林艺术精髓的思考和作法，无论在实践上还是理论上，都有很强的意义。

国务院最近公布了《关于加强城市基础设施建设的意见》（以下简称《意见》），其中与风景园林关系密切的内容是：推动生态园林城市建设，加大社区公园、街头游园、郊野公园、绿道绿廊等规划建设力度，到2015年，确保老城区人均公园绿地面积不低于5m^2、公园绿地服务半径覆盖率不低于60%。实现这些要求，肯定会大大改善我国城市的基本面貌和生态条件，功德无量。当然，实施的难度也很大，风景园林界要加油了！

在这个《意见》中，还要求各地编制城市排水防涝设施规划，加快雨污分流管网改造与排水防涝设施建设，目标是用10年左右时间建成较完善的城市排水防涝、防洪工程体系。刚巧本期《基于流域概念的城市生态排水体系构建——以溆浦县城区生态排水体系构建为例建》一文，涉及城市防洪，文中强调的是顺应自然，因势利导，很不错。近些年我国在城市水务上出现了许多怪现象：不管地形主要靠设计和建造泵站来管理水流；挖山头做"洪水调蓄池"；把地面水集中起来后在地下打个大洞来蓄洪（所谓深邃）；在严重缺水的地区建设"湿地公园"；有人甚至提出把全国的城市绿地统统下挖20cm用于调洪，实际上是要毁掉我国过去城市绿地建设的全部成果……总之，城市治水的花样不断翻新，虽然孤立地看有所成效，却都有共同的特性：越花钱越好。很多绿地成为实质上的牺牲品，应该引起风景园林系统的警惕。

过去，我们行业的工作基本上是在"建设用地"的范围内，所以大家很少关注"非建设用地"。《绿色基础设施视角下的非建设用地规划策略》一文，首次把行业的眼光系统地触及"非建设用地"领域，很有意思，值得关注。

Phase 10, 2013

The 9th China (Beijing) International Garden Expo is undoubtedly with the highest standards among all sessions, and its impact is certainly the greatest. This month we take "The 9th China (Beijing) International Garden Expo" as the theme and invite authors to introduce the good works from Beijing, Jiangsu, Shanghai, Hangzhou, Wuhan, Ji'nan and other places, for our readers' appreciation or working or touring reference. Previously we introduced "South China Garden", and we will continue to publish articles on this topic.

The relationship between art and science has always been a vexed issue. My view is: the depth of art is human nature, and the art is an expression of humanity, while science must rely on thinking, and thinking, particularly deep thinking, is also an aspect of humanity, the only aspect that makes people confident about the truth (it should be relative truth). Therefore, the art includes thinking, but not equally higher than thinking, if using the distance to truth as the standard. Although sometimes the artistic imagination is closer to the truth, but this kind of truth has not been proven, it cannot be deemed to have been established. In this regards, treating all arts with class analysis is a systemic absurdity with tremendous uproars that attempts to achieve the negative to the "generality" through the mechanical understanding of the "particularity". In his paper of *Expressing with Poetic Grace, Carrying the Spirit and Blending New Technique – On Beijing Garden Design in the 9th China (Beijing) International Garden Expo*, Mr. Xia Chenggang introduces his thinking and practice of inheriting and developing the essence of Chinese royal garden art in the design and construction of "Beijing Garden" in the 9th Garden Expo, which has a big significance in practice and theory.

The State Council has recently released the Views on *Strengthening the Construction of Urban Infrastructure*, and the contents closely related with landscape architecture are: to promote the construction of ecological garden cities, and to strengthen the planning and construction of community parks, street parks, country parks, and green corridors, to ensure that the per capita green area in the old city is not less than $5m^2$ and the park green space service radius coverage not less than 60% in 2015. Fulfilling these requirements will considerably improve the basic outlook and ecological conditions of Chinese cities, which would have boundless benefits. Of course, the difficulty of implementation is also great, and our landscape architecture professionals should work harder!

In this Views, local governments are also required to prepare urban drainage and waterlogging prevention facilities planning, and accelerate the transformation of rain and sewage diversion pipes and networks and the construction of flood drainage and waterlogging prevention facilities, and the aim is to build a better urban drainage, waterlogging prevention, and flood control systems in about 10 years. It happens that we have the paper of *The Build of Urban Ecological Drainage System Based on the Watershed Conception – A Case Study of the Construction of the Urban Ecological Drainage System of Xupu County* this month, touching upon the issue of urban flood control, and it is very good that the paper highlights the harmony with nature and making the best use of the circumstances. In recent years many strange phenomenon have appeared in urban water affairs of Chinese cities: mainly relying on the design and construction of pumping stations to manage the flow regardless of the terrains; digging hills as the "flood control pools"; collecting all ground surface water and storing floodwater in underground holes (the so-called deep); building wetland parks in areas of serious water shortage; someone even proposes to dig the whole country's green space 20cm lower to control the flood, which is in fact to destroy all the achievements of urban green space construction in the past years; and so on. In short, the urban flood control patterns are ever-changing pattern, which seem effective in isolation, but they all have a common characteristic: the more costly the better. Many green space systems become the substantial victims. Our landscape architecture industry should be vigilant of this.

In the past, our industry's work was basically within the scope of "construction land", so we paid little attention to "non-construction land". The paper of *Strategy Study on the Non-constructive Land Planning from Green Infrastructure Perspective* puts our industry's focus on the field of "non-constructive land" for the first time, and it is very interesting and worthy of attention.

2013年11期

2013年10月的"主编心语"上谈到了科学和艺术的关系，谈到艺术想象有时更加靠近真理。出于对杨丽萍的舞技之倾慕，9月底我在保利剧院欣赏了她的收官大作——舞剧《孔雀》。让我惊异的是，作为舞蹈艺术家，她对宇宙的终极思考，对生命的赞美，对艺术本质的理解，远远超过我们中的许多名人大师。《孔雀》力图告诉人们：是神灵（上天？规律？）和时间无情地主宰着宇宙；生命诞生于土地，有了生命才有了美善和丑恶；生命是那么美好，却摆脱不了丑；丑虽然也艳羡美，但是在群美的环绕中它还是本性难移，而美在群丑的围攻下，其前景或是毁灭，或是升华；对于美善与丑恶之争，神灵总是冷酷地旁观，时间照样无情地运转；生命或许灭亡，神灵还是神灵，直至时间戛然而止。

当然，上述理念是无法全部用所谓严格的"科学"手段来验证的，这恰恰证明了科学的局限性。但是，这些理念并不深奥，凡是以谦虚的态度观察和体验世界的人，凡是不把自己头脑中的东西看作高于一切的人，其思想几乎都可以走到这一步。于是，一个念头总是在我头脑中徘徊：我们自以为把大量生命体按照我们理解的所谓规律（包括形式美的规律）进行排列组合，就构成了LA学科的核心或精华，是顺天行道，还是逆天行为？我们是否应该想想，为何大自然不是这样地显示这些规律？

人，与孔雀还不一样，人是生物中的异数，有能力掌握一定的规律，所以，我不是自然生态主义者，我觉得大自然应该为人留出一块空间，让人类发挥自己的优长，同时需要注意，丑恶也是人类的创造。终于，我又回到思路的起点：既然不可能让人类来消灭人类的创造性，能不能给它留出一些空间，让它充分显露和发挥，从而在人天之间（实际上还是人与人之间）达成一种交易，不要去毁掉更大空间的大自然？能否（比如说）把大地至少一半还给万物，交回自然；再划出一块地面（比如毛主席常说的1%~3%，最多不超过5%）交给城市，交给人类，让人类中的一些人可以自由发挥？其余的土地，则让它处于中性地位，即供第一产业（农林牧渔）和人类游憩所使用，以体现上天对人的好生之德。

我在20世纪50年代上小学、初中的时候，教室墙上总爱贴着不少名人语录，其中一句是："不要等待大自然的恩赐，向自然索取是我们的任务！——米丘林"。这位米丘林是个苏联的果树育种家，70后的一代就可能已没听说过此人了，其本人可能是个老实人，不过被打着科学旗号的某些政治野心家（诸如苏联靠把摩尔根遗传学打成资产阶级反动学说，当上农业科学院院长的李森科之流）利用而已。小时候觉得这句话很鼓舞人，老了回头看看，凡是总想无限索取（不管是向别人，还是向自然）的人，他们的结局归根结底应该打上一个大问号。

谨用上述想法作为本期主题《亚洲文化景观》的推介语。

Phase 11, 2013

In "Chief Editor's Words" of October of 2013, we talked about the relationship between science and art and that the artistic imagination sometimes comes much closer to the truth. Out of the admiration of Yang Liping's dancing skills, I enjoyed her ending masterpiece—the dance drama "Peacock"—in Poly Theatre at the end of September. What surprised me is that, as a dance artist, her ultimate thinking of the cosmos, the praise of life, and the understanding of the nature of art, are much higher than many celebrities and masters. The "Peacock" was trying to tell people: it is the divinities (Gods? Rules?) and time that master the universe mercilessly; life was born in the land, and then there are beauty and virtue and evil and ugliness; life is so beautiful, but it could not break away from evil and ugliness; although the evil and ugliness envy beauty and virtue, it is hard to change their nature even in the surrounding beauty and virtue, and the beauty and virtue face the prospects of destruction or sublimation under the siege of evil and ugliness; in the fighting between the beauty and virtue and the evil and ugliness, the divinities always coldly stand by, and the time still runs inexorably; life may perish, but the divinities are still there, until the time halts.

Of course, the above-mentioned ideas could not all be verified by the so-called strict "science" methods, which justifies the limitations of science. However, these ideas are not so hard to understand, and for people who observe and experience the world with humility and do not regard something in the minds of above all else, their thoughts almost can all come to this step. Thus, a thought always lingers in my mind: we think that we put a lot of living things for permutations and combinations in accordance with our understanding of the so-called laws (including the law of formal beauty) and that will constitute the core or essence of the landscape architecture discipline, does it go along with or against the natural law? Should we think about that why the nature does not show the laws that way?

People are not the same as peacocks. People are the anomaly among the creatures and have the ability to master certain rules. So, I am not the natural ecology advocate, and I think the nature should leave a space for people to play their superiority, and we also need to pay attention that the evil and ugliness are also human creation. Finally, I went back to the starting point of thinking: since it is impossible for human beings to destroy their creativity, why not leave them some space to show and play, resulting in a deal between human and heaven (in fact still between humans) and not to destroy the larger space of the nature? Can (for example) at least half of the earth be returned to the nature, and some parts of land (like 1%~3%, or 5% to the most) to the city, to the human beings, so that some of them can have the free space to play? The rest of the land can stay in the neutral position, which is for the primary sector (agriculture, forestry, animal husbandry and fishery) and human recreational use, to reflect the virtue of divinities towards human beings.

In the 1950s when I was in my elementary and junior high schools, there were always many quotations of celebrities and masters posted on classroom walls, and one said: "Do not wait for the gift of nature, and our mission is to claim from nature.—Michurin." Michurin was the fruit tree breeder of the former USSR, and those who were born in 1970s might have never heard of this man. He might be an honest man, but was exploited by certain political careerists (those like Lysenko who took the position of the President of USSR Academy of Agricultural Sciences by labeling the Morgan genetics as the bourgeois reactionary doctrine). The quotation sounded inspiring for a child, but now looking back after years, those always with unlimited claims and demands (whether to nature or not) should be marked with a big question mark.

I would like to take these ideas as the introductory remarks for the theme topic of this month, "Asian Cultural Landscape".

2013年12期

2013年11月14日，《南方都市报》胡晴舫专栏短文的题目是《人造风景线》，写道："当我们想到风景线，是否一定青山绿水，城市能否是一种自然？……如果看见的是一片灰色的巴黎屋顶或伊斯坦堡高高低低的楼房建筑，是否也算一道漂亮风景线？城市一定不是自然环境的一部分吗？"

由此我联想到近来IT界也有人提出：虚拟世界也是人们头脑之外的，所以也是客观世界的一部分。正如我在2013年第9期主编心语中指出的：这种思维的最大危险"在于人类和自然以及现实之间被大数据隔离，由于逃避了事实检验这一关，人性中'恶'的潘多拉盒子可能被随意打开。"

所谓主观与客观，是有了人（或许包括外星人）才出现的对立面。何谓客观事物？即不以人的意志为转移的事物。城市、楼房、电子游戏之类，本来就是按照人的意志创造出来的，怎能算作环境自然的一部分？人的思想如果仅仅停留在头脑中不外化，没有什么危害，这也是进步国家取消了思想罪的重要原因。但思想一旦外化，就必然利用自然材料，所以这种产物的属性已经不是纯客观。当然，人造物的主观成分的权重，是有区别的，我想，书本的主观权重最大，其次是艺术品特别是耗材极少的纯艺术品；权重最小的至少包括矿产品和初级农产品。

在总是理不清人与自然关系的西方世界总是有些人喜欢把"自然"这个概念搅浑，这种作法混淆了主客观的区别，为主观权力的任意扩张，为人类对自然的贪婪剥夺打开了思想通道。比如，城市和建筑需要大量的自然和社会资源，它不是纯艺术，不适于单纯从纯艺术的角度去评价，建筑和城市过分地扩张必然影响人与自然的平衡以及人与人之间关系的平衡。最近一个什么劳什子的"世界高层都市建筑学会"把"2013年度全球最佳高层建筑奖"颁给了央视的"大裤衩子"，我们不讲别的，仅就巨大悬臂所造成的巨大浪费这一条就足以证明这个奖励的荒谬。美国为何不把建在9·11恐怖袭击遗址上540m高的世贸大楼也搞成一个撅屁股哈腰的违背自然规律的地标？那不是更具有世界高层都市建筑学会评委会所谓"作为结构工程学上的杰作，央视大楼是那些希望打破摩天大楼既定设计条框的人的一个实体教学课。建筑的设计挑战传统，同时也是对紧密合作和研究并最终达成成果的验证和奖励"的作用吗？

风景园林虽然也不适于仅用艺术标准评判，但有所不同，它所利用的主要材料（植物）的本性还是"上帝"给的，其目的是尽量在"上帝"给出的条件中为人创造最好的环境，所以有人称之为第二自然（或许有一天人类通过转基因之类方法进一步剥夺了"上帝"的权利范围，但是否就如美术作品之类算作纯人工产物，还需要探讨）。正因如此，风景园林才获得了国家的特别重视，成为一级学科，成为在中国改革开放大潮中越来越被重视的生态建设的一支生力军。

胡晴舫先生最后道出了令风景园林人深省的思绪："关于城市蚕食大自然的反省，好比人类对科技的使用，当我们逐渐习惯了科技编写之后的自然、城市改造之后的环境，我们对地球自然的依赖难道真的少了吗？还是只是我们忘记了？人类对自然的向往，如同动物的求生直觉。当站在画廊里观赏画家的风景画，画中看似没有一个'人'，然而那个'人'就站在画的前面，而他的欲望、他的生活方式却像无形的手，改变了他自己眼前的风景线。"

Phase 12, 2013

Hu Qingfang's essay in the November 14 edition of *Southern Metropolis Daily* was titled *Artificial Landscape*, which wrote: "When we think of landscape, should it necessarily be green mountains and clear water, or city can also be a kind of nature?... Can a stretch of grey roofs in Paris or the jagged buildings in Istanbul be also be seen as a beautiful landscape? Is city definitely not a part of the natural environment?"

From this I connect in mind that someone from the IT sector also suggested recently: the virtual world is also out of people's minds, so it should be part of the objective world. Just as I mentioned in the September issue of our journal, the biggest danger of this kind of thinking "lies in the separation of human with nature and reality by the big data, and because of the escape from the point of reality check, the Pandora's Box of 'evil' in human nature may be freely opened."

The so-called subjective and objective become opposites only with the appearance of human beings (perhaps including the aliens). What are the objective things? Those things that do not change with human will. Things like city, building, and video game are actually created in accordance with the will of people, how can they be counted as part of the natural environment? People's thoughts are harmless if they just stay within mind, which is also the main reason why progressive countries cancelled the ideology penalty. However, once the thoughts come outside of mind, they will inevitably use natural materials, so the properties of the consequences are not purely objective. Of course, the weights of subjective components in the artificial things are different, and I think the subjectivity of books weigh the most, followed by artworks, especially pure artworks with very few material costs, and things with minimal subjectivity at least include minerals and primary agricultural products.

In the Western world where the relationship between man and nature is always in disarray, there are always some people who like to put the concept of "nature" muddy. This approach confuses the distinction between objective and subjective, and opens the ideological channel for the arbitrary expansion of subjective powers and human greedy plundering of the nature. For example, city and building need a lot of natural and social resources, and they are not pure art, so it is unsuitable to evaluate simply from the pure artistic perspective, and the excessive expansion of building and city will inevitably affect the balance between human and nature as well as the balance between humans. Recently a so-called "World High-rise Urban Architecture Society" conferred "2013 Global Best High- rise Architecture Award" on CCTV's "big pants". If nothing else, only the huge waste of the giant cantilever is enough to prove the absurdity of this award. Why has the United States not make the 540m-high World Trade Center building on the relics site of the 911 terrorist attacks into a landmark which sticks up ass, bends the back, and runs against the natural law? Should that more have the so-called functions as the judging panel of "World High-rise Urban Architecture Society" said, "As a masterpiece of structural engineering, the CCTV building is an object lesson for those who wish to break the framesof skyscraper design. The architectural design challenges tradition, and also the verification and reward to close cooperation and research and the eventual results?"

Although it is not suitable to judge landscape architecture only with artistic standards, but it is different, since the properties of main materials (plants) it uses are actually given by the "God", and the aim is to create the best environment for people by making best use of the conditions given by the "God", so it is called the second nature. (Perhaps one day humans further deprive the power range of the "God" through transgenic methods, but it still need to be explored whether it could be regarded as pure artificial works like the fine arts.) For this reason, landscape architecture has gained a special attention of the country and became the first-level discipline and an increasingly emphasized new force for ecological construction in the reform and opening-up of China.

Mr. Hu Qingfang finally revealed the thoughts that trigger the profound thinking of landscape architecture professionals: "The reflection on urban encroachment of nature is like the human use of technology. When we become accustomed to the technology-transformed nature and the urban- developed environment, does our dependence upon the earth and nature really become less? Or we just forget it? Human yearning for nature is like the survival instincts of animals. When standing in the gallery watching the landscape painting, there is seemingly no 'person', but the 'person' is standing in front of the painting, and his desire and his lifestyle are like invisible hands changing the landscape before his own eyes".

2014年

2007年　2008年　2009年　2010年　2011年　2012年　2013年　　　2015年　2016年

2014年1期

2013年12月12~13日，中央城镇化工作会议在北京举行。我的记忆中，过去研究城市建设问题的会议最高级别也就是达到国务院，可见此次会议之重要。新华社发表的会议公告总长近3600字，涉及风景园林行业的大约800~900字，其中最主要的有：

"高度重视生态安全，扩大森林、湖泊、湿地等绿色生态空间比重，增强水源涵养能力和环境容量；不断改善环境质量，减少主要污染物排放总量，控制开发强度，增强抵御和减缓自然灾害能力，提高历史文物保护水平。"

"要坚持生态文明，着力推进绿色发展、循环发展、低碳发展，尽可能减少对自然的干扰和损害，节约集约利用土地、水、能源等资源。要传承文化，发展有历史记忆、地域特色、民族特点的美丽城镇。"

"要以人为本，推进以人为核心的城镇化，提高城镇人口素质和居民生活质量。"

"按照促进生产空间集约高效、生活空间宜居适度、生态空间山清水秀的总体要求，形成生产、生活、生态空间的合理结构。减少工业用地，适当增加生活用地特别是居住用地，切实保护耕地、园地、菜地等农业空间，划定生态红线。"

"根据区域自然条件，科学设置开发强度，尽快把每个城市，特别是特大城市开发边界划定，把城市放在大自然中，把绿水青山保留给城市居民。"

"要体现尊重自然、顺应自然、天人合一的理念，依托现有山水脉络等独特风光，让城市融入大自然，让居民望得见山、看得见水、记得住乡愁；要融入现代元素，更要保护和弘扬传统优秀文化，延续城市历史文脉；要融入让群众生活更舒适的理念，体现在每一个细节中……在促进城乡一体化发展中，要注意保留村庄原始风貌，慎砍树、不填湖、少拆房，尽可能在原有村庄形态上改善居民生活条件。"

"城市规划要由扩张性规划逐步转向限定城市边界、优化空间结构的规划。城市规划要保持连续性，不能政府一换届、规划就换届。编制空间规划和城市规划要多听取群众意见、尊重专家意见，形成后要通过立法形式确定下来，使之具有法律权威性。"

上述文字约占文件总长1/5。要将这些内容落实到行动中，教育科研必须走在前面。会议文件一开始就强调我国必须"城镇化目标正确、方向对头，走出一条新路，"又指出我国需要"培养一批专家型的城市管理干部，用科学态度、先进理念、专业知识建设和管理城市。"我想，我们学科今后很长时间主要将围绕着这些问题开展工作。

我特别欣赏文件所说的"推进城镇化必须从我国社会主义初级阶段基本国情出发，遵循规律，因势利导"。从基本国情出发，就不要动辄洋人如何如何；遵循规律，就要尊重科学，而不是观念第一，理念为先；因势利导，就必须实事求是，强化调研，而不是拍脑袋决策，感情用事。其实，所有这一切的基础都是科学研究。我们急需开展的科研工作很多，例如何谓生态文明；何谓美丽城镇；如何科学地提高城镇土地利用效率，制定科学的城镇建成区人口密度指标；上述指标与融入让群众生活更舒适的理念是什么关系；如何制定生活空间宜居适度的考核标准；风景园林如何参与增强抵御和减缓自然灾害能力，节约资源；风景园林师在提高城镇人口素质和居民生活质量中如何发挥作用；如何评价城市的生产、生活、生态空间的结构合理性；如何划定大环境的生态红线；什么样的城市与大自然的关系是科学合理的；现代化与乡愁是什么关系；如何具体解决融入现代元素与保护和弘扬传统优秀文化、延续城市历史文脉之关系；如此等等。许多人问我如何申报课题？什么文章有发表价值？循着这些方面开展研究不就是答案嘛！

Phase 1, 2014

The Central Urbanization Work Conference was held in Beijing in December 12-13, 2013, and it was very important and high-leveled. Xinhua News Agency published the meeting announcement in the total length of nearly 3,600 words, among which about one-fifth are about landscape architecture industry, and the most important include:

"To attach great importance to ecological security, expand the portion of forests, lakes, wetlands and other green ecological spaces, and enhance the ability of water conservation and environmental capacity; continually improve environmental quality and reduce emissions of major pollutants, control development intensity, enhance ability to withstand and mitigate natural disasters, and improve historical heritage protection."

"To insist on ecological civilization, put efforts to promote green development, recycle development, and low-carbon development, minimize disturbance and damage to the nature, and make economical and intensive use of land, water, energy and other resources. To carry on cultural heritage, and develop beautiful towns with historical memories, geographical features and ethnic characteristics."

"To form a rational structure of production, life and ecological spaces. To reduce industrial land, appropriately increase living space, especially residential land, practically protect arable land, garden, vegetable land and other agricultural space, and designate ecological red line."

"To reflect the idea of respect for nature, harmony with nature, and human- universe unity, rely on the existing unique sceneries and landscape contexts, and blend the city into the nature; to integrate with modern elements, and also protect and promote traditional excellent cultures, for the continuation of the historical contexts of cities; to integrate with the idea of making people's lives more comfortable… In promoting the development of urban-rural integration, to pay attention to keep the original style of the village, and improve the living conditions of residents in the original form of village as far as possible."

"Urban planning should gradually shift from the expansion planning to the planning that delimitates city boundaries and optimizes spatial structures. Urban planning should keep the continuity, instead of changing the planning along with the change of the government. The spatial planning and urban planning should listen to people's opinions, respect experts' comments, and be confirmed through the legislation to empower them with legal authority."

To implement these contents into actions, education and scientific research should go in the front. The file stressed in the beginning that China should "take the correct urbanization goal and the right direction, and make a new way", and also pointed out that China needs to "train a group of expert-type urban administrators, and build and manage the city with the scientific attitude, advanced concepts, and professional knowledge." I think our discipline will carry out the work around these issues for a long time in the future.

I especially appreciate that the document said, "promoting urbanization should start from the basic national conditions of the primary stage of socialism of China, follow the laws, and make the best use of the circumstances." In fact, the basis for all these is scientific research. We have a lot of research work need to be carried out urgently, for example, what is ecological civilization; what is beautiful town; how to scientifically improve the efficiency of urban land use; what is the relationship between these indicators with the idea of making people's lives more comfortable; how to set the assessment criteria for the modest living space; how landscape architecture functions in enhancing the abilities to withstand and mitigate against natural disasters and conserving resources; what kind of roles landscape architects play in improving the quality of the urban population and the quality of life for residents; how to evaluate the structural rationality of urban production, living and ecological spaces; how to delimitate the ecological red lines for the larger environment; what kind of relationship between city and nature is scientific and reasonable; what is the relationship between modernization and nostalgia; how to substantially well deal with the relationship between integrating with modern elements, protecting and promoting the traditional outstanding cultures, and continuing the historical and cultural contexts of cities; and so on. Many people ask me how to apply for research projects and what kind of papers are easier to publish, and doing research on these topics is rightly the answer.

2014 年 2 期

这两年我们编辑关于遗产的主题比较集中，原因很多，其中之一是在我们的头脑里有一种想法：与前些年对遗产的破坏相比，与国内至今对遗产的认识相比，我们说的还不算太多。风景园林的任务是正确处理天人关系，艺术创作不是我们唯一的任务，艺术界的动态仅仅是我们的参照系之一，我们没必要处处与艺术界攀比。

本期主题"文化景观与遗产保护"里，我首先推荐《历史公园保护与发展策略》一文。英国邱园的经验证明，与时间和空间关系最为密切的园林是不可能一蹴而就的，也许只有习惯于从固定视点欣赏画面的西方人才能产生"最终效果"之类的理念，但他们一直在探索改革。历史上中国园林就是活态的，一直在修改变动，绝不是某个"大师"设计后就固定了。因此对风景园林的评价不能抄袭源于西方石头遗产的方法，对园林的保护也不能抄袭源于对几何形体式园林的博物馆式保护。在评价的基础上分类分级，特别是区分博物馆型和活态型，应该是我国风景园林文化保护改革的重点方向。至于如何保护数量不多的处于世界顶级的中国古典园林，还需要中国人和世界的精英们共同认真探讨，园林这种自然与文化、物质与非物质遗产的综合体应该探寻自己的保护理论。十分明显，我国在文化遗产保护的理论和政策上又再次落后了，所以如何处理好行政主导与跨行政行业之间的关系，应该列入改革的内容。

谈到法制建设，本期《我国城镇园林标准建设概况研究》首次对国内外行业标准体系建设、组织编制以及管理情况进行了全面总结，并提出了若干建议。从文章中我们还感觉到，除了城镇园林，如果还能对"风景名胜区"、"风景园林综合"的标准体系同样进行一次系统整理，那就更好了。

近年来，我受益最大的本专业的国外学术文章有两篇，一篇是2012年第一期的《可持续之美》（作者Elizabeth K. Meyer），另一篇是本期Arnold. R. Alanen的《为何保护文化景观》，两篇都是美国人写的，也都道出了所谓LA的核心困境，美国精英的洞察力和聪明，普通美国人的真诚坦率，由此可见一斑。前一篇的内容这里就不再重复，我们从后一篇至少可以知道，原来所谓"景观建筑（LA在中文的一种翻译）"可以不懂"建筑（指中文的房屋）"，知道源于地理学的landscape在地理界的尴尬，知道源于固定视点、固定画面的landscape（对此，翻阅任何一本20世纪50年代以前的英文词典即可明了）在面对现实中不断转换的"空间环境"时是多么束手无策。写到这里，我突然悟道：所谓景观都市主义把"图形"的意义捧到那么高，用所谓"非自然式景观"的图形来对付大地，原来根源在这里！其实，我们中国的老祖宗根本就没有这些烦恼，我们何必一定要去主动重蹈一次覆辙？

综上所述，风景园林必须建立自己的坐标体系，这个体系是这门学科的唯一代表，而其他体系只能在这个体系中寻找自己的合适位置，更不要去妄想在风景园林界充当主导。这个任务，可以也应当由中国的风景园林人来解决！

Phase 2, 2014

Our journal has had many themes on heritage in recent years, and one of the many reasons is that in our minds there is an idea: compared with the destructions of heritage in previous years and the understanding of heritage so far in China, we have not talked too much about it. Landscape architecture is to correctly handle the relationship between human and heaven, and artistic creation is not our only task. The trends of arts community is just one of our frame of reference, and we do not need comparisons with the arts everywhere.

In this month's theme "cultural landscape and heritage protection", the first one I recommend is *Protection and Development Strategy of Historic Park*. The experience of the Kew Garden of the United Kingdom proved that landscape architecture that has the closest relationship with time and space cannot be accomplished at one stroke. Perhaps only Westerners who are accustomed to appreciate the picture from a fixed viewpoint will have the ideas like the "ultimate effect", but they have been exploring reforms. Historically Chinese garden is in a living state, always being modified and changed, by no means a fixed design by some "master". So the evaluation of landscape architecture should not copy from the methods originating from the Western stone heritage, nor could the protection of landscape architecture copy from the museum-style protection of geometric gardens. It would be the major direction of Chinese landscape architecture cultural protection reform to make classification and grading based on evaluation, especially the distinguishing of the museum type and the living-state type. As to the protection of the rare and world-class Chinese classical gardens, Chinese and world elites need to explore together seriously. Landscape architecture is a complex of nature and culture, material and non-material heritage, and its own protection theories should be explored. Obviously, China is left behind again in the theory and policy of cultural heritage protection, so how to handle the relationship between government-led administration and cross-department and cross-industry interaction should be included in the reform.

On legislative establishment, this month's paper *Study on the General Construction Situation of Chinese Urban Landscape Standards* made a comprehensive summary of the domestic and foreign industry standard system construction, organization, and management for the first time, and put forward a number of proposals. We also feel from the paper that, in addition to urban landscape, it would be better if the standards systems of "scenic sites" and "integrated landscape architecture" could also be sorted out in the same way.

In recent years, there are two foreign academic papers that I enjoy the greatest benefit, one is Elizabeth K. Meyer's paper of *Sustaining Beauty* (in the January 2012 edition of our journal), and the other is this month's *Why Cultural Landscape Preservation?* by Arnold R. Alanen. The two papers are both written by Americans and both revealed the core awkward situation of landscape architecture, and the insight and intelligence of American elites and the sincerity and honesty of ordinary American people are evident. I will not repeat the contents of the former paper, while we can at least know from the latter one that the so-called "scenic building (a Chinese translation of LA)" can have nothing to do with "architecture (referring to buildings in Chinese)", "landscape" that originated from geography has an awkward situation in the geographical circle, "landscape" that originated from fixed viewpoint and fixed image (seen in English dictionaries before 1950s) is helpless in the ever-changing "spatial environment" in the reality. Now I suddenly realize that: the so-called landscape urbanism puts the meaning of "graphics" up to a high level and uses the so-called "unnatural style landscape" graphics to deal with the land, and the original root is here. In fact, our Chinese ancestors simply did not have these troubles, why should we have to repeat the mistakes?

In summary, landscape architecture should establish its own coordinate system, and this system is the only representative of this discipline, while other systems could only find their proper places in this system, never to play the dominant role in the landscape architecture circle. This task can and should be solved by Chinese landscape architecture professionals.

2014年3期

本期主题"女性与风景园林"是由我国当下几位女性骨干刘晖、杜春兰、金荷仙等发起的。能够与"三八妇女节"呼应,赶在3月出版,多亏了几位发起人的努力,在此表示感谢。细读主题文章,有一个问题引起我的关注:风景园林是否更偏于女性特质?

重庆大学杜春兰教授直言:"女性美与自然美原本相似,而女性更是城市中自然的缩影"。天津大学张天洁博士《管窥女性主义视角下的风景园林史英文文献研究》中有这样一段话:"美国风景园林师协会在成立之初便拒绝了传统的园艺和花园设计协会,而建立了同建筑学和土木工程等职业的联系,这样规避了被认为女性化的自然和园艺领域。"这段话道出了一个实情:在对待人与自然的关系上,landscape architecture走出了一步有问题的棋。毫无疑问,在人类史上女性是园林的最初开拓者。我想,在我国人居环境学科群的3个一级学科中,风景园林显然更偏于阴性,如果把风景园林也改造成阳性为主,这个学科群就失去了平衡。然而历史经历了一个大螺旋,近数千年来,男性逐渐成了风景园林的主角,直到最近,我国各地学校都在反映风景园林学生中女生数量和成绩普遍超越男生,这是否意味着风景园林史上一个新的大时代的到来?

林箐教授明确指出:"女性没有男性那么雄心勃勃",我觉得这是好事。女性比男性更细心和敏感,也更热爱生命,把处理人类与自然关系这一个最大问题(另一个最大问题是人与人之间的关系)主要交给女性引领,比交给那些不知天高地厚,动不动就想显示自己"伟大"力量的愣头青男人,可能更靠谱。虽说"人定胜天"之类的话语已经没有了市场,实际上很多信奉科学主义的人内心里还是以为人的能力是无敌的。在自然面前,人类是自以为高出一头,还是应该持有谦卑的心态,这个问题并没有根本解决。本来,在对待看不清楚的问题时,采取小心的态度,是真正的负责态度,当初邓小平提出改革要"摸着石头过河"就是一个典范。顶层设计,只有在基本上做到全盘掌握,了然于胸的前提下,才可以实行。这也是我对人类现在就企图大规模规划地球表面感到深深忧虑的基本原因。

最近,美的集团的基金会捐资在广东佛山顺德区建设一个公园,邀请了建筑、规划、园林3家设计院出方案,建筑院的方案明显以建筑群的景观效果为引领,规划院的方案则以表达规划师的场地控制能力为意图,只有园林院的方案在追逐"天人合一"的境界,似乎验证了我的上述观察。

清华大学沈洁同学的《从哲学美学看中西方传统园林美的差异》表现出很高的哲学和美学修养,我读后的主要感想是人类对"自然"的解读一定要慎重。而武汉大学魏合义同学的《从ASLA年会议题看美国LA研究的发展趋势》正切合了许多师生的现实需求。从以上2篇文章中我们可以体会到中国的博士研究生成长的脉搏,令人欣慰。不过我有个感想:我们的教授和博士何时能够多了解些中国的实际和历史,不再读了美国的"圣贤书"便"下车伊始就哇啦哇啦(毛泽东语)"?那时我国就真的自立于世界民族之林了!比如鼓吹城市农业时应该想想:中国的城市哪里去找地来搞农业?农业进城还是不是走紧缩高效城市的路线?中国城市环境下生产的农产品还能吃吗?

在实践方面,丁朝华研究员的《树木移植成活的新理论、新技术和新方法》和任梦非高级工程师的《我国西北地区大型景观湖储水防渗技术措施探讨——以新疆昌吉市滨湖河中央公园项目为例》价值很高,特予推荐。

Phase 3, 2014

The theme of this month "Female and Landscape Architecture" is initiated by several prominent females in the landscape architecture industry, Liu Hui, Du Chunlan, Jin Hexian, and others. Thanks to these promoters' contributions, the journal could be published in March and in concert with the Women's Day. Reading through the papers, a question causes my concern: whether landscape architecture is more biased toward femininity?

Professor Du Chunlan of Chongqing University puts it bluntly: "Female beauty and natural beauty are originally similar, and women are even the microcosm of nature in city." There is a passage in the paper of Dr. Zhang Tianjie of Tianjin University, *A Brief Review of the English Literature Studies on the History of Landscape Architecture from a Feminist Perspective*, that "in its early days of establishment, the American Society of Landscape Architects rejected traditional gardening and garden design associations and set up links with the professions like architecture and civil engineering instead, which avoided the natural and horticultural fields that were considered feminine." This passage tells a truth: in dealing with the relationship between human and nature, landscape architecture took a problematic step. There is no doubt that women were the pioneers of gardens in human history. I think in the three first-level disciplines in the human habitat environment discipline group of China, landscape architecture is more feminine, and if transforming it more masculine, the discipline group will lose balance. However, history has experienced a large spiral, in thousands of years, men have gradually become the protagonist of landscape architecture. Until recently, many universities report that the numbers and academic performances of female students are better than that of male students in general. Does this mean the arrival of a new era in the history of landscape architecture?

Professor Lin Qing clearly states: "Women are not as ambitious as men." I think this is a good thing. Women are more attentive and sensitive than men, and they love life more. It is more reliable to let the big issue of dealing with the relationship between human and nature (another one is the interpersonal relationship) led by women rather than by harum-scarum men who have an exaggerated opinion of their abilities and frequently want to display their "great" powers. Although there has been no market for these sort of discourses like "man can conquer nature", in the heart of scientism believers they still think that the human ability is invincible. In the face of nature, the question whether human should regard themselves higher than nature or they should hold humility is not fundamentally solved. Actually, when dealing with problems that are not clear, it is truly responsible to take the careful attitude, and it is an example when Deng Xiaoping proposed that reforms should go ahead by "crossing the river by feeling the stones" is an example. Top-level design is possible only on the premise of full grasp and understanding. This is also the basic causes that I am deeply worried that people attempt to make large-scale planning of the surface of the Earth.

Recently, the foundation of Midea Group donated to build a park in Shunde District of Guangdong Foshan, and invited three institutes of architecture, planning and landscape architecture to make the plan. The architecture institute's plan was obviously guided by the visual effects of the building complex, the planning institute's plan aimed to express the planner's ability of site control, and only the landscape architecture institute's plan chased the realm of "human-universe unity", which seems to validate my above observations.

The Comparison of Traditional Chinese Gardens and Western Gardens from the Perspective of Philosophy and Aesthetics by Dr. Shen Jie of Tsinghua University shows high philosophical and aesthetic accomplishments, and the main impression after my reading is that human interpretation of "nature" must be careful. *A Discussion on the Landscape Architecture Research Trends in the USA Based on the Analysis of the ASLA Annual Meeting Topics* by Wei Heyi of Wuhan University is in line with the real needs of many teachers and students. From the two papers we can understand the pulse of the growth of Chinese doctoral students and it is gratifying. But I have a feeling: when can our professors and doctoral students more understand the actual situation and history of China, instead of making impractical or improper decisions just with foreign concepts or doctrines? At that time China will really stand on the its own among the nations of the world. For example, when advocating urban agriculture we should think: where to find land for agriculture in the cities of China? Should urban agriculture follow the route of condensed high-efficient city? Are the agricultural products from the urban environments of China edible?

In practice, *New Theory and Techniques for the Survival of Transplanted Trees* by Researcher Ding Zhaohua and *Discussion on Water Storage and Seepage Control Technologies for Large-scale Landscape Lakes in Northwestern China – Case Study of Lakeside River Central Park Project of Changji, Xinjiang* by Senior Engineer Ren Mengfei are of high value, and I especially recommend them to our readers.

2014 年 4 期

科学与理念，本质上是两回事，科学追求客观规律，理念崇尚个人头脑里的冲动。理念当然可以带动实践的演变，但理念应该通过科学的检验才知道具有多大的真实度，科学也需要通过实践的检验才可知道是否具有"正能量"，因为科学已经常常与政治、资本的利益纠结在一起。当现代西方设计学进入我国的设计领域时，带来更多的是理念，"理念当头"似乎已经成为设计界的圭臬，反之很多设计师把科学视为妨碍理念的障碍和累赘，并以能够打破这个障碍为荣，不幸的是很多似懂非懂的官员和老板也对此趋之若鹜，才得让"理念"这个幽灵在大地上到处游荡，让"北京三鸟"之类的东西得以成为现实。

湿地问题，实质是一个科学问题，不是视觉景观问题。人类居住自古是优选高燥但有水源保障的地点，而湿地卑湿，百虫杂存，百病丛生，并不很适于人类生活。我们现在重视湿地，是科学告诉了我们湿地在地球生态系统中的重要作用。在城市的生态系统中，湿地也具有必要的作用，所以城市当然可以拥有湿地，其中有些可以适当开放供民众接近自然，成为湿地公园，但切忌搞成湿地游乐场，如果把鸟都吓跑了，湿地就失去了一半意义。另外，要区分湿地景观和湿地公园，湿地景观是一种景观样式，许多小型湿地景观其实没有多少生态作用，更多是一种摆设或教材。在人类视觉记忆中，湿地是无数景观类型中的一种，并非特别出色，所以至今所谓湿地设计，也就那么几种模式，我也看不出还能有多大"改进"的余地，除非大量增加人工景观，而这是与湿地建设的本质相矛盾的。在昆明，房地产商搞了个以湿地为卖点的地产项目，有同仁问我怎么样，我劝他如果自己住就别买，就是基于上述想法。

从学科来讲，真正重视湿地，就应该重视科学研究。遗憾的是，至今我们的绝大多数湿地设计，基于的都是理念，而不是自然规律与科研数据。我还没有见过一个设计可以拿出有关水流、水量、水质、生物种类、生物量、生物体内各种有害物质的含量、淤泥性质与数量等的数据，特别是它们与时间、地点、水层等的关系，并在设计中包含有各种必要的后续措施。我不知道一些"设计"所谓一天可以净化多少水有什么依据，反正只要敢吹，就不愁没有市场。如果我们自封为商人，这样做似乎还符合"在商言商"之道，但我们现在自豪的是"一级学科"，这样下去可怎么是好？

所以，当成玉宁教授介绍澳大利亚蒙纳什大学对土壤水文和湿地的研究有相当水平的科学依据时，我特别感兴趣，为此请他组织了这期的主题。当然，有的作者可能不了解我的心结，有的文章可以看作是为向中国介绍自己而作，但至少是在抛砖引玉，对改造我们自己的工作还是大有裨益。其中，我积极推荐《生态型景观——利用雨水生态过滤系统为城市提供多重环境功能与效益》这篇文章，因为它暗示了一种我们可以学习的学科途径。

Phase 4, 2014

Science and concept are substantially two different things that science is in the pursuit of the objective laws while concept advocates personal mind impulses. Of course, concept can drive the evolution of practice, but concept should be tested scientifically to get its fidelity, and science should also pass the test of practice to know whether it has "the positive energy", since science has often been intertwined with the political and capital interests. When modern Western design entered China, more concepts were brought, and "putting the concept first" seems to have become a standard of the design world, while many designers regard science as obstacles and burdens of concept and are proud of breaking it. Unfortunately, many half-comprehending officials and bosses scramble for this, which made the ghost of "concept" wandering around China, and made the buildings of three birds (the bird's nest – National Stadium, the bird's egg – National Centre for the Performing Arts, and the bird's legs – CCTV Headquarters Building) become a reality in Beijing.

Wetland, in essence, is a scientific issue rather than a visual landscape issue. Since ancient times humans have been inhabited dry but water resources secured areas, while wetlands are wet with many insects and diseases, not very suitable for human life. We now pay attention to wetlands, since science tells us the important role of wetlands in the ecosystem of the Earth. In the urban ecosystem, wetlands also have the necessary effect, so the city can certainly have wetlands, and some of them may be appropriately open for the public to get close to the nature and become wetland parks, but they should not be made into wetland playgrounds, otherwise the wetlands will lose the meaning if the birds are scared away. And wetland landscape and wetland park should also be distinguished. Wetland landscape is a kind of landscape style, and many small scale wetland landscapes are actually of little ecological functions, more of decorations or teaching materials. In the human visual memory, wetland is one of the many landscape types, not particularly good, so up to now the so-called wetland designs are still within the several models, and I cannot see much room for "improvements", unless a greatly increased number of artificial landscapes, which is contradictory to the essence of wetland construction. In Kunming, real estate developers take wetland landscape as the selling point, one of my colleagues asked me how about this, and I advised him not to buy if he wanted to live there, which was based on the above ideas.

For the discipline, if wetlands are really attached importance, scientific research should be emphasized. Unfortunately, so far the vast majority of our wetland designs are based on concepts, not the natural laws and scientific data. I have not seen any design that can come up with the data of water flow, water quantity, water quality, biological species, biomass, contents of harmful substances in the living bodies, the nature and quantity of silt, etc., especially their relations with time, location, and aqueous layers, as well as a variety of necessary follow-up measures in the design. I do not know what is the basis of the so-called volume of purified water a day some "design" boasts, but anyway, as long as one dares to boast, there is a market for it. If we are self-styled businessmen, that would be in line with the way of "business is business", but we are proud of the "first-level discipline", and it will not be good if it continues to go this way.

So, when Professor Cheng Yuning introduced that the soil and wetland study in Monash University in Australia has a considerable level of scientific evidence, I was particularly interested and invited him to organize for this month's theme. Of course, some authors may not understand what I think, so some articles can be seen as the introduction of themselves to China, but at least they set the initiative and serve as a modest spur, which is very good to transform our own work. Among them, I recommend the paper of *Ecological Landscape – Stormwater Biofiltration for Delivering Multiple Benefits to Cities*, because it implies a disciplinary way that we can learn.

2014年5期

本期的内容十分丰富，相信对理论更感兴趣的教师和博士们会深有体会。我听到很多战斗在实践第一线的同仁们反映，杂志似乎越来越难懂了，因此我们也很矛盾：我们晋升为一级学科的好处，已经在逐步体现；但是一级学科是社会赋予我们的重任和对我们的企盼，我们自身的学术水平不要说和其他学科的学术尖端水平相比，就是和兄弟学科比较也差一大截。刊物如何走下去，望读者多给我们一些反馈。

2014年4月16日《中国社会科学报》头条标题是"计算社会科学：计算思维与人文灵魂相融合"，读来发觉记者张清俐很有水平，基本清晰准确地反映出事物的全貌，可以感到有能力投入计算社会科学的人应该是人群中的精英。不过，读完掩卷沉思，发现该领域的一些主张还需要进一步思考：

"计算社会科学是社会科学与自然科学交叉的典型代表。"

风景园林学科的同仁大都主张我们是社会科学与自然科学交叉的典型代表。那么，计算社会科学与风景园林，到底谁更典型呢？这个问题从细节上是纠缠不清的，还需要从更高层次审视。对人与自然的关系，有两类看法，一类是认为人的能力无限，最终可以认识一切奥秘，从而掌控自然和自己，其中又可分为"人是上帝"与"人获自由"两派；一类是认为人与自然始终是矛盾的双方，人不能胜天，其中又分为"天人合一"与"天人不相胜"两种看法。我认为，"人获自由"和"天人合一"二者相较，它们的内涵游移度都太大，非常灵活，难以证伪，可暂且不表；仅就"人是上帝"与"天人不相胜"相比，哪个更符合辩证法和中国传统哲学的精神，则是不言而喻的。其实，作者的标题很聪明，它采用了"计算思维与人文灵魂"，避开了"自然科学与社会科学"的提法。我个人倾向于认为社会科学只是人文认知的一个组成部分。

"通过全面分析反映历史行为的数据，原本难以捉摸的人类社会活动变得可被量化、解析和洞见，甚至有可能被预知、得到预先处理。"

科学，是一种求真的方法，曾有过无往不胜的光辉记录，但科学方法的局限性，也是毋庸置疑的。科学可以应用于那些有大量重复性的社会问题研究，而那些主要发生在人类社会领域的再现性很差的现象，例如艺术家的独创，科学基本上是无能为力的，除非科学可以疯狂到再造出秦始皇、苏格拉底、耶稣、李白、贝多芬、"第二次世界大战"……

"传统社会科学一般通过问卷调查的方式收集数据，以这种方式收集的数据往往不具有时间上的连续性，以此对连续的、动态的社会过程进行推断的准确性有限……正是通过（大量）数据挖掘，社会科学家可以处理非线性、有噪声、概念模糊的数据，分析数据质量，从而聚焦于社会过程和关系，分析复杂社会系统。""大数据提出跨学科的协作与训练，学科间统一的理论指导是当前最大的挑战。"

这段话，对我们学科的科学研究工作有很好的警示作用。从科学角度看，我们的多数科研还处于初级水平，有些甚至不入门，不知道如何明确目的，如何进行试验设计，如何分析试验结果，只知道如何缩小题目减少争议，如何立项，如何符合论文格式，如何通过评审之类，这正是前些年我国所谓学术评估、学科评估、教学评估的最大问题，是学术不端的集中表现，浪费了大量科研教育经费。但对于这段话中隐含的对大数据时代的过高预期，我还是担忧，当一个大数据组成的虚拟世界把人类和真实世界隔开后，"实践是检验真理的唯一标准"这条保护了人类几千年的求真法则，还能不能被人们遵循，这的确是个问题。

Phase 5, 2014

The journal is very rich in content this month, which I believe teachers and PhD candidates interested in theories would fully understand. I heard a lot of colleagues in the practical fields saying the journal is becoming incomprehensible, and therefore it is a contradiction: the benefits of landscape architecture uplifted as the first-level discipline are gradually reflected, but the title of first-level discipline also means the great responsibilities and expectations from the society, and the academic level of the discipline is incomparable with the cutting edge or even the brother disciplines. We welcome readers to feedback us on how to run the journal well.

The headline of the April 16 of 2014 edition of *Social Sciences in China* read "Computational Social Sciences: The Blending of Computational Thinking and Humanistic Soul", and I find that reporter Zhang Qingli is with a very high standard and has clearly and accurately drawn the whole picture, and I can feel that those in computational social sciences should be the elites. However, with meditation after reading, I think some ideas in this field needs further consideration.

"Computational social sciences is the typical representative of the intersection of social sciences and natural sciences."

Colleagues in landscape architecture would mostly advocate that our discipline is the typical representative of the intersection of social sciences and natural sciences. Which is more typical? This issue is tangled from the details and need to be reviewed from a higher level. On the relationship between human and nature, there are two types of views, one considers human have infinite capacity and can eventually understand all mysteries, and thus control nature and themselves, which can be divided into two factions of "human as God" and "human with freedom"; the other considers human and nature are always contradictory sides, people cannot triumph, which can be divided into two factions of "human-nature unity" and "human and nature cannot conquer each other". I think the connotations of "human with freedom" and "human-nature unity", if compared with each other, are too wavering, very flexible and difficult to falsification, and we can just leave them. Simply comparing "human as God" and "human and nature cannot conquer each other", it is self-evident which is more in line with dialectics and the spirit of traditional Chinese philosophy. In fact, the author takes the title cleverly, using "computational thinking and humanistic soul" to avoid the wording of "natural sciences and social sciences". Personally, I tend to think that social sciences are only a part of the humanistic cognitions.

"Through a comprehensive analysis of the data reflecting the historical behaviors, the originally elusive human social activities can be quantified, analyzed and understood, and even predicted, and treated beforehand."

Science is a method of seeking truth and enjoys the invincible brilliant record, but the limitations of the scientific method is also beyond doubt. Scientific research can be applied to those repetitive social problems, while for the phenomena with poor reproducibility in human social fields, such as the artist's originality, science is basically powerless, unless science would crazily reproduce the First Emperor of Qin, Socrates, Jesus, Li Bai, Beethoven, World War II…

"The traditional social sciences generally collect data through questionnaires, and the data collected in this way often do not have continuity in time, and the accuracy of the conclusion of the continuous and dynamic social processes based on the data is limited… through (large volume) data mining, social scientists can handle non-linear, noisy, fuzzy concept of data, analyze data quality, and therefore focuses on the social processes and relationships, and analyze complex social systems." "Big data present the interdisciplinary collaboration and training, and the unified interdisciplinary theoretical guidance is currently the biggest challenge."

This passage have a good warning for the scientific research of our discipline. From a scientific point of view, most of our research is still at the primary level, and some even without a start, not knowing how to clarify purpose, how to conduct experimental design, how to analyze the results, only knowing how to narrow the topic and reduce disputes, how to set a project, how to meet the paper format, how to pass the assessment, and the like, which is exactly the biggest problem of the so- called academic assessment, disciplinary assessment, and teaching assessment a few years ago. It was the concentrated show of academic misconduct and wasted a lot of research and education funding. But regarding the too high expectations implied in this passage, I am still worried that, after human and the real world are separated by the virtual world composed of big data, it is really a problem if "practice is the sole criterion for testing truth" that has protected the human race for thousands of years could be followed.

2014年6期

本期发表了清华大学杨锐教授的《论"境"与"境其地"》，这是杨教授多年积累、5年思考结果的结晶。我个人完全同意文章的主旨："境"为风景园林学的核心范畴。此"境"已不再是纯物理、纯天然、纯客观的环境（environment），更不是可以由人任意捏造的对象（object），而是人与自然经过长期互动平衡达到的（以及由此推想还可以较长时间持续下去的）理想境界，这个"境"既是物构造的境，也是人的境，是人类对自然原生物境的提升，并由此同时获得名词与动词的双重含义，用西文方式来表达是jing，或者更应该是大写的JING。

Jing范畴的提出，如能得到社会的普遍承认，必将极大地提升我们学科的地位，同时空前地提升本学界的视野高度，园、囿、林、苑、绿化、景观、环艺、市容、场所、风景、开敞空间、第二自然等多种出现于学科领域内的概念，将汇于一体。也许，将来环境保护、环境卫生、环境地理、环境文化、社会心理、城市生态、人类生态等都会集合在jing的大旗下，那就是大写的JING！

我一直有一个心结：中国人，即使不谈自己历史上的成就，已经进行了人类史上规模最大的地表景物改造工作将近20年，难道真的就拿不出自己的理论？难道仍然是毛主席所批评的"言必称希腊"？现在终于得到一个答案："NO"！

为什么境的范畴是中国人提出？这是因为只有中国人有这样的历史文化积淀。我们知道，大多数古代文明关心的是人与神的关系，只有中国人提出了天的概念，关心人与天的关系。当然，历史上对于天也有被神化的理解，但在这个问题上，对中国和世界影响最大的是老子，他的天就是宇宙，是非人格化的，是自然的。在老子那里宇宙的法则被称为道，道是不以人的意志为转移的。人类是天地所生，自然是人类的父母，人可以认知父母，可以对父母提出要求，但人不能作父母的主人，疯狂掠夺父母的人为天下大众所不齿，这就是中国人对人天关系的基本观念。所以，中国人一贯关注"境"，通过"境"感受父母恩泽，感知天地。

但是，我还认为在全球一体化的大背景下，源于中国的理论体系不应纯粹由中国人提出。基于中国哲学和历史的所谓"中国理论"不应是中国私有的，而应该是世界的，所以具有深厚中国文化修养同时又有留学经历并大量参与国际学术活动的学者，是最有可能作出这种贡献的。其实，20世纪20～40年代留学的人最具有这种资质，可惜他们回国后时运不佳，动乱频仍，国力衰弱，难以获得大量的实践机会和理论探索的时间。

很多人以为风景园林只是个实操的学科，对哲学范畴之类进行研究有什么用？此言谬矣。就人天二极而言，学术探讨范畴无非三大类：天与天、天与人、人与人，前者是自然科学，后者是社会科学，中间就是JING学。风景园林作为处理人天关系的文化，虽然不同的人可能会用不同的方式对待自然，但大自然在相同问题上不会用不同的方式对待不同的人，这就是风景园林共性的最大公约数之所在。风景园林里的文化内容虽然可能大量涉及人与人的关系，但处理人人关系不是风景园林文化本身的核心。认识到这点非常重要，类似逻辑上偷换概念，用次要矛盾掩盖主要矛盾，搅乱风景园林文化的核心关怀之类的东西在全世界都很多,我们如何不被迷惑?如何获得火眼金睛?必须具有思想高度。我认为,"境"范畴的提出,就具备了这种高度。

Phase 6, 2014

The paper of *On "Jing" and Its Development in Landscape Architecture* of Professor Yang Rui from Tsinghua University is published in the current issue, and it is Professor Yang's accumulation over the years and the result of his five-year thinking. Personally, I totally agree with the purport of this paper: "Jing" is the core part of the landscape architecture study. This "Jing" is no longer a purely physical, natural and objective environment, nor any object that can be fabricated by anyone, but the ideal state after a long period of human-nature interactions (and thus presumably also can be sustained longer) instead. It is the physical jing and also the human jing, the human improvement of the natural original physical environment, and therefore it acquires the noun and verb meanings, spelt as jing, or JING in the uppercase.

If universally recognized by the society, the proposal of the category of jing will greatly enhance the status of our discipline and unprecedentedly raise the academic vision of the discipline, the various concepts appeared in the discipline, such as garden, park, forest, yard, greening, landscape, environment art, city appearance, place, scenery, open space, and second nature, will be integrated into one. Perhaps in the future environmental protection, environmental sanitation, environmental geography, environmental culture, social psychology, urban ecology, and human ecology will all be included in jing. And that should be a big thing.

I always have a knot: Chinese people, not talking about their achievements in the history, have been transforming the surface scenery in the largest scale of human history for nearly two decades, is it true that they cannot put forward their own theory? Is it still "There is nowhere but Greece", as Chairman Mao criticized? Now we finally get an answer: "NO"!

Why is the category of jing proposed by Chinese? It is because only the Chinese people have such a historical and cultural heritage. We know that most ancient civilizations concern about the relationship between man and God, while only the Chinese people put forward the concept of heaven and care about the relationship between man and heaven. Of course, there was also the deified understanding of heaven in history, but it is Lao Tzu who gave China and the world the greatest impact, and his heaven is the universe, non-personified and natural. According to Lao Tzu, the law of the universe is known as Taoism, and Taoism does not change with people's will. Human beings are born from heaven and earth, and the nature is the parents of human beings, human beings can recognize parents and request parents, but they are not the masters of their parents, and plundering parents is despised by the public, and this is the basic idea of the Chinese people about the human-heaven relationship. So the Chinese people have always been concerned about environment and feel parents' love and perceive heaven and earth through "jing".

However, I also believe that in the context of globalization, the theoretical system originated from China should not be made solely by the Chinese people. The so- called "Chinese theories" based on Chinese philosophy and history should not be privately-owned by Chinese, but of the world instead. So scholars who have profound Chinese culture accomplishments, study abroad experience and active involvement in international academic activities are most likely to make this contribution. In fact, most of those who studies abroad in 1920s to 1940s had this kind of qualification, but unfortunately they had bad luck when they return the war-devastated and weakened China and it was difficult to get a lot of opportunities for practice and theoretical exploration.

Many people think that landscape architecture is just a practical discipline, what are the use of philosophical studies? It is wrong. Regarding human and heaven, the academic research areas are no more than three kinds: heaven and heaven, heaven and human, human and human; the former is the natural science, the latter is the social science, and the middle is the study of JING. As to landscape architecture – the culture dealing with the human-heaven relationship, although different people may use different ways to treat the nature, but for the same question the nature does not treat different people with different means, and this is the greatest common divisor of landscape architecture commonalities. The cultural contents of landscape architecture might involve a large number of human-human relationships, but dealing with human-human relationships is not the core of landscape architecture culture itself. It is very important to recognize this point. Recently someone moved out the "similar class analysis" to treat the traditional landscape culture and continuously repeated the theory of dividing landscape culture into the so-called "small feet and big feet", and this is to cover the principal contradiction with the secondary contradiction, disrupting the core caring of landscape culture, logically disguised replacement of concept. There are a lot of wrong things around the world such as covering the principal contradiction with the secondary contradiction, disrupting the core caring of landscape culture, covertly exchanging concepts, how can we not be confused? How can we have penetrating eyes insight? You must have thought highly. I think the proposal of the category of "jing" has such a height.

2014 年 7 期

绿城集团在苏州建设了一个住宅区——桃花源，意在传承苏州园林。日前开了一个"南北园林"对话，由苏州大学主办。我参观了一下这个"桃花源"小区，又听了孟兆祯院士和苏州大学曹林娣教授的发言，收获很大，更深刻地感到用西方理论看待中国园林，是路线错误。比如孟先生对同里"退思园"曲廊由依墙到独立作出分析：九为最大，由九曲到九窗，九窗分嵌九字"清·风·明·月·不·须·一·钱·买"，却以和"退思草堂"相对立的"闹红一舸"为结点和转折点，这个处理揭示了园主人表面退而思过，内心却极为扭曲和不满的复杂心态。又如曹教授批评桃花源铺地：厅的北边铺海棠纹，南边铺冰裂纹，整个搞颠倒了！让我震惊，体会到中国文化讲求从联系中看待和认知事物，竟然达到如此细致的程度。又比如，一处墙角，用墙隔开，墙上开了一个贝叶形小门，里面种了一丛竹，点了一块石，曹教授点评：这里面和佛教没有关系，干嘛用贝叶？

我们做设计时想过这些吗？比如，我们为某商人设计了一个"文人园"，想过这个主人的真实心态吗？要知道，真正的艺术是人性和人心的展示，表现技艺只是手段，现代流行的对形式和手段的孤立追求只是表面热闹而已。最近审查了一个城市广告规划，热衷于把仅存的少数代表城市文化的空间也用广告攻占，长远来看，对于自诩是文化人的设计者，只能说被卖了还在帮别人数钱。又如我们设计了一个漏窗，把边缘做成长卷样，可是看出去却只是一株芭蕉，谁见过用长卷画一张芭蕉呢？所以，让风景园林师决定总平面，让建筑师设计建筑，让园艺师在平面图上配置植物，对一个讲究照顾四面八方关系的中国园林来说，是根本不可取的。

我由此想到何为"格物"。西方对一个事物的认知是从将其与四周联系的隔断入手，尽量减少其他因素对其的干扰，或者只留下一两个外因，以求得对事物本身自有规律或对少数外因作用效果的认知，如果还是认知不透，就再次将事物分割，分别深入其中。为了达到对某个事物绝对真理的认识，就必须彻底地割断各种联系和变化，而这在现实中基本做不到，只能用脑子去追寻，这就是毛泽东将"孤立、静止、片面"的认识方法误称为"形而上学"方法的原因。（我在本科时曾将其视为毛主席根本不懂"形而上学"的概念，暗中取笑。）中国人却是反其道而行之，老祖宗们是从事物的变化及其与周围事物的联系之中来观察事物，认识事物，这就是所谓"格物"，离开了联系与变化，中国人就不能认识事物。两种思维方法各有优劣。西方的确可以做到越来越深入、越精细、越准确，但一直这样下去，其世界观就会成为一堆几乎互不相干的碎片的堆积，顶多是若干"图层"的堆叠，像一部电影片，所谓的连续动作，其实是可以打断、分割、剪接、插入的，并不是真实的联系，于是，每个人都有了发明创造的自由，胡编乱造成为不少人出格成名的途径。直至系统论的出现，才开始解决这个问题，但是按照系统论（包括其他新理论）的方法，用数学真正复原世界原本的联系，何其困难，想必有数理根底的人可以想象得到。中国方法的弱点，除了难以深入，还在于过于强调一切联系，不会量化，难分轻重，所以会将一些细微的、表面的联系也强调起来，所谓"风水"里的许多理论就是典型。毛主席特别强调了"抓主要矛盾"，但是如何科学地、动态地抓住主要矛盾，还是需要我们继续探索的重大任务。

Phase 7, 2014

The real estate developer Greentown built a residential community – Taohuayuan (The Peach Garden) – in Suzhou, hoping to carry on the heritage of Suzhou garden, and recently a northern-southern garden dialogue organized by Suzhou University was held there. After visiting the community and listening to the speeches of Academician Meng Zhaozhen and Professor Cao Lindi from Suzhou University, I gained a lot and more deeply felt that it was a wrong route to treat Chinese gardens with Western theories. For example, Academician Meng analyzed the winding corridor of Tuisiyuan (The Retreat and Reflection Garden): nine means the biggest and the most in Chinese thinking, and from nine twists to nine windows, with nine Chinese characters– "qing (clear), feng (wind), ming (bright), yue (moon), bu (no), xu (need), yi (one), qian (money), mai (buy)" – inscribed in the nine windows, and with Naohongyike (The Ship-shaped Building)– opposite to Tuisicaotang (The Retreat and Reflection Cottage) – as the ending and turning point, the design revealed that the garden owner seemed retreating and reflecting, but he had extremely distorted and discontent complicated mentality. Another example is Professor Cao's criticism upon the pavement in Taohuayuan (The Peach Garden): it was totally upside down to make the crabapple-shaped pavement in the north part of the hall and the broken-ice-shaped pavement in the south. I was shocked to feel that the Chinese culture emphasizing the treatment and cognition of things through connections reached such a detailed level. In another example, a corner was fenced by a wall, a Corypha umbraculifera leaf shape door was opened in the wall, and a clump of bamboo and a rock were put in the corner, and Professor Cao criticized: there was no relationship between Buddhism, and what was the use of the Corypha umbraculifera leaf?

Have we ever thought about these in design? For example, when we design a "literary garden" for a businessman, have we thought about the true state of mind of the garden owner? We should know that, the true art is the display of human personality and mind, the display skill is only a means, and the presently popular isolated pursuit of forms and skills is just superficial. Recently I reviewed a city advertising planning, which was keen to put advertisements on the very small number of remaining space that could represent the city culture; in the long run the designer who called himself as intellectual would only be hooked in his own lies. In another example, someone designed an ornamental perforated window, with the edge in the long scroll shape, but there was a Chinese banana tree in sight, who had ever seen a Chinese banana tree painted in a long scroll? So, for Chinese garden looking after various relationships, it is simply unwise to let the landscape designer determine the general plan, let the architect design the building, and let the horticulturist configure plants on the plan respectively.

I thus think of what the "nature of things" is. The Western cognition of a thing starts from the separation of its contact with the surrounding, minimizing the interference with other factors, or leaving only a few external factors, in order to achieve the cognition of their own laws or the effects under the few external factors; if there is still no thorough cognition, the thing will again be divided to get in deeper respectively. To achieve the absolute true cognition of a thing, the various connections and changes should be completely cut off, but it is basically impossible in reality and can only be in pursuit in brain, which is the reason that Mao Zedong mistakenly called the "isolated, static, and one-sided" cognitive method as the "metaphysical" method. (In my undergraduate years, I secretly made fun of that Chairman Mao did not understand the "metaphysical" concept at all.) However, the Chinese people go the opposite, and our ancestors observed and cognized things through the changes of things and their connections with the surroundings, and this is the so-called "nature of things". Without connections and changes, the Chinese people cannot understand things.

The tow ways of thinking have their advantages and disadvantages. The Western one can indeed be more and more in-depth, sophisticated, and accurate. But if being like this all the way down, their view of the world will become a pile of debris almost independent of each other, and at most the stack of several "image layers"; just like the film, the so-called continuous action can actually be broken, split, spliced, inserted, not the real contact, so that everyone has the freedom for inventions, and fabrications become a way to gain fame for a lot of people. Until the emergence of the system theory, it began to solve this problem, but it is quite difficult to truly recover the original connections of the world with mathematics according to the system theory (including other new theories), and I am sure those with mathematical basis can imagine that. The weakness of the Chinese one, in addition to the difficulty to be in-depth, also lies in too much emphasis on all connections, not being able to be quantitative, being hard to identify the weighing points, so that some minor and superficial connections are also emphasized, and many theories in "feng shui" are very typical. Mao Zedong particular emphasized "grasping the principal contradiction", but how to grasp the principal contradiction scientifically and dynamically is still the big task for us to explore continuously.

2014年8期

全球化,本应在平等公正的基础上综合全人类的文化优秀成果,绝不是只对欧美文化的顶礼膜拜。基于"两希(希腊、希伯来)文化"的欧美文化,如今已发展出严重依赖于对自然资源和不发达国家疯狂掠夺的金融资本体制,同时,它又与数字技术结合起来,发展成"两虚(虚拟经济、虚拟世界)文化",这一套东西是否可以将人类带入世界大同,我是很怀疑的。

把中国的文化纳入西洋的系统,值得警惕。这样做的前提,是默认中国一切都落后于西方。中国在近二三百年的确在政治、军事、科技、经济等方面落后了,当然有其传统文化的基因,从人类的长远利益看,这种基因是好还是坏,还不能仅因为不重视经济强权和发展速度而盖棺论定,不过在人天关系的思维上,中国人还是远远高于欧美。对很多真正懂得中国园林的人来说,汉字"园林"不等于源于篱笆圈子里种植东西的"garden",中国园林很早就摆脱了花园或药圃,走上了山水崇拜,进而步入关心人的心灵以及天人关系的境地。同理,中国也不迟于4世纪(东晋)就第一个喊出了"风景"这个词汇,即人可以感知(包括各种感觉和知觉)的被人(包括心灵)介入的自然。根据陈俊愉院士在本刊的回忆,1928年留美,在20世纪50~70年代主管中国城市规划的程世抚先生,于70年代后期就提出中国建立"风景区",从而成为如今中国式国家公园"风景名胜区"和被称为"风景园林"学科的滥觞。如果顺着这个逻辑观察下去,西方人的思维完全是走了一个大圈,至今未走上正路。

西方由于将自然与人工完全对立的思维传统,便遇到自然美的难题,所以landscape与garden是完全不同的起源。在16~17世纪的西方风景画中,就有不少描绘自然的恐怖的作品。至于地理学引入landscape,最初是借用,其研究方法也是自然科学的老思路——用自然规律解释地表的形象(早期的地理学只研究由于地质构造形成的地貌-topography的成因,不关心由生态和社会原因造成同属外貌的植被等)。一旦介入人工因素,地理学就不再是纯自然科学,所以至今地理学界对于landscape还是争议不断。

形式逻辑告诉我们,一个概念的内涵越简单,其外延就越大。许多人以为一个概念里加进去的"内涵"越多,其包括的东西(也就是外延)就越多,这完全是一种对逻辑学的无知。landscape将学科外延扩大,主要靠将内涵简化到只管视觉,特别是日本人将其翻译成景观,又把内涵中的land抛掉,使景观的外延比landscape更加扩大,以致诱使西方人现在想发展出scape这种学科,我想这是许多人特别偏爱"景观"一词的主要原因。西方理性文化自来轻视那种用外表掩盖实质的思想,当代由于金融资本的利益,才使重视感觉刺激的各种流派兴盛起来,scape如何又具有生态意义,至今无人阐述,是故意还是疏忽?然而,当另一些人再想把综合感觉的、特别是涉及文化和精神内涵加进landscape之中时,其外延必然缩小。西方人将这样一种内涵和外延飘忽不定的极易被偷换内涵的概念作为一种学科,实在令人扼腕叹息。

Phase 8, 2014

Globalization should integrate in the outstanding achievements of mankind on the basis of equality and fairness, by no means only paying homage to European and American culture. The European and American culture based on the "Greek and Hebrew cultures" has now developed the financial capital structure heavily dependent on the crazy plundering of natural resources and underdeveloped countries, while it has also combined with digital technology and developed into "the double virtual (virtual economy and virtual worlds) culture", and I am very skeptical whether this series of things could bring mankind into the universal brotherhood.

It is worthy of attention to put the Chinese culture into the Western system. The premise of doing this is to acquiesce in that all of China lag behind the West. China has indeed been lagging behind in the last three hundred years in the political, military, technological, economic and other aspects. Of course, it has its own traditional culture gene, and from the long-term interests of mankind, that the gene is good or bad could not be judged only by not paying attention to economic power and development speed, while as to the human-heaven relationship, Chinese people are still much higher than the European and American ones. For many people who truly understand Chinese garden, the Chinese character "*yuanlin*" (garden) is not equal to the *garden* of planting within the fences. Chinese garden broke away from garden or medicine garden very long ago, began the worship of landscape of mountains and waters, and then then entered into the realm of caring about human mind and the human-heaven relationship. Similarly, China had raised the word of "*fengjing*" (landscape) no later than the fourth century (Eastern Jin Dynasty), namely the nature that was intervened (including through mind) and could be perceived (including senses and perceptions) by people. According to Academician Chen Junyu's memories in our journal, Mr. Cheng Shifu, who studied in U.S. in 1928 and took charge of urban plan of China in 1950s to 1970s, suggested the establishment of "scenic site" in China in 1970s, which became the origins of the Chinese-style national parks "famous scenic sites" and the "landscape architecture" discipline. If observing along this logic, the Western thinking has completely gone in a big circle and has not been on the right path.

Due to the Western thinking tradition of put the nature and the artificial in completely opposite positions, they encounter the problem of natural, so landscape and garden have completely different origins. In the Western paintings of 16th to 17th centuries, there were a lot of works depicting the horror of nature. As to the introduction of geography study into landscape, originally it was borrowing, and its research means was also the old ideas of natural sciences – explaining the image of the earth surface with the laws of nature (early geography studies only researched on the geological causes of topography, and did not care about vegetation that was part of appearance and caused by ecological and social reasons). Once artificial factors being involved, geography is no longer a pure natural science, so up to now the geology circle still has continuous disputes upon landscape.

The formal logic tells that the simpler the connotation of a concept is, the more its extension is. Many people think that the more "connotation" they put into a concept, the more things (namely the extension) it will include, which is a complete ignorance of logic. *Landscape* expands the extension of the discipline, mainly by simplifying the connotation into just visual sense. And especially the Japanese translated *landscape* as *jingguan* (scenery) and cut off *land* in the connotation, and made the extension of *jingguan* even larger than that of *landscape*, thus tricking the Westerner into thinking of establishing the discipline of *scape*, and I think this is the main reason many people especially prefer the word "*jingguan*". The Western rational culture originally despises the ideas that cove the substance with appearance, and it is the interest of contemporary financial capital that make various genres paying attention to sensory stimulations flourishing. On how scape has ecological significance, no one has explained, is it intentional or negligent? However, when others want again to put sensations – especially those related to culture and spiritual connotations – into *landscape*, its extension is bound to shrink. It is really a pity that the Westerners take the concept, which has erratic connotation and extension and whose connotation could be easily changed, as a discipline.

2014年9期

有点巧合，本期有4篇涉及"看"的文章。

《"言说"与"实践"——〈园冶〉与江南园林互文性诠释》一文揭示了一个重要问题：外国园林主要是看的，他们特别重视一眼看去地面上的画面图形，除了英国风景园有些改变，到了现代景观，这一特点又回来了。而中国园林基本是读的，须逐字逐段读下去，直至豁然，所谓"文章是案头之山水，山水是地上之文章"。过去我们行业的教学和研究主要用西方"看"的体系解读中国园林，有些不着边际。

《景观空间视觉吸引机制实验与解析》涉及中外差别的另一个问题，即"境与景"。过去，境的问题一直被空间概念笼罩着，冯纪忠先生的旷奥理论，继承柳宗元，融合空间论，是中国景园理论体系构建过程中的重要一步；孙筱祥先生的"生境、画境、意境"理论，理出了境的概貌；最近杨锐教授的《论"境"与"境其地"》，终于比较完整地结构出了"境"理论的体系。而景物，一直是西方体系的核心，其理论成熟度远比中国大得多。但是，若将景物作为风景园林理论的核心，我们就只有跟在建筑、美术、雕塑等后面跑的份。

《基于可视性图解与视域分析的园林空间造景研究——以重庆市川剧艺术中心为例》同样涉及如何借鉴兄弟学科成果的问题。文章运用Depth map软件，通过可视性图解和视域分析，确定园林空间中的景观敏感点，同时将景物按视觉与行为的影响程度进行分类，进一步制定园林空间的造景对策。这种方法对迅速提高我们的视觉科研水平和提升景观设计效果，非常有用。

西方人对视觉的各种马太效应（即累积优势，强者越强）做了大量研究，的确有用、有效，使他们在制造超乎自然赐予的视觉对象方面，获得强大的能力，并发展出一套设计学体系。不过，如果我们把眼光放长远来观察，所有马太效应都是暂时的，没有长久性的，因为没人会因为这几种效应而盯死一个东西永远不动。这说明还有更强大、更深刻的原理在管理动物的视觉。其实生态学和系统学已经证明，这个世界不会有什么东西一直具有强大的吸引力，否则世界的一切物质和能量都会被这种东西吸引进去，从而毁灭世界。我想，即使有这种东西，为了人类的长远利益，也应该尽量抵制这种东西，而不是盲目追随之。《道德经》曰："天之道，损有余而补不足；人之道，则不然，损不足以奉有余。"老子的这种观察，远比《圣经》深刻。

《中小尺度生态界面的图式语言及应用》一文一开始就介绍了"景观的语言"概念。景观从画面转变到语言，似乎与中国文化更接近了，虽然，语言还不等于文章。Wiki解释语言：Language is the human capacity for acquiring and using complex systems of communication, and a language is any specific example of such a system. The scientific study of language is called linguistics."景观语言"的概念可能源于"建筑语言"。建筑是全部由人操控而成的，所以可以在某种程度上视为人类的一种表达；但是，与建筑不同，景观除了表达人的意图，还含有大量大自然想表达的内容（如自然规律），尺度越大，后者的权重就越大。因此，风景园林套用建筑语言要特别小心，本文从中小尺度景观入手，是得当的。此外，本文还提出了一种可供实践操作的设计方法，我想特别对初学者，还是不错的。关键是学习者应逐渐深入理解这些图式语言背后所隐含的规律和道理，就不会把景观设计当作随意的语音拼凑，而是"读得懂"的文章。

Phase 9, 2014

It is a coincidence that there are four papers related to "seeing" in this issue.

The paper of *"Speaking" and "Practice" — Intertextuality of Yuan Ye and Jiangnan Gardens* reveals an important problem: foreign gardens are mainly for seeing, in that they pay special attention to the graphics on the ground at a glance, and except some changes in British gardens, this feature is back in the modern landscape. However, Chinese gardens are basically for reading, verbatim and paragraph by paragraph until sudden comprehending, which is the so-called "articles are landscape on desk, and landscape is articles on the ground". It is far-fetched that Chinese gardens were interpreted with the Western "seeing" system in our LA teaching and research in the past.

Experiment and Analysis of Visual Attraction Mechanism in Landscape Space relates to another problem of Chinese and foreign differences, namely "jing (realm) and jing (scenery)". In the past, the realm issue had been covered by the concept of space. It has been an important step in the construction of Chinese landscape garden theoretical system that Mr. Feng Jizhong's Kuang'ao Theory inherited from Liu Zongyuan and blended with the space theory; Mr. Sun Xiaoxiang's theory of "habitat environment, picturesque realm and conception" rimmed an overview of jing; and lately Professor Yang Rui's *On "Jing" and Its Development in Landscape Architecture* finally structured a relatively complete system of Jing Theory. Scenery has always been the core of the Western system, and related theories are much more mature than Chinese ones. However, if taking scenery as the core of landscape architecture theory, we can only follow behind architecture, fine arts, and sculpture.

Research on Landscape Space Scenery Creation Based on Visibility Graphics and Viewshed Analysis — With the Example of Chuan Opera Art Center in Chongqing also involves how to learn from the outcome of brother disciplines. In this paper, with the use of Depth map software and through visibility illustrations and viewshed analysis, the landscape sensitive points in the landscape space were determined, sceneries were classified according to the degrees of visual and behavioral impacts, and the landscape creation measures for the landscape space were further made. This method is very useful in rapidly increasing our level of visual scientific research and landscape design effects.

The Westerners has made a lot of research on the Matthew Effect (namely cumulative advantage, the strong getting stronger) of vision, which is very useful and effective and gives them powerful capabilities in creating supernatural visual effects, and a set of design system is developed. However, if looking from a long term, all of the Matthew Effect are temporary, since no one would just stick to something just because of several visual effects. This shows that there are more powerful and profound principles in the management of animal vision. In fact, ecology and systematics have proven that there is nothing in this world that has a lasting powerful attraction, otherwise all matters and energy of the world would be drawn into this kind of thing and the world would be destroyed. I think that even if there is such a thing, for long-term interests, humans should also be possible to resist this kind of thing, rather than blindly follow it. As said in *Tao Te Ching*, "Nature's motion decreases those who have more than they need and increases those who need more than they have; it is not so with man. Man decreases those who need more than they have and increases those who have more than they need." Lao Tz's observation was profounder than the *Bible*.

The Pattern Language and Its Application in Meso- and Micro-scale Eco-interface introduced the concept of "landscape language" first. Landscape changing from graphics to language seems much closer to the Chinese culture, even though language is not equal to article. As explained in Wiki, "Language is the human capacity for acquiring and using complex systems of communication, and a language is any specific example of such a system. The scientific study of language is called linguistics." The concept of "landscape language" might originate from the "architectural language". Architecture is all made through human manipulation, so it can be seen as an expression of human beings in a way. However, different from architecture, landscape, in addition to expressing human intention of the landscape, also contains a lot of content that the nature wants to express (such as natural laws), and the larger the scale, the greater the weight of the latter. Therefore, we should be especially careful to apply architectural language on landscape architecture. It was proper that the paper started from small-scale landscape. In addition, this paper proposes a design method for practical operation, and I think it is good especially for beginners. It is essential that learners should gradually deepen the understanding of the laws and truths behind the patterns language and then they will regard landscape design as a "comprehendible" article rather than a casual phonetic patchwork.

2014年10期

本刊的创始人之一，曾任主编17年的余树勋先生不幸于2013年辞世，他为本刊和中国风景园林界作出了巨大的贡献，特别是他知识之渊博，人格之高古，可谓今人之楷模。为此本期编发了一组"纪念余树勋先生"的专题约稿，作为余主编逝世一周年之际的怀念。

许多所谓的现代科学方法都根基于远古智慧，离开这些智慧的启示，只剩下方法，很可能本末倒置。例如现在很多人在利用"德尔菲法"，得到一些结论便自以为是放之四海而皆准的真理，四处宣扬，不顾这些结论的条件。其实呢，古希腊德尔菲（Delphi）的阿波罗神庙前殿的神谕是"认识你自己"，它是借神祇之口教诲凡人："认识你自己，噢，人哪，你不是神"，可用阿波罗神庙中的另外2条箴言："凡事不可过分"和"自恃者必毁"为证。

到文艺复兴时期，人文主义者费其诺回溯到这条神谕时，却对此进行了反解："认识你自己，噢，人哪，你这穿着凡人外衣的神的子孙"。的确，单纯从人类文化的繁荣上看，文艺复兴开启了一个伟大的时代，但是从人天关系上看，当人们忘记了"凡事不可过分"和"自恃者必毁"之时，有可能正在开启一扇危机的大门！

太过于放纵自己，是这个时代人类的标记之一。人性的解放是否没有边界？至今人们不敢讨论，我感到不是好事。快vs慢，逐新vs顾旧，刺激vs平淡，自豪vs谦虚，炫耀vs自律，前者总是容易抓住听众，获得成功，赢得粉丝。最近又有人拿那个"文化大革命"中交白卷的"英雄"，现在成为上市公司的某老板说事，谓之"是金子总能放出光彩"，这让我们又一次听到"唯成功论"的喧嚣。其实，"文化大革命"中此人是政治的工具，现在此人是金融的工具，从这方面看没什么本质差别。

本期的主题"地理设计"，集中介绍了世界上与我们学科发展密切相关的最重要新潮流。在此我想说几点感想：（1）地理设计的最大优点是紧紧追随科技的发展，这是偏重于文化思考的人居环境界相对薄弱的环节，将其作成主题，可以呼唤风景园林界内的重视；（2）地理是个外延几乎包罗一切地表和大气的概念，所以要特别小心，不要搞成让人类干预一切。我还是期望人类干预不超过大地的1/3，将一半以上的大地还给上帝，见2014年第5期《关于中国风景园林的地位、属性与理论研究》；（3）地理设计要注意给纯人性留出表现的空间，地球上适当地留出这样一点地盘还是可能的；（4）可将"地理设计"望文生义地理解为遵从地学规律针对地理对象的设计，这个中国人的翻译比起日本人将landscape翻译成景观，确切得多。

本期《弹性理念，乡村重塑中的风景园林思考》是篇好文章，观察深刻，目标明确，为中国风景园林发现了一片很大的领域和重要的机遇。我要特别感谢杨滨章教授对该文的认真审阅和指导。

Phase 10, 2014

Mr. Yu Shuxun, one of the founders and the 17-year chief editor or our journal, unfortunately passed away last year. He made a great contribution to our journal and the Chinese landscape architecture community, and is known as the model for his profound knowledge and elegant personality. For this, we compile and publish a set of commemorative articles this month, in memory of the first anniversary of Mr. Yu's death.

Many so-called modern scientific methods are rooted in ancient wisdom, and it is possible poles apart if only the methods are there without the revelation of the wisdom. For example, many people are using the "Delphi method" to get some conclusions, and opinionatedly take them as the universal truth and preach everywhere, regardless of the conditions of these conclusions. Actually, the oracle in the front hall of the Temple of Apollo in Delphi of ancient Greece is "Know Yourself", which is teaching the mortals through the mouth of the gods: "Know yourself, oh man, you are not the God." It could be proved with another two mottos in the Temple – "Nothing should be overdone" and "Self-conceited ones will ruin".

In the Renaissance period, the humanist Marsilio Ficino made a reverse explanation on this oracle: "Know yourself, oh man, thou the sons of God, in the coats of the mortals." Indeed, just from the perspective of human culture prosperity, the Renaissance opened a great era, but from the perspective of human-heaven relationship, when people forget "Nothing should be overdone" and "Self-conceited ones will ruin", it might open the door of crisis.

Too much self-indulgence is one of the marks of humans in this era. Is there no boundaries for the liberation of humanity? So far people dare not to discuss, and I think this is not good. Fast vs slow, chasing for the new vs attending to the old, stimulating vs simple, pride vs humility, show-off vs self-discipline, the formers always easily grasp the audience, succeed, and win fans. Recently, some people talked about somebody, the "hero" who handed in an unanswered examination paper in the Cultural Revolution and now the boss of a listed company, and called it "The gold always glitters", which makes us once again hear the uproar of "the theory of only success". In fact, this man was the political tool in the Cultural Revolution, and is the financial tool now, and there is no essential indifference between these two.

This issue's theme "GeoDesign" focuses on the introduction of important new trends of the world closely related to the development of our discipline. Here I would like to express a few thoughts: 1)The biggest advantage of GeoDesign is closely following the development of science and technology. It is the relatively weak part of human habitat community that lays emphasis on cultural thinking, and making it as a theme can call the community's attention; 2)Geography is a concept with the extension that almost covers all the earth surface and the atmosphere, so we should be especially careful not to let human intervene into everything. I still hope that the human intervention does not exceed the one- third of the land, and more than of the land is returned to the God (seen in *On the Status, Attribute, and the Theoretical Research of Chinese Landscape Architecture*, May 2014 edition of this journal); 3)GeoDesign should leave some space for the expression of pure humanity, and it is still possible to set aside such room on the earth; 4) "Dili sheji (GeoDesign)" can be literally interpreted as design for geographic objects by complying with the geographic laws, and this Chinese translation is much more precise than Japanese translating "landscape" as "jingguan".

Resilience Concept for Thinking of Landscape Architecture in Rural Remolding in this issue is an excellent paper, with insightful observation and clear target, and it discovers a vast area and important opportunities for Chinese landscape architecture. And I would like to thank Professor Yang Binzhang for his careful review and guidance of this paper.

2014 年 11 期

无机界是他组织的。生命是什么？是自组织的出现。自然生物的妙处，在于他们可以自行组织自己物种的形态、位置、相互关系和生命长短，而这一切，都不曾改变无机界的规律，更不会影响生物界的最基本"价值观"（原则）：物种自己的寿命延续和物种总量的逐渐丰富。

只有人的出现，改变了自然生物界的基本"价值观"。人类把太阳照耀的地方遮蔽起来，把天然降水流经的轨迹进行改道，把深藏地下的矿藏挖掘出来并对之实施化学处理，把物种丰富的草原和森林改变成单一作物或者城镇工厂；为了满足自己的物欲，人类肆意地对自己和别的物种狂妄地进行他组织活动。

是否只有等到科学发展了，生产力提高到某种水平，人们才能达到更高的认识呢？不一定。中国的古人从另外一条渠道早就达到了这个认识高度，那就是如实地把万物包括人类自己看作是自然的一分子。在这种思维影响下，以及对物质的有机和无机的区分还不清晰的条件下，把大地山川也视为富有生命的自然而予以尊重，当然无可厚非，从中可以感受到远古"万物有灵论"的影子，从大方向讲，至少也可以避免人类因狂妄导致的迷航。

正因如此，清华大学袁琳等的《古代山水画中的地域人居环境与地景设计理念——宋〈蜀川胜概图〉（成都平原段）为例》一文引起了我特别的注意。文章以北宋李公麟"蜀川胜概图（成都平原段）"为媒介，解读了古代成都地区人居环境艺术表现，阐释了古代山水长卷中蕴含的"地景设计"理念。文章总结道："古代山水长卷是将山水生命气象与画家心灵境界融合并符号化的产物，这种符号化的创造过程体现着无所不在的生命精神，画卷整体在一个巨大而有序的世界中提供了一种人类和谐存在的宇宙性视角。"讲得多么好！该文把艺术视角的大尺度地景与生态科学的生命现象直接挂钩起来，并将其上推到至少北宋时期，而那时，西欧还处于中世纪的前期，即黑暗时期，西欧的艺术家，那时只关心人与上帝的关系。

蜀川胜概图描绘的地方，我很熟悉，可以说除了岷江源，基本都去过。回想起来，就是李公麟画笔的表达：整块地域就像一个大有机体，犹如一气呵成，脉络关联，相互扶持，气势回荡，万物生发，天人无瑕，生机盎然。由此我联想到：大尺度的生态规律和生态内涵的艺术表达，是否天生就是同体的？古语云："万法无常"，无常就是永动，那么古人如何理解这个无常呢？他们肯定不是像我们一样从物理化学规律去理解物质的变化，他们会不会把宇宙看成一个生命体（即一切皆是有机体）？

重庆大学毛华松的《西湖文化的演进历程及其历史意义——〈永乐大典·六模湖〉中的西湖文献统计分析》提出了中国一个非常重要的建城模式"山—湖—城"，从今天所谓雨洪管理来看，极为科学，此种模式正好让湖面在水源和城市之间起到调蓄作用。因为中国大体西高东低，所以城市的西湖的水利作用最大，因而历史上保留下来的西湖最多，而相依的景观虽然是次生产物，但也是幸运产物。

南京林业大学赵亮等的《图解"风"与"水"——流体景观元素模拟与分析法在风景园林规划设计中的应用》把注意力集中到流体力学对风景园林的影响，这一点非常重要，将来一定会应用到城市绿地系统规划中和城市小气候分析与设计中。我更重视文中对各类CFD软件在风景园林规划设计中的应用从软件特点、主要用途、面向对象、在风景园林中应用局限性4个方面进行的比较，可以警示对电脑软件迷信的不良倾向，希望能对近年来的大量此类文章有所启示。

Phase 11, 2014

The inorganic world is other-organization. What is life? The emergence of self- organization. The beauty of the natural biology lies in that they can self-organize the forms, locations, inter-relationships and lifespans of their species, and also these did not change the functions of laws of the inorganic world, neither did it affect the basic "values" (principles) of the biosphere: the continuation of lifespans of species, and the gradual enrichment of total species.

Only the emergence of human beings changed the fundamental "values" of the natural biosphere. People cover up where the sun shines, divert the natural trajectories of rainfall flows, dig out mineral resources from deep underground and implement chemical treatment on them, change rich-species grasslands and forests into a single crop or urban factories. In order to satisfy their material desires, people can wanton themselves and other species with arrogant other-organization activities.

If only with the scientific development and productivity increased to a certain level, will people have such understanding? Not necessarily. Chinese ancients already reached this height of understanding of channels from another channel, and that is regarding all thing–including people themselves–as part of the nature. Under the influence of this thinking, as well as under the conditions of unclear distinction between organic and inorganic things, it is understandable that mountains and rivers were also considered to be full of life in nature and respected. You can feel the presence of the ancient "animism", but in the general direction, at least it avoided the drift off course by human arrogance.

For this reason, *Regional Habitat Settlements and Large-scale Landscape Design Concept in Ancient Shan-shui Painting: A Case of the Long Scroll of The Shu River (Chengdu Region Section) in the Song Dynasty* by Yuan Lin of Tsinghua University caught my special attention. Taking the Long Scroll of the Shu River (Chengdu Region Section) by Li Gonglin of Song Dynasty as the medium, the paper analyzed the artistic expression of the habitat environment of the ancient Chengdu area, and explained the "large-scale landscape design" concept included in the ancient shan-shui long scroll. The paper concluded: "The ancient landscape long scrolls in China are symbolic products combining landscape images with spiritual realm of the painters, which express ubiquitous living spirit and construct an ordered world in which people can live in harmony with nature, and offer a cosmic perspective." Well put! The paper directly linked the large-scale landscape in the artistic perspective with the phenomenon of life in ecological sciences, and pushed up to at least the Northern Song Dynasty, and at that time, the Western Europe was still in the early Middle Ages, namely the Dark Ages, and Western artists then only cared about the relationship of human with the God.

I am very familiar with the area the Long Scroll depicted, and have been to every places except the source of Min River. In retrospect, it is the expression with Li Gonglin's brush: the entire area is like a large organism, seemingly at one stretch, with context correlation, complementing each other, in an imposing manner, with prosperous living things, and vitality. Thus I think: Whether large-scale ecological laws, and the artistic expression of ecological connotation, are born in one body? The old saying goes, "Everything is impermanent," and impermanence is perpetual, then how did the ancients understand this impermanence? They certainly did not understand the changes of materials with physical and chemical laws like us, did they regard the universe as a living body (i.e., everything is an organism)?

The Evolving Procedure and Historical Significance of the West Lake – A Statistic Analysis on Literatures about the West Lake in The Yongle Canon: Six Pattern Lake by Mao Huasong of Chongqing University put forward a very important Chinese urban construction mode "mountain-lake-city". It is very scientific from the perspective of presently-called stormwater management, and this mode precisely let lake play the role of regulation and storage between water and city. Because China mainland is roughly high in the west and low in the east, and the water conservancy role of west lakes of the cities functioned great, so west lakes are mostly preserved from the history, and the scenery dependent on the lake is a secondary product, and also a lucky product.

Graphic Wind and Water: Application of Fluid Landscape Elements Simulation and Analysis in Landscape Planning and Design by Zhao Liang of Nanjing Forestry University focused on the influence of hydrodynamics on landscape architecture, which is very important and in the future will be applied to the urban green space system planning and urban microclimate analysis and design. I pay more attention to the papers introduction of the application of various types of CFD software in landscape planning and design, and the comparison from the four aspects of software characteristics, main purposes, oriented objects, and application limitations, which can alert the negative tendencies of computer software superstitions, hoping to give some inspirations for a large number of such articles in recent years.

2014 年 12 期

近日在"汉网"上看到一篇佳作《治理信息洪水建设精神家园》(原作者子曰师说·刘吉山),其中讲道:"面对以互联网为代表的、史无前例的信息浪潮冲击,随波逐流者有之,兴风作浪者有之,被拍在沙滩上者亦有之。这真像人类文明的又一场大洪水。如果长期任由网民在这个虚拟世界里随心所欲,那么,负面的、虚假的、有害的信息也势必借助新技术的推动,爆炸式传播,最终给现实世界带来灾难性影响。对于这样一把双刃剑,应尽早采取理性措施,我们不能等它伤了人再讨论如何驾驭。就像习近平主席在首届世界互联网大会的贺词所说,'集思广益、凝聚共识、贡献创见,推动互联网更好造福人类'。"

问题是治水有堵和导之争,堵是与规律对抗,导是顺应规律而为。大禹通过总结经验发现与规律对抗是错误的,这可能是中国人自古尊重自然和规律的思维方式的源头。现代的许多规划设计不少是与自然规律对抗的,如在缺水的荒原造林,在多雨地区把城市绿地下挖20cm储水,在分水岭上挖大湖,在滩涂上堆大土山,在平缓的内河河岸搞人造沙滩,在可能爆发山洪的山溪旁造休闲景点,在远大于自然安定角的坡面上种树来稳固土壤,在河曲的冲击岸建房或在淤积岸建港,在主航道上立桥墩,在过洪的水道沙洲上种植树丛,在没有阳光的楼层搞生态农场……而这一切,无非打着景观(背后是经济利益)的旗号。

我觉得,过去以为风景园林似乎超脱经济利益的态度是有问题的,但近来兴起的"景观资本主义"可能问题更大,为唯利是图打开大门。资本这东西显然是具有两面性的,对待它,应该持有像上文对待信息一样的两点论态度。我对"唯×"主义的各种形态(如唯美主义、唯新主义)皆持怀疑态度,"唯×"就是"只有",就是极端,就是不可分,这不符合辩证法。

在20世纪60年代毛主席提倡"一分为二"并反对"合二而一",当时我很不理解,甚至私下做了批驳,现在想来毛主席的思维更为犀利深刻。关键问题在于毛主席提倡无限可分,导致了斗争扩大化,没有看到可能(例如极限状态,或者在一定条件下)有不可分的(或分也没有意义的)终极的"一",而这个量词"一"却是"二性同体"的,我们必须把数量的"一"与性质的"一"区别开来。佛家用"不二"来指代"唯一",可能就有了这种意头,在极限处,"一分为二"与"合二而一"没有区别。

从这点看,所谓"一生二"是有问题的,如果一和二都是指性质,即在阴阳两性之上还有无性,就无法阐述世界为何会启动,或只能假设另外还有个上帝在变戏法。所以,"无极生太极,太极生两仪(即阴阳)",只是祖先们的一种推想,不必奉为圭臬。但太极图所示的境界,就是那个"二性同体"的、不纯粹的、终极的"一"。当然,阴阳只是我们祖先的一种比喻性界定,从现代物理出发,可以将其改称为正负。零,就是正负二者的交接点或界面。当人们还不知道或看不到"负"的时候,就以为从零中生出了有,提出所谓"无中生有"。

再进一步思考,既然没有纯粹的"一",那么西方思维体系的核心模式:把一切还原到唯一的原点,再从原点推导出一切规则,真的是绝对真理吗?我们有必要把这种思维奉为楷模吗?这就是我为什么坚信基于中国思维的中国风景园林理论和实践,必将高居世界之林巅的根本原因。

Phase 12, 2014

Recently I read on the web a good post titled "To Manage the Information Flood and Build the Spiritual Home" (originally posted by Liu Jishan), which says: "Facing the impact of the unprecedented wave of information represented by the Internet, there are the ones going adrift, the ones making waves, and the ones being submersed. This is really like another great flood of human civilization. If letting the web users do as they wish in this virtual world for long, the negative, false, and harmful information will be bound to explosively spread with the promotion of new technologies, and ultimately result in the catastrophic impact on the real world. For such a double-edged sword, rational measures should be taken as soon as possible, and we cannot wait for it to hurt people and then discuss how to manage it. As said in President Xi Jinping's congratulatory message to the First World Internet Conference, "we should brainstorm, reach consensus, and give thoughtful contribution to promote the Internet better for the benefit of mankind'".

The problem is that there is the dispute of dredging and blocking in flood control, and blocking is the confrontation of the natural laws while dredging follows the natural laws. Yu the Great found by summing up experiences that confronting with the natural laws is wrong, and this may be the source of the Chinese people's way of thinking of respecting the natural laws. In many modern planning and design projects, we find that many of them are against the natural laws, such as the afforestation in the dry wasteland, lowering the urban green space at 20cm for the storage of water in rainy areas, digging a large lake on the top of watershed, piling up a large dirt hill on the beach, creating man-made beach along the placid inland river banks, building leisure attractions along brooks with the hidden danger of the outbreak of mountain torrents, planting trees to stabilize the soil on the slopes with an angle much larger than the natural, building houses on the flood impact bank in the river bend or the harbor on the silting bank, setting up bridge piers in the main channel, planting trees in the flood waterways sandbar, building the ecological farm on the floors without sunlight… all these are nothing more than the name of landscape architecture (while the economic interests are the background).

I think that the past attitude of landscape architecture seemingly transcending economic benefits is a problem, but the recent rise of "landscape capitalism" may be a bigger problem–opening the door for mercenary. The capital obviously has two sides, and treating it should hold the same two-point attitude as the one for the above- mentioned information flood. I am skeptical about the "only-x" doctrines in various forms (such as aestheticism, and the only-new doctrine); the "only-x" is "only", extreme, un- detachable, not in conformity with the dialectics.

In the 1960s Chairman Mao advocated "dividing one into two" and opposed "combining two into one", and I did not understand and refuted this privately, but now I think his thinking is far sharper and profounder. The key problem was that Mao advocated infinite divisibility, which led Bigger Struggle, and he did not see the possibility (such as under extreme or certain conditions) of indivisibility (or no sense if divided) of the ultimate "one". This "one" has two meanings, and we should differentiate the number "one" and the property "one". The Buddhism taking "non-dual" to refer to the "only one" may have this meaning, and in extremity "dividing one into two" and "combining two into one" have no difference.

From this point of view, the so-called "two out of one" is problematic. If one and two both refer to the property, there are *yin* (female), *yang* (male) and asexuality, and that cannot explain how the world were started, or we can only assume that the God performs tricks. So, "*tai chi* comes out of nothing, and *tai chi* gives birth the meters (*yin* and *yang*)" was just one of the ancestors' suppositions, and we should not look up that as the standard. But the realm shown in the *tai chi* diagram is that "dually co-bodied", non-pure, ultimate "one". Of course, *yin* and *yang* are only a figurative definition by our ancestors, and from the viewpoint of modern physics, it can be changed into positive and negative. Zero is the transition point or interface of the positive and negative. When people do not know or do not see the "negative", they think there is something coming out of zero and propose the so-called "creating something out of nothing".

Thinking further, since there is no purely one, then the core model of Western thinking system: to restore everything to a unique origin, and then all the rules derive from the origin, is really the absolute truth? Should we put this kind of thinking as a paragon? This is the fundamental reason why I believe the Chinese landscape architecture theories and practices based on Chinese thinking are bound to hold the high summit of landscape architecture of the world.

2015年

2007年　2008年　2009年　2010年　2011年　2012年　2013年　2014年　　2016年

2015 年 1 期

如果将地球看作是一个有机体，她的细胞是什么？我觉得应该是小流域。一个小流域是一个生态过程基本完整的地理细胞。20世纪七八十年代，中国水利界曾经掀起过一个小流域规划浪潮，后来似乎无疾而终，可能是当时只看到小流域工程的水土保持作用，还有工程量小，不需要很多钱，甚至国家不需花钱的意义。现在我们从生态角度观察，会发现小流域更大的意义，本期重庆大学赵珂《以小流域为单元的城市水空间体系生态规划方法——以州河小流域内的达州市经开区为例》一文，把小流域理念引入城市，让我们对城市生态问题有了更深一步的理解。

城市是人的聚集地，我们在城市中所做的一切应该以为人考虑为主，在我看来这是天经地义的事，即便是保护环境、保护生态，实事求是地讲，最终目的还是为人。地球上的城市总面积，不到地面面积的百分之一，我们没有必要在这里还为苍蝇、蚊子、老鼠、老虎考虑，我想这是小孩都明白的道理，可是很多专家却另有主张，完全不管人的需要，视每年都要登场的禽流感、登革热之类为无物，在城市里推广湿地、原始林，抽象地提倡物种多样性。

这里我看到的还是一种西方逻辑：把事物还原到"原点"（例如物种多样性），再从"原点"出发强加于一切。追逐并尊重原理，本是一种善举，从这里可以发现，如果将其奉为无条件的"唯一"，同样会出现问题。

我们风景园林工作者，头脑里一定要有人，你的作品是人用的，只有想到人，风景园林才能达到"善境"，因为善，是与人为善。而且，如果深一步想想，人是有好坏之分的，我们应该为好人服务，这才是真善。具体到当下，我觉得可以将"为法治服务"也作为设计原则之一。过去我们曾经批判过19世纪奥斯曼改造巴黎是为资产阶级镇压工人起义服务，现在想来，其实奥斯曼也不无道理，我们不是很拥护城市中的摄像头？现代乱世中，最倒霉的肯定是善良的老百姓。清华同衡规划院贾培义《城市公共开放空间的防卫性景观设计研究》一文，开始正式涉及这个问题，很及时。据我所知，很多园林管理部门都对安全问题有所考虑，积累了不少经验，可惜无人总结，也没有被学界和领导重视。例如，现在一些景观喜欢到处摆设鹅卵石，这不是为恐怖分子提供武器？诸如此类问题还有很多。

感谢华南理工大学袁晓梅教授利用其访美机会，为本期组织了"康复花园"主题文章。袁教授本来是研究声景的，到美国后发现康复花园是当下的热门。我觉得这一组文章最突出的集中点，是对人的热爱与关怀。虽然至今我们还不了解康复花园的科学原理（例如生化过程），但出于对人的关怀，花园还是回到医院。19世纪中叶，医学和风景园林作为科学，几乎是同时起步，后来医学专门化发展迅速，实践前线的临床医生反而深受实验室里的科学家们控制，"他们信奉这样一种错误的观念，即患者对医疗环境的感知不会影响人类的免疫系统。"（见《花园重归美国高科技医疗场所》一文）。这里所揭露的事实，值得至今还完全执迷于欧美体系的各界人士，包括学者、专家、官员等各种涉及国家民族前途的人士，做一次深刻反省。中国的前途，绝不是亦步亦趋地追随这个体系能够解决的。

Phase 1, 2015

If taking the earth as an organism, what is its cell? I think it should be a small watershed. A small watershed is a geographic cell with the basically complete ecological process. In the 1970s and 1980s, the Chinese water conservancy sector set off a wave of small watershed planning, but later it seemed to die a natural death. It might be the time only seeing the soil and water conservation function of small watershed projects, as well as small quantities, without much costs, or even no cost from the government. Now observing from an ecological point of view, we will find the greater sense of small watershed. In *The Method of City Water Space System Planning Based on Watershed: A Case of Dazhou Economic and Technological Development Zone in Zhou Watershed* by Zhao Ke from Chongqing University, the concept of small watershed is introduced into city, and thus we could have a deeper understanding of urban ecological issues.

The city is a gathering place for people, and everything we do in city should be considered for human, and in my opinion this seems to be a matter of course. Even the protection of the environment, and ecological protection, realistically speaking, the ultimate goal is human. The total urban area on the planet is less than one percent of the ground area, and here we do not need to consider for flies, mosquitos, mice and tigers, which I think is the truth even the children are aware of, but many experts had other ideas, completely regardless of human needs, taking the yearly-coming bird flu and dengue fever as nothing, to promote wetlands and primeval forests in city, and promote species diversity abstractly.

Here I see also a Western logic: to restore things to the "origin" (such as species diversity), and then impose on everything based on the "origin". Chasing and respecting the principles is originally a good thing, but if it is regarded as an unconditional "only", problems also occur.

For our landscape architecture professionals, we should have human in mind, our works are for human, and only thinking of human can landscape architecture achieve the "good", since the "good" means good for human. Moreover, if thinking further, human are good or bad, we should serve the good ones, and that is the real good thing. Specifically to the moment, I think we can take "the service to the rule of law" as one of the design principles. In the past we have criticized that the transformation of Paris by Haussmann in the 19th century served the bourgeois repression over workers' uprising, and now we think Haussmann's transformation actually made sense. Don't we quite support the city's cameras? In the modern troubled times, good people are certainly the most unfortunate. *Research on the Security Landscape Design of Urban Public Open Space* by Jia Peiyi of THUPDI formally addresses this issue, which is very timely. As far as I know, a lot of landscape management departments have considered the security issues and have accumulated a lot of experience, but unfortunately no summary has been made and no attention has been paid by the academic circle or the leadership. For example, now some landscapes like to have a lot of pebbles as decorations, which is not provided as terrorists' weapon? There are many similar problems.

I would like to express my gratitude to Professor Yuan Xiaomei of South China University of Technology for organize a series of "healing garden" theme papers during her visit to the United States. Professor Yuan originally researched on soundscape, and found that healing garden is the current hot topic when she was in the United States. I think the most prominent point of this series of papers is the love and care for human. Although so far we do not know the scientific principles of healing garden (e.g.the biochemical processes), but for the care of people, gardens return to hospitals. In the mid-19th century, the medical science and landscape architecture, as sciences, started almost at the same time, then the specialization of medical science developed rapidly, practical clinicians are influenced by laboratory scientists. "The medical theories embraced the mistaken notion that the human immune system was unaffected by the patient's perception of the medical environment." (In the paper of *The Return of the Garden to High-Tech Medical Facilities in the United States*) The fact revealed here requires a deep introspection of those who are still totally obsessed with the Western system, including scholars, experts and governments officials who are involved in the nation's future. China's future could not be worked out by blindly following this system at all.

2015 年 2 期

1972年，一本由"罗马俱乐部"智库策划的名为《增长的极限》的书出版。该书的作者预言：如果采取"放任模式"，人类的文明将于2015～2030年开始走向崩溃。今年已到了2015年，澳大利亚墨尔本大学的一项研究表明，虽然到2010年全球还尚未出现崩溃的迹象，但对数据的分析显示，这本书中的预测具有相当高的精准度。把书中预言的曲线与40年来实际发展轨迹相比，人口、工业经济的曲线拟合度非常高，而且一个转折在开始显现。对于污染和资源消耗，由于世界对环保的重视，已经不属于"放任模式"，恶化的速度有所减缓，但大趋势仍然近似。还有一个值得注意的倾向，由于金融资本的扩张（包括各国大量印刷钞票），近几年服务行业的增长率大大超过了书中的预测。

再从社会人文方面看，2015年年初恐怖分子对巴黎《查理周刊》的袭击事件引起对"自由的边界"的讨论。争论围绕着启蒙的意义、信仰的地位、不公与暴力的关系、普世价值、文明的冲突等方面展开，热闹得很。我想与其纠结这些，不如从更高的立场，审视一下人在世界和历史中的地位。

我不知道机械思维的危害到底有多深，但从很多例子可以看出来。比如所谓"绝对真理"，依从这种思路，人最后只有完全服从绝对真理。其实这种想法本身已经否定了人的地位。对于无神论者来说，必须承认人是宇宙的异化产物，理论体系中才能有人的地位。据说普列汉诺夫曾经问列宁：既然革命胜利是必然的，我们何必还要为之奋斗？这个问题的根本纠结亦在于此。中国《易经》的命理，既有天命，也有人运，这在"绝对真理"信众看来乃是悖论，不可理解。

人类的"异物宿命"和人类的"长寿愿景"的矛盾（其中也暗含着个人与人类的矛盾关系，因为强调个性就是强调异物），是人类这个物种本身的根本矛盾，也是人性的核心问题。解决这个问题的基本路线，应该是"不过分张扬的人类"。

不过分张扬的人类的概念是：以兼顾人类合理的生活质量与延续人类整体生命为目的，努力维护地球质素，自觉约束人们的欲望，放弃对速度和实力的竞争，实行审慎（可持续）发展策略的人类。这种人类社会必然是和平、民主、守法、自律、尊天悯人、有计划发展的社会。必须指出，自律是自觉地尊重规律，既不是自以为是，更不是自觉地被操控。

因此，教育的首要目标是育人，育出不过分张扬的自觉人类，而不是培养只懂得人类"发展"所需知识的工具。

若人类果真已经到达了重大转折点，我们对此的准备是不足的，大量传媒宣扬的仍然是放纵和任性，根本的原因是这样才符合资本快速增长的需要。

风景园林学，从根本上是属于"不过分张扬的人类"的学科，有些奇怪的是，它却在社会飞速发展的历史阶段才得到飞速发展。这是否意味着，当下我们学科中有很多东西是属于"张扬的人类"的？例如，所谓景观资本，所谓视觉刺激。当然，在保护学生未来和大量从业者饭碗的前提下，我们不需要发动一场清除这些"不良因素"的革命。但是让我们心里清楚这类东西的本质，眼前利用它而不是迷信它，并准备好一旦不合适就立即抛弃它，还是有必要吧。

最后，要特别感谢中国城市规划设计研究院风景园林规划研究所贾建中所长，为组织本期"国家公园"的主题文章所付出的努力！

Phase 2, 2015

In 1972 a book titled *Limits to Growth* by a think tank "Club of Rome" was published, and the author predicted that our civilization would begin to collapse in 2015-2030 if taking the "laissez-faire model". 2015 comes this year. As shown in the study by the University of Melbourne, Australia, although there had not yet been signs of collapse around the world until 2010, the book has a very high prediction as revealed by the data analysis. Comparing the predicted curve in the book with the actual development trajectory in the past forty years, the population curve and the industrial and economic curve fit very well, and a turning point is starting to show. The world's attention to pollution, resource consumption, and environmental protection is no longer in the "laissez-faire model", the pace of deterioration has slowed, but the trend is still approximate. A noteworthy tendency is that the growth rate of the service sector in recent years is much higher than predicted in the book due to the expansion of financial capital (including countries printing a large number of money).

Then from the humanities aspect, the terrorists' attack to the Paris *Charlie Hebdo* at the beginning of the year arouses the discussion on the "border of freedom". The debates are on the meaning of enlightenment, the status of faith, the relationship between injustice and violence, universal values, and the clash of civilizations, and it is very boisterous. I think we should review the human position in the world and history from a higher standing, rather than tangled in these debates.

I do not know how deep the harm of mechanical thinking is, but it can be seen in many examples. For example, the so-called "absolute truth", and in this idea, human will only finally fully obey the absolute truth. In fact, this idea itself has denied the status of human. For atheists, it must be admitted that human is the product of cosmic alienation and then there is the status of human in the theoretical system. Plekhanov once asked Lenin: "Since the victory of the revolution is inevitable, the why should we strive for it?" This is also the tangled root of this problem. Tangled root of this problem is also in this. The numerology in the Chinese *Yi Jing* (The Book of Changes) includes both the mandate of heaven and human fortune, which seems to be a paradox in the "absolute truth" believers and is incomprehensible.

The human contradiction of "alienation destiny" and "longevity vision" (which also implies the contradictory relationship between the individual and mankind, because individualism is to emphasize alienation), is the fundamental contradiction of the human species in itself, and also the core issue of humanity. The basic way to solve this problem should be the "not overly assertive mankind".

The concept of the not overly assertive mankind means: to balance the reasonable quality of human life and the continuation of human life as a whole, try to maintain the quality of the earth, consciously restrain human desire, abandon the competition for speed and strength, and implement the prudent (sustainable) development strategy. This human society must be a society with peace, democracy, law, self-discipline, respect for nature and care for human, and a planned development. It must be noted that self-discipline is to respect the law self-consciously, neither self-righteous, nor consciously manipulated.

Therefore, the primary goal of education is to educate people, bringing out the not overly assertive conscious human, instead of those tools only knowing the knowledge for the needs of human "development".

If human has truly reached a major turning point, we are not ready for this, and the indulgence and self-willed are still propagated by a large number of media, and the fundamental reason is that this is consistent with the rapid growth of capital needs.

Landscape architecture study is fundamentally a subject of the "not overly assertive mankind", but it is strange that it gets rapid development until in the historical stage of rapid development of society. Does this mean, a lot of things in our discipline at this moment belong to the "overly assertive mankind"? For example, the so-called landscape capital, and the so-called visual stimulation. Of course, on the premise of protecting students' future and jobs of a lot of practitioners, we do not need to launch a revolution to purge these "adverse factors". But it is still necessary for us to understand the nature of such things in mind, use it at this moment instead of believing it blindly, and get ready to abandon it immediately once it is inappropriate.

At last, we would like to express our gratitude for Jia Jianzhong, Director of Landscape Architecture Planning Institute of China Academy of Urban Planning and Design. He makes tremendous contribution to organize the "national park" theme papers of this issue!

2015年3期

羊年（2015年）春节，我总的感受是，人类最终不会把社会交给经济学家。经济发展的重要性仅仅存在于人类社会的初期（青少年期），人类一旦进入中年，更多的应该是考虑维系各种关系的平衡（不是说维系现状，因为现状是一种极端不合理的不平衡），而不是忙于打破它。我对大形势的判断是：21世纪是人类从青年向成年转折的时代，吊诡的是，在这个时期我们却要负担起改造现存关系的艰难任务。

近来原油价格的暴跌，让那些唯经济发展论的人重新振奋了精神：看哪，哪里有什么资源危机，按照20世纪70年代某些人的估计，到21世纪20年代就该石油枯竭了，他们简直是杞人忧天！在人类还没有能力看穿数十千米厚的地壳之前，资源的储量本来就是连误差都难以确定的估计。但无论如何，地球的总量是有限的，这点不需要杞人忧天，需要的只是对后代的关心。更重要的是环境质量问题，虽然我对所谓温室气体理论充满着狐疑，但有一点是肯定的，大气中二氧化碳增多的根本原因是人类烧掉了大量原来藏在地下的各种碳氢燃料，将之归罪于牛放屁和有机物腐败之类的说法本身就有问题，因为那本来就属于大自然的碳循环。

中国人口众多，使得中国人不可能像人少的国家那样去消耗资源。中国人必须探寻一条能够国富民丰但又节约资源的道路，而这条路正是全球人口共同走向富裕的正路，这就是中国模式的根本意义。春节前有个在欧洲留学的学生来看我，聊起这几年的经历，他一脸对欧美文明羡慕的神情，不过3个小时后，当他明白这个世界的主要矛盾之一是发达国家对不发达国家的剥夺，以及欧美模式不可能成为全人类的生活模式后，神情变得严肃起来。我相信，今后他必将重新审视他所看到的一切，进入一个新的精神层次。

那些肤浅并炫耀奢华和浪费的文化令人讨厌。曾经打着"为广州科技文化发展服务"旗号建设起来的广州购书中心，经过4个月的闭馆改造在春节前重新开放。我一进去，发现全变了，装修现代了，空间丰富了，有了某种"高大上"气派了，但一半柜台不再卖书，变成打着高雅旗号的售卖所谓文化商品和休闲的场所，即使是卖书的地方，乍一看似乎励志、成功和儿童教育的书很多，其实基本是在进行自私教育，科技书籍只占大约5%的面积，需要大学问才能编纂和查询的大百科一类辞书被赶到了负一层最偏远的角落，整个书城几乎成为时尚、通俗、娱乐眼球一类文化的一统天下。这种终于追上了全球时髦潮流的文化环境能够培养出可以引领全人类方向的民族吗？当然，问题的根本不在书店，而在于教育出了大问题。

其实中国人只要有自强精神，完全没有必要自卑。不久前媒体报道了对抗埃博拉的解放军援塞拉利昂医疗队：虽然中国医院接诊病人最多，但我们一个感染的都没有。这就是我们的底气。一开始西方人主导援塞，想给我们评级，挂个什么灯，不到绿灯不能开业。李进队长直接把他们挡出去了，他们有什么资格给我们评级！人，只有自信，才能以平等的心态分辨他应该学习什么，他应该摈弃什么，才能强大起来。在国内还充斥着崇洋精神的时候拼命宣扬民族自谦，到底为了什么？

2015年北京电视台《环球春晚》，最后上场的西城男孩尚恩（Shane filan）唱出：I

am strong, when I am on your shoulders. You raise me up to more than I can be.（站在你的肩上，我才强大。你的提升使我超越了自我。）我真诚地希望，将来有一天外国朋友们会诚挚地对中国人说这句话。

Phase 3, 2015

In the Chinese New Year of the Goat, I always feel that the human will not hand over the society to economists in the end. The importance of economic development exists only in the early years (adolescence) of human society. Once entering the middle age, the human should more consider to maintain the balance of relationships (not to say maintaining the status quo, because the status quo is extremely irrationally unbalanced), rather than being busy breaking it. I judge the bigger picture as: the 21st century is the transitional stage for the human from adolescence to adulthood, and the paradox is that in this period we have to shoulder the difficult task of remolding the existing relationships.

Recently the crude oil prices plummeted, and that re-inspires those who only believe in the economic development theories: behold, there is no resources crisis; according to some estimates in 1970s, the oil would dry up in 2020s, and they are simply unfounded! Before people have the ability to see through tens of kilometers of the crust, the reserve of resources is only an estimate that even errors cannot be determined. Nevertheless, the total amount of the earth is limited, which does not require alarmist, just need to care for future generations. The issue of environmental quality is more important. Although I doubt the so-called greenhouse gas theory, one thing is for sure: the root cause of the increase of carbon dioxide in the atmosphere is that the human burned large volumes of various hydrocarbon fuels originally hidden in underground; those blaming the cow fart and the decomposition of organic matters are talking nonsense, because that belongs to the carbon cycle of the nature.

China's large population makes the Chinese people cannot consume national resources like countries with smaller populations. The Chinese people should explore a road to wealth and prosperity with resource conservation as well, and this road is the right one for the whole world toward common prosperity and also the fundamental meaning of the Chinese model. A student studying in Europe came to see me before the Spring Festival, and had an envious look at the Western civilization when talking about his experience in recent years. But three hours later when he knew that one of the major contradictions of the world is the developed countries' deprivation of the underdeveloped countries, and the Western pattern cannot become the lifestyle of all mankind, his expression turned serious. I believe that in the future he will re-examine everything he see and enter a new spiritual level.

Those superficial, luxury-flaunting and wasteful cultures are annoying. Guangzhou Book Center, originally developed to serve the scientific and cultural development of Guangzhou, reopened before the Spring Festival after four months of renovation. When I stepped in, gosh, all have changed, modern decor, abundant space, and with some kind of "splendid and posh" style, but half of the counters no longer sold books and became places for so-called cultural goods and leisure with the elegant banner, while in other book-selling counters, inspirational, success and children educational books were abundant at first glance, but in fact they were books educating selfishness, scientific books accounted for only about 5% of the whole area, dictionaries and encyclopedias with profound knowledge were piled up at the most remote corner of the basement, and the entire bookstore almost became a place for fashionable, popular, and entertainment cultures. Can this cultural environment that finally caught up with the global fashion trends bring up the people that can lead the whole mankind toward the right direction? Of course, the fundamental problem is not in the bookstore, and it is the education that has a big problem.

In fact, as long as the Chinese people have self-reliance, self-abasement is totally unnecessary. Not long ago, the media reported the Chinese PLA medical team to Sierra Leone against Ebola: the Chinese team's hospital received the most patients, but we do not have any infection. This is our confidence. At first the Westerners dominated the medical aid and wanted to give ratings to the Chinese team, hanging lights and the hospital could not be open without the green light.

The Chinese team leader Li Jin directly block them out, and they were qualified to give a rating! Only people with confidence can have the mentality of equality to distinguish what to learn and what to abandon and then they will become powerful. Why should people advocate national self-deprecating when China is still filled with the worship of foreign things?

In this year Beijing TV "Global Spring Festival", Shane Filan (Westlife) in the last play sang: "I am strong, when I am on your shoulders. You raise me up to more than I can be." I really hope that someday in the future foreign friends will say this sincerely to the Chinese people.

2015年4期

2015年第1期"康复花园"主题，因故延迟到本期才得以全部发表。这个主题的组织工作，得到了美国弗吉尼亚大学建筑学院William Stone Weedon荣休教授，"设计与健康中心"联席主席Reuben M. Rainey，以及Dirtworks景观公司总裁David Kamp的大力支持。他们参与了主题策划，精心挑选了景观案例，并协助论文邀请。Reuben M. Rainey教授还对论文的英文稿件进行了编辑与修订，特此致谢！

战略和策略，总的看它们应该是一致的，但是偏偏有表面看来相反的情况发生，比较著名的是"用正义的战争消灭非正义战争，从而实现和平"的命题，战略目标是和平，实现策略却是战争。相类似的是发展与和谐的关系，发展的动力其实是差距，是不和谐，发展的目标是社会以及人天关系的和谐。为了消除世界上发达与不发达的不和谐，不发达就必须加速发展，这就必然带来高速发展的很多弊病，如资源消耗、环境污染、基尼系数加大、国际关系紧张等。我们怎么看这类问题，是否全都怪在不发达国家身上？有些考验很多大V级人物的智商。

2015年3月13日美国宾夕法尼亚大学设计学院在北京召开了"宾夕法尼亚大学中国景观研讨会"，充分展现了美国学术的敏感性和开拓精神，很值得我们学习。我感觉宾大是一片好心，想帮助中国找到解决发展中很多问题的途径和办法，并作为以后很多发展中国家的借鉴。会议在沃顿商业中心召开，暗示着其中必有很多商机。对于世界上风景园林近半个多世纪的发展，除了20世纪中叶设计上的"哈佛革命"，影响更大的是世纪之交的生态运动，其主导却是宾大，"设计跟随自然"和"景观都市主义"，是在宾大打下的基础。有人哀叹好像制高点都被美国人占完了，其实不然，这也很有些考验我国业界特别是一些大咖的智商。

本期的主题是"棕地再生"，我要特别感谢清华大学的郑晓笛博士，她为组织这个主题付出了大量心血。棕地，已成为风景园林在生态运动中很惹人瞩目的对象。我在20世纪70年代后期主持过一次某重点污染区大规模环境本底检测，包括大气、水体、土壤和食品，在分析结果时，我就感到土壤污染远比空气和水体麻烦，当时只想到如何从化学角度处理，根本没想到园林也可以加入其中。老实说，即使想到，也不敢提倡，因为风景园林的处理办法，从整个地球化学角度看，终究还是浅层次的，打个比方，把土壤中那些不可降解的污染物视为一个个的小恶魔，如何将其抓出来并集中起来关进牢房才是根本解决办法。然而，棕地终于成为本学科的一颗明星，这让我想到的是中国文化或许还是太保守，这种思想作风不改变，是根本无能加入世界文化发展大潮的。郑晓笛在《棕地再生的风景园林学探索》一文中提出了"棕色土方"的概念，十分有助于我们重新整理对付土壤污染的思路。

说到土方，其实就是地形改造，这方面中国人的积累远比其他民族深厚。西方传统的土方工程大体分为3类：平整土地，挖方池（或运河），堆馒头山（或土坝、漫坡）。如何理解自然地形系统，理解山水关系，理解地形与小气候，特别是地形审美，那绝对是中国人的长项。

Phase 4, 2015

The papers with the theme of "Healing Garden", originally planned for the January edition, were delayed for some reason and are fully published this month. The organization of this theme has been strongly supported by Professor Emeritus William Stone Weedon, School of Architecture of the University of Virginia, Professor Reuben M. Rainey, Co-chair of the Design and Health Center, and Mr. David Kamp, President of Dirtworks. They participated in the theme planning, carefully selected landscape cases, and helped to call for papers. We are also grateful that Professor Rainey even edited and revised the English papers.

Strategy and tactic, in a general view, should be the same, but there are some apparently opposite situations happen, and the more famous example is the so-called "eliminating the unjust war with the just war to achieve peace", in which the peace is the strategic goal while the war is the realizing tactic. Of course, it is worth considering if it is really the case. It is similar to the relationship between development and harmony. The driving force of development is in fact the gap, which is unharmonious, and the goal of development is the social and human-nature harmony. In order to eliminate the undeveloped and unharmonious of the world, the undeveloped should accelerate the development, which would inevitably bring about a lot maladies in the rapid development, such as resource consumption, environmental pollution, the Gini coefficient increase, and the tension of international relations. How can we look at these issues? Are the undeveloped countries the ones to be blamed? Can we all whether in developed countries who are all to blame? This is really a test for the so-called celebrities' IQ.

In March 13, the School of Design of the University of Pennsylvania held "Penn- China Landscape Symposium" in Beijing, which fully shows the sensitive and pioneering spirit of the American academia and is worth our learning. I feel Penn is well-intentioned and would like to help China to find the ways and means to solve the problems in the development and make it reference for many developing countries, and of course, there will inevitably contain a lot of business opportunities. As to the development of landscape architecture of the world over the past nearly half a century, a lot of people noticed only the "Harvard Revolution" in the mid-nineteenth century, but in fact the more influential ecological movement during the turn of the last and present centuries was led by Penn, which also laid the foundation for "design with nature" and "landscape urbanism".

The theme of this month is "Brownfield Regeneration". Dr. ZHENG Xiaodi from Tsinghua University did a lot for the organization of this theme, and I would like to express our gratitude for his contribution. Brownfield has become the striking object in the ecological movement of landscape architecture. In late 1970s I once presided over a large scale environmental background detection of a seriously contaminated area, including air, water, soil and food. In the analysis of the results, I felt that soil pollution is far more serious than air and water problems. At that time we were only thinking how to deal with from a chemical way and did not expect that landscape architecture could also be added in. Honestly, even with that thought, we would not dare to advocate it, as the landscape architecture approach, from the entire geochemical perspective, was superficial after all. In analogy, if regarding those non-biodegradable contaminants in soil as little devils, how to get them out and put them together in jail is the fundamental solution. However, brownfield finally become a star of the discipline, which makes me think that the Chinese culture is probably still too conservative, and if this style of thinking not changed, it can never join the world cultural development tide. Dr. Zheng's concept of "brown earth-work" proposed in her paper *Landscape Research on Brownfield Regeneration* is very helpful for us to rearrange our ideas to deal with soil contamination.

The earthwork is in fact terraforming, in which the Chinese people accumulated much more knowledge than others. The Western traditional earthworks are roughly in three categories: land leveling, pool (or canal) excavating, hill (or dam, gentle slopes) heaping. But it is definitely the advantage of the Chinese people to understand the natural terrain system, the relationship between mountain and water, the terrain and microclimate, and especially the terrain aesthetics.

2015 年 5 期

中国有着世界上悠久的连续性的风景园林传统，近十多年中国进行了世界上规模最大的landscaping实践，从而拥有了世界上规模最大的风景园林教育事业，获得了数量最多的IFLA学生设计竞赛奖项，直至拥有了数家世界上资本雄厚的园林工程上市公司，但是，为何在世界上的LA界，门调最高的还不是中国人？这个问题值得中国风景园林界认真思考。希望有这方面的探讨文章投稿给本刊。其中有些属于技术性问题：如何能让国家尽速为风景园林立法，如何能让本行业代表进入国家各级最高权力机关——人民代表大会，如何能让国家为风景园林学科建设设置一笔专款，如何能让世界最高级的学术论坛固化到中国，如何建设一本不受盈利拖累的纯学术刊物，如此等等。

《城市滨河地区生态修复与生境营造——以烟台市鱼鸟河为例》一文展示的大量实景照片，无论是河岸形态、植物群落，还是画面意蕴，乃至人群活动，都浸蘸着浓浓的中国味道，又是具体的烟台场所。作为《中国园林》的主编，我当然希望有更多的这类设计文章。

对于中国园林，我们已有的认识比起很多重视民族文化的国家还远不能达到令人满意的水平。本期《徽州水口园林中的廊桥》让我获益匪浅，进一步加深了我对中国园林如何综合处理景物、功能、思想、社会以及人与自然的关系的理解。我过去曾经非常讨厌"封建迷信风水"，经过多年去粗取精，包括自己的思想变迁，我对传统文化的很多东西逐渐服膺，只恨自己过去所知太少，正因所知甚少，也增加了我进一步进行批判吸收的难度。

在世界性的学风恶化大环境下，实践性学科的研究，一定要紧密结合实践，从实践中来，到实践中去。仅从文献综述中找题目，找依据，很不可靠，因为有大量荒诞的论文混杂其中。这类荒诞论文或凭着编审不严，或凭着花钱买位，或凭着人际关系，跻身进了所谓科技刊物，一旦进入，又有更多的人盲目引用，于是谬种流传。特别对于学生，更要强调此点，因为他们根本无法辨别这类文章中的真假。

在本期的主题"风景园林专业教育"中，同济大学的刘滨谊教授撰文并特别关注到风景园林的时间维度，这是一个哲学和美学界潜在的新动向，刘教授一向对动向非常敏感，这是时代的需要。华南理工大学的袁晓梅教授为本刊组织了"声景"专题，应约者非常踊跃，稿件质量普遍较高，编成了本期的"声景"栏目，以更完善地反映世界上声景研究的动态。在此，我向袁晓梅教授表示感谢。

Phase 5, 2015

Now that China has the world's old continuous landscape tradition, has conducted the world's largest landscaping practices over the past decade, and hereby has the world's largest landscape architecture education, has won the largest number of IFLA Student Design Contest awards, and has several financially-abundant listed landscape engineering companies, why are the ones with the highest tone in the worlds LA circle not the Chinese people? It is worth the serious thinking of the Chinese landscape architecture circle. We welcome the papers one this topic. There are also some technical problems: how the national legislation can be established for landscape architecture as soon as possible, how the representatives of the industry can enter the highest authorities of various levels – the People's Congress, how the country can set up a special fund for the construction of landscape architecture, how the world's most advanced academic forums can stay longer in China, how to run academic journals not being dragged by profits, and so on.

Urban Riverside Area Ecological Restoration and Habitat Creation – Based on the Fish-birds River of Yantai presented a large number of real photos, and the riverbank terrains, plant communities, scenery implication, and even human activities are all of strong Chinese flavors, as well as specific places of Yantai. From the perspective of our journal *Chinese Landscape Architecture*, I certainly hope to have more articles of this kind.

As to the Chinese garden, the knowledge we have is far from satisfaction comparing with the countries paying attention to their national cultures. I benefit a lot from *On Lounge Bridge of Shuikou Gardens in Huizhou*, and it further deepened my understanding of Chinese garden in the integrated approach to handling the relationships among scenery, function, thinking, society, man and nature. I used to be very sick of the "feudal superstition feng-shui", but after years of rejecting the dross and assimilating the fine essence, as well as the changes of my own thinking, I gradually adhere to the traditional culture, and hate myself for knowing little in the past, and it is the poor understanding that also further increase the difficulty of my absorbing the traditional culture with criticism.

Under the global situation of academic and study deterioration, the study of practical disciplines should work closely with practices, from practice and to practice. It is quite unreliable to look for topics or basis only from literature and review, because there are a lot of absurd papers mixed in. These absurd papers ascended into the so-called scientific journals through loose editors, money, or relationships. And once inside, there are more blind citations, so absurd theories spread. Especially for students, this point should be emphasized, because they cannot distinguish between the true and the false in these papers.

About the theme of this month "Landscape Architecture Education", Professor Liu Binyi of Tongji University wrote the paper and paid particular attention to the time dimension of landscape architecture, which is a potential new trend in the philosophical and aesthetic fields. He has always been very sensitive to new trends, and it is the need of the times. Professor Yuan Xiaomei of South China University of Technology organized the theme of "Soundscape", contributors were very enthusiastic, and the qualities of papers were generally high, so we set up the special column on "Soundscape", in order to better reflect the trends and progress of world soundscape research. I would like to express our gratitude to Professor Yuan Xiaomei.

2015 年 6 期

西方社会的特点之一是始终保持着创新的态势，创新的先锋是提出新理念，有了能够打响的新理念一般就保证了它可以打遍天下，成为全球化的一员干将。依我看，这种路数保证了创新，却不一定保证正确。可惜很多人把二者混为一谈，以为创新就是对的，一些比较审慎的人虽然不相信创新就好，会先审视一下其理念如何，但也会以为新理念的方向正确就保证了创新的正确。

果真如此吗？让我们拿出一个例子剖析一下。

城市对雨水的态度，除了下雨时吸水，需要时释放2类，还有一类是当城市土地已经水饱和但还在继续下雨时，要保证城市不被水淹（当然需要根据社会发展和需要提出具体指标）和保证必需保护的城市植被（如大树、古树、名木和重要的群落）的生存。

在思考人类对地球影响后果的问题时，负责的人提出了"低影响开发"（LID）的理念，其核心思想是人类的活动最好能保持场地的基本特性不变，这样就可以把人类对大气、水体、上下游等环境的影响控制在最低限度，总体上维护地球的生态条件，实现人类对环境的友好。应该说，这是一个能够综合平衡人类利益和地球生态平衡的好建议。

在西方特别是美澳地区，城市绿地可以达到城市用地的2/3左右，城市人均用地是我国的2倍左右的条件下，将这个原则应用到城市，可以基本解决雨洪问题，因为雨水污染的浓度较小，单位面积绿地的雨洪径流负荷增加不大。

在我国城市绿地很少的条件下，让少量绿地去负荷整个城市用地的原生态作用，只是一种幻想。简单说，如果城市只有1/3的绿地，原则上就只能维系原城市用地1/3左右的生态作用（当然可以通过优秀的城市规划设计和绿地管理来合理地提高其效益）。30%左右的绿地去负担100%的径流，意味着绿地土壤中要多留2倍的雨水，这些雨水的污染浓度也要增加2倍，那么很多本地植物可能就承受不了，所谓"提倡本地植物"又成了一句空话。

被人为破坏的自然，应该由人为措施去解决，不应靠加重自然的负担来解决。这种思想，可称之为人类环境保护的"自作自受"原则。以城市土地的雨水贮藏作用为例，要想保持城市建成前后周边大地域雨水分配的状况基本不变，那些硬化地面的雨水洁净和下渗问题主要应该用人工措施解决，而不是像现在一些人坐在屋子里狂想出来的全都让城市绿地去解决，甚至提出"城市绿地一律下降20cm"这样的害国害民的"提议"。别的不说，我们知道中国现有的城市绿地负荷已经非常沉重：改善气候，优化大气，保护水质，居民游憩，市民健身，老人儿童，美化景观，旅游经营，文物保护，生物多样性……这些作用都被抛到了何处？特别是绿地负担着城市重要的防灾功能，我们知道大震之后几乎必有大雨，如果我们城市的绿地皆成了泽国，让我们的避灾群众到哪里去？救灾活动还如何进行？

归根结底，中国共产党一贯提倡的"实事求是"传统才是真正的科学之道。面对一时非常时髦的"雨水花园"，北京市明确提出"市政道路初期雨水绝对不能在绿地内利用"，就是很好的例子，详见本期《北京推进集雨型城市绿地建设的研究与实践》一文。

对于全国都在关注的"海绵城市"课题，由于中国的地域广大，情况多变，其难度和复杂程度是可想而知的。在此我要特别感谢中国城市建设研究院王磐岩副院长勇敢地承担起组织本期主题的重担。有着党中央的关怀和领导，相信中国城市的雨洪处理一定能沿着实事求是的道路很快达到世界的前列。

Phase 6, 2015

One of the characteristics of the Western society is maintaining a momentum of innovation. The vanguard of innovation is to raise new ideas, and the resounding news ideas generally can play all over the world and become a member of the go-getters of globalization. In my opinion, this approach ensures innovation but does not guarantee correctness. Unfortunately, many people confuse the two and think that innovation is just right, and some of the more cautious people who do not believe innovation is right will first look at how the idea is, but will also think that the right direction of new ideas ensures the correctness of innovation.

Really? Let's come up with an example and make an analysis.

Apart from absorbing water in rain and release water upon need, the city's attitude toward rainwater is to ensure the city is not flooded (of course specific indicators should be put forward according to social development and needs) and the survival of urban vegetation (e.g. big, ancient or famous trees and important communities) when urban land is already saturated with water while the rain continues.

When thinking about the consequences of human impacts upon the earth, the conscientious ones put forward the "low-impact development" (LID) concept, and its core idea is that human activities should better maintain the same basic characteristics of the site, to minimize the human impact upon the environments of atmosphere, water, upstream and downstream, and maintain the ecological conditions of the planet in general and achieve human friendly to the environment. It should be said that this is a good proposal that can comprehensively balance human interests and the ecological balance of the earth.

In the West, particularly the United States and Australia, green areas of the city can reach about two-thirds of urban land, and urban land per capita is about double of the Chinese one. Under this condition, applying this principle to city can basically solve the problem of rainwater, because the concentration of rain pollution may be small, and the increase of rainwater runoff load per unit green area is not big.

In the situation of less urban green space in China, it is an illusion to make the small amount of green space to shoulder the original ecological effects of all urban land. Simply speaking, if the city only one-third of green space, in principle, it can only sustain about one-third of the original ecological role of urban land (of course its effectiveness can be reasonably improved through excellent urban planning and design and green space management). In this case, the 30% green space to shoulder the burden of about 100% runoff, it means twice amount of rainwater need to stay in the green space soil, and the contamination of rainwater increases twice, so many native plants may not afford it and the so-called "promoting local plants" has become an empty talk.

The man-made sabotage upon nature should be resolved with man-made measures, rather than by increasing the burden upon nature. This idea can be called the "self-inflicted" principle of human environmental protection. Taking the rainwater storage effect of urban land as the example, in order to keep the situation of substantially constant rainwater distribution in the large surrounding area, the issue of rainwater cleaning and infiltration on the harden ground should be solved mainly with man-made measures, rather than now some people sitting in the house and thinking out of fantasy to solve all the problems with urban green space, or even putting forward the harmful to the nation and people "proposal" of "sinking all urban green space at 20cm". If nothing else, we know that the existing Chinese urban green space has a very heavy burden: improving the climate, optimizing the atmosphere, protecting water quality, for residents recreation, public fitness, the elderly and children, landscaping, tourism management, heritage conservation, biodiversity… where have these functions been put? In particular green space shoulders the burden of the city's important function of disaster prevention. We know there will almost be heavy rain after earthquake, and if all urban green space becomes swamps, where could the people go to avoid disasters? How can the disaster relief activities proceed?

Ultimately, the Chinese tradition of "seeking truth from facts" is real scientific approach. On the momentarily very fashionable "rain garden", the Beijing government clearly stated that "the initial rainwater on municipal roads must not be used in green space", which is a good example and can be further read in the paper of *Promoting the Research and Practice of Rainwater Harvesting Green Space Construction in Beijing*.

For the "sponge city" topic with nationwide concerns, due to China's vast area and changing situations, its difficulty and complexity can be imagined. Here I would like to especially express my gratitude to Vice President Wang Panyan of China Urban Construction and Design Institute for organizing the theme papers. With the care and leadership of the central government, I believe that China's urban stormwater treatment will quickly reach the forefront of the world from the "seeking the truth from facts" approach.

2015 年 7 期

"数字景观"作为最高端科技水平的代表，已经第二次成为本刊的主题。对于这个问题，我总是有些话不吐不快。

从"数字xx"到"智慧xx"再到"云计算"和"大数据"，也就是从IT（internet technique）到DT（data technique），技术的进展实在太快了，数字技术对人类的影响，必将远远超过机械电器时代的技术，因为后者只是人类体力和技能的替代者，而前者是人类头脑的替代者。论体力和天生技能，人不是生物中的佼佼者，只有人脑，才是人之所以为人的标志。以此观之，把迄今的技术进步分为1.0~4.0有些贪小，其实只有两大时代，我将之标志为T-I和T-II。

对于人脑的理性思维方式，也可以分为两大类：一类是单线或分枝状的、逻辑的、因果的；一类是网状的、博弈的、统计的。大体上前者是西方的优势，后者是东方的优势，我们当下教育的根本问题之一，是让年轻人都以为只有符合逻辑才是真理的唯一标准或是唯一的真实，看不到偶然、干扰和宇宙负面的作用，很容易走上一统化的思路。当T-II发展到大数据时，其技术也从以钻牛角尖为主转为以海纳百川为主，即着眼于数据关联性，而非数据精确性，从而算法趋于简化。《大数据时代》作者舍恩伯格（Viktor Mayer-Schonberger）明确提出大数据的简单算法比小数据的复杂算法更有效，"更具有宏观视野和东方哲学思维"。当然，随着技术的进步，大数据也会逐渐走上在必须时把计算精确化之路，这就是东西方文化交流的深化之路。在我看来，把一切都精确量化实在没有必要，我国很多科研经费都被浪费在这上面。

一统化的根子是一元论。对万事要追到根，追到起点、最早、顶尖、头头，追到元素、原子、原理、原初，是人类的天性，但谁也没有希腊人和希伯来人彻底。西文archi-这个词根的意思是原初，它孕育出了档案（archives）、考古（archaeology）、工匠领头人（architect）、原型（archetype）等一批概念。而对原点和原理孜孜不倦的探求，也成为从苏格拉底到海德格尔的西方哲学的基本特征，以为有了原点，再有了逻辑，就可以接近绝对真理。但世界果真是这样的吗？我呼吁中国人不要盲从。

对于人之所以为人，还是有两种基本看法：第一种认为人只不过是大自然的一种产物，特殊的异化产物，一种某种程度上的"异数"，人终究还要回归于大自然，人应该把大自然的权力尽量还给大自然，自己只保留一块足够的领域即可；第二种认为人因为有着特殊的（例如受到上帝的授权，或者能够发现规律体系）能力应该与大自然平等，甚至人权高于大自然的权力（虽然权力这东西纯粹是被人发明的），所以人一定能战胜自然，掌控自然，塑造出一个自己可以任意驰骋的自由的世界。T-II时代的来临，更为后一种思想增添了强有力的武器。问题是未来人是掌控技术？还是被技术掌控？一些人信誓旦旦地保证人总是可以控制技术，即使如此，这里的人是全体还是几个精英？难道未来真的会是另外一种奴隶制吗？

观察一下最新的热点"地球脑"（grobal brain）。地球脑是以大数据、云计算，以及无数拥有聪明大脑的人类个体，借助于互联网结合成的具有神经系统特征的自组织巨型网络，可以帮助决策甚至指挥人类。既然全球都被一个脑袋统帅，还要多样性干嘛？

感谢东南大学成玉宁教授组织了本期主题。

Phase 7, 2015

"Digital Landscape" as the representative of high-level technology has become the theme topic of our journal for the second time. For this topic, I always have something to speak out.

From the "digital xx" to "smart xx" and then to "cloud computing" and "big data", namely from IT (internet technique) to DT (data technique), technical progress is too fast, the effect of digital technology on humans will far exceed that of the technical machinery and electrical era, because the latter is just the replacement of human strength and skills, while the former is the replacement of human mind. Humans are not outstanding in terms of physical strength and innate skills, and it is the brain that makes people distinctive. In this view, it is a bit petty to divide the to-date technical progress into 1.0- 4.0, and in fact there are only two generations which I tag it as TI and T-II.

The rational thinking of human brain can also be divided into two categories: one is single or branched, logical, and causal; another is reticular, gaming, and statistical. In general the former is the advantage of the West, while the latter is the advantage of the East. One of the fundamental problems of our present education is to make young people think that only meeting the logic is the sole criterion of truth or the only true, neglecting the effect of accident, interference and adversities of the universe, which is apt to generate the idea of unification. When the T-II develops to big data, its technology is mainly converted from a dead end to absorbing and diversity, focusing on the relevance of data, rather than data accuracy, which tends to simplify the algorithm. Viktor Mayer-Schönberger, the author of *Big Data*, clearly put that the simple algorithm of big data is more efficient than the complex algorithms of small data, "a more macro perspective and Eastern philosophy thinking". Of course, as technology advances, big data will gradually embarked on the road of exact calculation when necessary, and this is the way to deepen the cultural exchanges between the East and the West. In my opinion, precise quantification of everything is really unnecessary, and many of our research funds have been wasted on it.

The root of unification is monism. To trace the root of everything, it is the human nature to trace the starting, the first, the top, the head, the element, the atomic, the theory, and the original, but no one does more completely than the Greeks and the Hebrews. The root "archi-" means original in the Western languages, and it gave birth to a number of concepts and words like archives, archeology, architect, and archetype. The tireless quest of the principle of origin has become the basic feature of the Western philosophy from Socrates to Heidegger who thought that they could be close to the absolute truth with the origin and the logic. But is the world really so? I advise the Chinese people no to follow blindly.

For the basic factors to be a human being, there are two basic views: The first is that man is nothing but a product of nature, a special product of alienation, the "anomaly" to a certain degree, people will eventually have to return to nature, people should try to put the power of nature back to nature, and they only retain a sufficient area; the second believes that man has the special (such as God-given, or being able to find the law system) capability and should be equal to the nature, human rights are even higher than the power of nature (although the thing of power is purely invented by man), so man will be able to overcome the nature, control nature, and create a whizzing free world. The coming of the T-II era more adds a powerful weapon to the latter ideology. The problem is that, are people controlling or controlled by technology in the future? Some people vowed that man can always control technology, but even so, does the man here mean all people or just some of the elites? Is the future actually another form of slavery?

Look at the latest hot spot "global brain". The global brain is a self-organizing giant network with neural system characters combining big data, cloud computing, as well as the smart brains of numerous individual persons via the Internet, and it can help make decisions or even directing humans. It seems the world has been commanded by a head, what is use of diversity?

Thanks to Professor Cheng Yuning of Southeast University for organizing the theme papers.

2015年8期

本刊有幸发表了刘家麒先生对孟兆祯先生《园衍》一书的意见和孟先生的回函，充分体现了老一辈学者对学术的严谨态度和无私精神。两位先生皆已年过八旬，是我的师辈，我自忖不如远矣。

学术讨论、议论和争辩，是学术发展的基本条件，近年来我们开了很多学术会议，但极少见到讨论和争辩，总是大家宣读一遍论文，就顺利结束，你好我好大家好，其实对探讨真理的意义不大，反而可能造成谬论的流传。这种风气也渗透到刊物稿件的评审过程中，如果编委审查严格了，有些作者反而会不满意。美国人统计过，就在近一两年里，中国的学术论文产量将超越美国，可是有意义的论文数量却与美国相差甚远。其原因当然很多，而缺乏好的学风，当是重要的一环。为了改变学风和有利于创立中国风景园林学派，我和国际著名杂志《*Landscape and Urban Planning*》的双主编之一象伟宁教授牵头开启了一个沙龙，命名"善境"。首届沙龙共约集了10位教授雅集于广东惠州南昆山，畅谈之外，凝聚众意成就了"雅集善境沙龙，开议中国学派"一文。

本期的论文，我郑重推荐《景观生态空间网络的图式语言及其应用》。我很欣赏此文的理论、方法和作风，相当正宗的科研方法，以及一种实事求是、尊重现状、接地气的路线，多样性也因此油然滋生。乡间大自然乃至城市绿地的生态，是人与自然互动的领域，将其参照语言规律分析为字、词、句和语法规则，以说明糅合了自然关系和人造关系的景观生态关系的相对客观而又生动的组合与流程的规律，虽然是初步的，但是可以试行和推行。这让我想到，人类已有了仿生学，基于自然科学的生态学研究也不妨试一试"仿人学"。

此外，《城乡规划体系中生态绿地的现状问题与对策》一文对国家级顶层设计非常重要。我的看法是，生态绿地这个词从字面上很难定义，是"专门为了区域生态划定的绿地"还是"具有生态效用的绿地"？按前者，不应占用城市建设用地指标，但可包括国土上各种自然保护区和指定用于生态为主的目的的绿地；按后者，似乎各种绿地都可以划进来，那就等于绿地了。所以原则上这不是个学术问题，应该由中央政府首先做出政策性决定，定了，大家照做就行了。我倾向于前者，因为内涵清晰，便于实行。

最后，我想谈谈《还君明珠——探索历史图像中的广州行商园林》的读后感。现在已经很难想象那群富可敌国甚至世界首富的十三行商人园林的奢华，它们基本皆毁于清末民初，大约从广州泮溪酒家或北园酒家的几个豪华间装修（是新中国成立初期集一些待拆建房屋配件而成）或可隐约领略当时的情景。虽然，作为商人园林肯定会走上中西合璧的路线，但那时，他们是以"上国"的心态来接受西方事物的，那时的中西合璧肯定与鸦片战争后的中西合璧有本质的区别。我想，现在的中国人，在看待中西合璧时，亦当与积贫积弱的旧时代有不同的心态，才合乎人道。

特别感谢北京林业大学董丽教授，为组织本期"城市生物多样性"的主题文章所付出的努力！

Phase 8, 2015

We had the pleasure to publish Mr. Liu Jiaqi's comments on *Yuan Yan* and Mr. Meng Zhaozhen's reply, which fully shows the rigorous academic attitude and selflessness of the older generation of scholars. They are both over eighty years old, of the generation of my teachers, and I think I am not a quarter as good as them.

Academic discussions, debates and arguments are the basic part conditions of academic development. In recent years, we have held a lot of conferences but rarely seen discussions and debates. We always read the papers and then made a successful conclusion, and everyone was fine, but it is actually of little significance in seeking the truth, conducive to the spread of fallacy instead. This trend also penetrates into the review of manuscripts for publishing, and some authors would not be satisfied if the editorial board members are strict. American statistics show that in coming years the number of China's academic papers will surpass that of the US, but the number of significant and substantial papers of China are far behind. There are, of course, a lot of reasons, and the lack of good style of study is definitely an important part. In order to change the style of study and be conducive to the creation of Chinese landscape school, Professor Xiang Weining, one of the dual chief editors of *Landscape and Urban Planning* (and the internationally renowned magazine), and I, opened an academic salon named "Fine Environment". The first session of the salon gathered ten professors in Nankun Mountain of Huizhou, Guangdong Province, where we talked and discussed, summarized our opinions and achieved the paper *Gathering for the Fine Environment Salon and Opening the Chinese School*.

Among the papers of this month, I strongly recommend *The Pattern Language and Application of Landscape Eco-space Network*. I appreciate the theories, methods and style of this paper, the quite authentic scientific method, and a no-nonsense, respect for the status quo, and down-to-earth way, and thus the diversity spontaneously breeds. The countryside natural or urban green space ecology is an area of nature- human interaction. Analyzing it as characters, words, sentences and grammar rules referring language laws to describe the relatively objective but vivid combinations and processes laws of natural-artificial-blended landscape ecological relationships is, although preliminary, pilot and feasible. It makes me think, we have had bionics, and the ecological study based on natural sciences may also try "imitation of human".

In addition, the paper of *Study on the Situation, Problems and Countermeasures of Ecological Green Space in Urban Planning System* is very important for the national top-level design. My view is that it is hard to define the term of ecological green space literally. Is it "the green space specially designated for regional ecology" or "the green space with ecological functionality"? According to the former, it should not occupy the city construction land, but may include a variety of nature reserves and the green space designated mainly for the purpose of ecology; according to the latter, it seems that all kinds of green space can be drawn in, and it is equal to green space. Therefore, in principle, this is not and academic question, the central government should first made policy decisions, and once set, we go on the line. I prefer the former, because it has clear connotation and is easy to implement.

Finally, I want to talk about my review on *Iconological Study on Hong Merchants' Gardens in the Historical Images*. It is hard to imagine how luxurious the gardens of this groups of the world's wealthy businessmen were, which were basically destroyed in late Qing Dynasty and the early period of the Republic of China. We may have some blurring taste of the scene from the decoration of the deluxe rooms (with the parts of the demolished houses during the early period of the founding of New China) of Guangzhou Panxi Restaurant or Beiyuan Restaurant. As merchants' gardens, they certainly followed the style of the combination of Chinese and Western elements, but at that time, they hold the mentality of "the central empire" to accept Western things, which was fundamentally different from the combination of Chinese and Western elements after the

Opium War. I think that the Chinese people now surely have a different mentality upon the combination of Chinese and Western elements comparing with the people of the poor and weak old ages, and it is human sympathy.

We would like to specially express our gratitude for Professor DONG Li of Beijing Forestry University. She makes tremendous contribution to organize the "Urban Biodiversity" theme papers of this issue!

2015 年 9 期

本期的主题是"城市绿量",衷心感谢华中农业大学姚崇怀副教授组织的这批稿件。

跟海绵一样,绿量这个概念也被某些人搞得一塌糊涂。大约十五六年前我和陈自新先生聊过,他提出绿量概念的初衷,从生态原理上说是想了解单位面积上光合作用效能的大小,但这东西因素太多,变化太大,很难有稳定的定量,所以才想到用单位面积上叶绿素的多少来替代。而用叶面积(注意,没有枝干)来替代叶绿素,正好国际上已经有人提出过LAI(叶面积指数)的概念,比较可行,不过已经是再一步的近似,因为不同叶子的绿色浓度不同、活性不同、叶面朝向不同、受光强度不同等,其光合效率当然也不同。这些陈自新都考虑过,我也大加赞赏,称赞他是真正的科学工作者。

后来有人提出用树冠体积(原始的表达是"植物茎叶所占据的空间体积")的大小来表达三维绿量,则距离研究植物生态作用的实质——光合作用效能的初衷越来越远,甚至很不可靠。如果陈自新先生还在世,是绝对不会这么干的。原因很简单,同样大小不同植物的树冠,其叶(以及叶绿素)的密度相差很大,而且,这种方法忽视了地被层和下层灌木的作用。把如此荒疏的方法直接推广,至少很不严谨,这里谈不上是学说问题,而是个思想方法和科学态度问题,即对表象和实质之关系的理解问题。

实事求是地讲,用三维绿量来替代二维绿量(绿地面积)本来是好意,也很适用于景观的量化。但在中国城市绿地还很少,总体绿量严重不足的时代,用垂直绿化和屋顶绿化来替代绿地面积,推行结果只能是更加促进城市绿地的减少。一个根本的问题是:当城市规划和相关法规的指标还处在平面的时代,不科学地强调三维绿量,后果到底是好还是不好?

自从"理念"这个东西传入中国,我们老祖宗的实事求是精神被很多人抛到了脑后。我以为,中国革命的成功,根本在于走的是实事求是的路线,而不是靠听起来高大上的理念。实事求是,就是要具体问题具体分析,要心中有数,要掌控变化,要了解条件。拿海绵这个理念来说,中国城市里有多少土地可以海绵?占1/3的房屋地基不能海绵,占1/5的道路下面不能海绵(除非这条道路不再承重),这就去掉了一半多城市用地;此外,很多市政工程、交通工程、基础设施、工业设备等的下面也不能海绵,原有的水面下面本来就是海绵……把这些都除掉,可以海绵化的城市土地可能仅剩下1/3左右。即使如此,还要考虑避灾、防险、市民活动等的需要,不能一概海绵化了之。总之,我们再不能不加思考地把某些西方人在人少地多条件下搞出来的"理念"捧到天上,奉为放之四海而皆准的真理,特别是大学里不能再由这类东西肆无忌惮地流传。而这类现象,正是我所收到的很多论文的通病。

我的学生告诉我,过去他很欣赏网络上的大V、大咖,觉得他们很能发现问题,说得很有道理。现在他明白了,面对中华民族承担的人类历史重任,中国现在需要的是如何解决问题,真正地解决问题。在这方面,所需要的知识、思想方法和人格修养,远不是仅靠读点书,发点议论,再加上胆大妄为就可以达到的。

Phase 9, 2015

The theme of this month is "urban green volume", and I want express our heartfelt gratitude to Associate Professor Yao Chonghuai of Huazhong Agricultural University for collecting the papers.

The concept of green volume, just like the concept of sponge city, is messed up by some people. When I talked with Mr. Chen Zixin about fifteen or sixteen years ago, he put forward that the original intention of the concept of green volume, in terms of the ecological principle, was to get the size of the effectiveness of photosynthesis per unit area, but there were too many factors and big changes, difficult to have stable quantity, so he just thought of using the quantity of chlorophyll per unit area instead. As to using the leaf area (please note, no branches) instead of chlorophyll, someone in the international community has just proposed the concept of LAI (leaf area index), more feasible, but more approximation as well, since the photosynthetic efficiency differs with different concentrations of different green leaves, different activities, different orientations, and different light intensities. Mr. Chen Zixin had considered all these matters, and I also much appreciated this and praised him as a real scientist.

Later, someone proposed to use the crown volume (the original expression was "the spatial volume occupied by plant stem") to express the amount of three- dimensional green volume, but it was getting away from the original intention of studying the ecological role of plant– the photosynthesis efficiency–and even unreliable. Mr. Chen would not do like this if he is still alive. The reason is simple that the densities of leaves (chlorophyll) of the same size canopies of different plants vary greatly, and this approach ignores the functions for groundcovers and sub-layer shrubs. It is, not rigorous, at least, to directly promote this out-of-practice method. We cannot say it as a doctrine issue, but a problem of the way of thinking and scientific attitude, namely the understanding of the relationship between appearance and substance.

Practically speaking, it would have been well-intentioned to replace the two- dimensional green volume (green area) with the three-dimensional green volume, also very suitable for landscape quantification. But in the era of limited urban green space and insufficient overall amount of green volume, using vertical greening and green roofs to substitute green space can only accelerate the reduction of urban green space. The fundamental question is: when the indicators of urban planning and relevant regulations are still in the planar era, the consequence of unscientifically emphasizing the three- dimensional green volume is good or bad in the end?

Since the "philosophy" (idea) came to China, the pragmatic spirit of Chinese ancestors are thrown away by many people. I think that the success of the Chinese revolution lies in following the pragmatic route, rather than the sounding-well philosophy. To be realistic, it is to analyze specific issues, know fairly well, take the changes under control, and understand the conditions. Talking about the concept of sponge city, how many areas of Chinese cities can be sponge? No for 1/3 of housing foundations, and no for 1/5 underground areas of roads (unless this road no longer load-bearing), which removes more than half of urban land; in addition, the underground areas of many municipal engineering, traffic engineering, infrastructure, and industrial equipment cannot be sponge, while the underground areas of original water surfaces are already sponge. After removing all these, only the rest 1/3 urban land can be sponge. Even so, taking disaster prevention and public events into consideration, these land cannot all be sponge. In short, we should never unthinkingly applaud the "idea"–the one some Westerners make under the conditions of fewer people and more land – to the skies and regard it as a universal truth again, and especially these type of things should no longer spread with impunity in universities. And this phenomenon is the common problem of many papers I have received.

My student tell me that in the past he admired the web opinion leaders, thinking that they were able to find problems and sounded very reasonable. Now he understands that, in the face of the human historical responsibilities shouldered by the Chinese nation, how to solve the problems and solve them really is what China needs now. In this regard, the required knowledge, thinking and personality cannot be achieved by just reading some books, proposing some discussion points, and being full of guts.

2015年10期

弹指一挥30年，转瞬间，本刊已达而立之龄，同时，也是中国风景园林学会（含其前身中国园林学会）的而立之年。为了纪念和祝贺，吴良镛、齐康、郑孝燮、孙筱祥、程绪珂、孟兆祯几位前辈为本刊写来题词，我们深表感谢。此外，还有很多前辈、同仁、朋友、作者和读者发来了贺文和稿件，寄托了他们的祝福和期望，我们将集成一期"增刊"发表。

而立，《辞海》释曰："《论语·为政》，'三十而立。'指年至三十，学有成就。"2011年，学科成为一级学科；2014年，根据全国科学技术名词审定委员会（经国务院授权，代表国家进行科技名词审定、公布的权威性机构，经其审定公布的科学技术名词具有权威性和约束力，全国各科研、教学、生产、经营以及新闻出版等单位应遵照使用）的安排，中国风景园林学会启动了《风景园林学名词》编写工作；2015年，《中国大百科全书》（第三版）的编辑工作启动，在这一版，风景园林学将独立成卷。中国的风景园林学科，确实已到了"学有成就"的阶段。

为此，本期以"《中国园林》创刊30周年纪念"为主题，从理论、教育、规划与设计、经济与管理、园林植物等方面较全面地回顾和总结了学科30年的发展路程，并对30年来的《中国园林》载文做了统计分析。

细品本期的论文，它们反映出我国风景园林学科的科技水平在迅速提高，仅题目中就出现了大量高层的或前卫的关键词或词组。我大致开列一下：设计语汇，4I体系，快速城镇化，生态文明构建策略，信息化集成管理模式，绿化废弃物循环利用，镇域生态服务评价，数值模拟，景观生态网络方法，模糊综合评价法，绿地热舒适度，空间格局特征，意境的循环演进模式……这只是当下一期的部分内容而已。想当初30年前，我作为中国首届硕士研究生刚刚毕业3年，这些词汇我都是闻所未闻，学科的进步不可谓不速，行业的作用不可谓不大，卅年的变化真真已是两重天！

下一步呢？我想首先还是把中国传统这一块搞好，比起发达国家，我们这一块还是不够深入，例如荷兰，不就是大坝和圩田么，可人家搞了几百年，还在深入，我们一些人对传统只是浅尝辄止，浮在表皮，就在那里哇啦哇啦，甚至数典忘祖，很不应该。同时，我们对世界上的各种发展也要深入下去，不再是介绍和初探，而是真正搞懂它，懂得其原理和推理，懂得其使用条件和适用范围，准确运用各种理论和方法，并能够分析和鉴定其成果的有效性。这就要求更深刻和准确的数、理、化、生知识，更重要的是实事求是做学问的态度和对立统一的辩证思维方法。这样，再过10年，我们肯定可以做到"四十而不惑"！

中国人一定要打破对现状的迷信，懂得一切都在变化。举例来说，国际通用影响因子来作为文章的重要评价标准，对此我们当然也要采用，但同时要看到，这个影响因子是不讲价值观的。比如"第二次世界大战"，2颗原子弹杀害的几十万人和中国被杀害的3 000多万人，孰重孰轻？你去查查影响因子，绝对是讨论前者的文章高，也就是说，表面客观下面有世俗的影响，甚至控制性作用。

Phase 10, 2015

In a flash of three decades, our journal have reached its adult life, and it is also that of the Chinese Society of Landscape Architecture (including its predecessor Chinese Garden Society). We are deeply grateful that other Mr. Wu Liangyong and other predecessors presented their inscriptions for commemorations and congratulations. In addition, a lot of seniors, colleagues, friends, writers and readers sent their congratulatory letters and papers to express their wishes and expectations, and we will collect all these into a supplementary issue of our journal.

According to the *Chinese Dictionary*, the thirty-year-old adult life means being independent with accomplishments in learning. In the year of 2011, the landscape architecture discipline became the national first-level discipline. Last year, as arranged by the China National Committee for Terms in Science and Technologies (CNCTST, established upon the approval of the State Council of the People's Republic of China, is the organization authorized by the State Council to examine, approve, and promulgate the scientific and technological terms on behalf of the Chinese government. The scientific and technological terms examined and approved by CNCTST are authoritative and have a binding force in China in 2014. They should be observed and used by all organizations including scientific research and educational institutions, production and management units, news and publishing circles, etc.), Chinese Society of Landscape Architecture launched the compiling of *Landscape Architecture Terms*. The editorial work of the *Encyclopedia of China* (third edition) has started, and in this edition, landscape architecture discipline will be a separate volume. The landscape architecture discipline of China has indeed entered the stage of "accomplishments in learning" in 2015.

For this reason, in this issue we have "The 30th Anniversary of *Chinese Landscape Architecture*" as the theme, make a more comprehensive review and summary of the thirty-year development of the discipline from the aspects of theory, education, planning and design, economics and management, plants and flowers, and have a statistical analysis on the papers published in *Chinese Landscape Architecture* in the thirty years.

The papers in this issue reflect the rapid increase of the science and technology level of China, and a large number of high-level or avant-garde keywords or terms have appeared in the titles of the papers, including design vocabulary, 4I system, rapid urbanization, ecological civilization construction strategy, information integration management, green waste recycling, township ecosystem services evaluation, numerical simulation, landscape ecological network, fuzzy comprehensive evaluation method, green thermal comfort, spatial pattern characteristics, cycling evolution model of artistic conception. They are only part of the contents of the current issue. Thirty years ago – just three years after graduated as the first class with Master's degree, these words were totally unheard to me. The progress of the discipline is quite fast, the role of the industry is so substantial, and the changes in thirty years have really made another world.

What's next? I think, above all, we should make a good job of the Chinese tradition, since we are not deep enough in this area, comparing to the developed countries. For example, the Netherlands has just dams and polders, as we thought, but they have been engaged in these for hundreds of years and are going deeper. Some of us are very wrong to spurt a hubbub of voices after only scratching the surface of the tradition, or even forgetting their origins and predecessors' contributions. At the same time, we should delve into the depths of all kinds of development of the world, no longer just introductions and preliminary explorations, but really getting to know and understanding the principles and reasoning, the application conditions and ranges, accurately using a variety of theories and methods and being able to analyze and identify the validity of results. This requires a more profound and accurate mathematics, physics, chemistry, biology knowledge and, more important, the realistic scholarly attitude and the unity-of-opposites dialectical thinking. Thus, in another ten years, we certainly can be "at thirty, one can begin to understand the world".

Chinese people must break the superstition on the status quo and know that everything is changing. For example, the impact factor (IF) is the internationally accepted evaluation criteria for academic papers, and of course we should also use this, but at the same time we should be aware that the impact factor does not include values. For example, in the World War II, the hundreds of thousands of people killed by the two atomic bombs, and the over thirty million Chinese killed by the Japanese invasion, which is more important? The impact factor of the papers on the former is definitely higher. That means the seemingly objectiveness has secular influence and even plays a controlling role.

2015年11期

本期"屋顶绿化"由北京园林科学研究院风景园林规划设计研究所韩丽莉所长组稿，在此深表谢意。

不久前我参加了东南大学举办的"中国第二届数字景观国际研讨会"。东大在数字景观这个尖端领域已经走在了世界前沿，获得了国际认可，这在中国风景园林界是可喜的开端。相信会有更多的中国学府基于中华文化对人天关系的精深博大的理解，秉承中国人优良的智商基因，勇敢地向世界LA的前沿进取，占领学术高峰。

近来本刊及其他刊物涉及数字化、数理模型等方面的稿件迅速增加，同样反映了我国学界的这种动向，可喜可贺。另一方面，由于在这方面对大多数单位还是初涉，严谨的科研程序、审查制度还正在建设，各种问题难免出现。我想在此对常见问题做个初步梳理，供大家参考。

1）不了解条件。不了解国外模型的适用条件，也不了解自己课题的真实条件，盲目搬用。所有模型都是在一定的现实或假定条件下做出来的，不是万能的，适用越宽泛的，越需要在使用之前根据现场条件做出校核，特别是对那些利用线性回归推导出来的模型。不幸的是，目前常见的模型大量都是些线性回归作品，我们就只有强化使用前的校核工作。过去作为科普，本刊发表过一些模型的介绍性文章，今后，对模型不经校核利用的文章，我们不再发表，除非已经证明它可以利用在文中的案例。还要提醒一句，对那些低水平杂志上发表的文章本刊一般不认可为依据。

2）不了解机制。不了解国外已有模型的原理，不了解它是为了解决什么问题，如何推演出来的，就无限推广。例如径流本身的大规律是S形曲线，某些径流模型是线性或二次指数曲线，显然研究的是很局部的问题，靠它来解决雨洪问题只能是误导。又如一些看来高大上的文章，细究起来很多概念的单位都错了，说明作者并不了解该概念的实质以及单位的意义。总之，机制、机制、再机制！中国的学科要上升到世界领先，就必须从对机制的了解抓起。

3）知识版权。除非公开发表的模型，其他从研究机构流出来的模型都可能有版权问题。所以，利用别人的模型（包括进行二次开发）必须获得版权，那些"基于……"的文章要注意了。其实，风景园林模型这类东西只不过刚刚起步，漏洞很多（或曰还相当粗放），大多数也不难，为何总是基于别人的东西进行研究？中国人难道没信心创建一套更好的模型体系吗？例如为何必须依靠GPS、GIS，从北斗起家不行吗？有些卫片或GIS信息是会误导的，如四川盆地是一片灰黄色，江苏省的树木覆盖率只有不到2%，某位大师已经规划了中国整个大地的东南部之类，已远离了事实真相。

4）惧怕数学，迷信图形。对于世界的本质（数？图形？物质？心灵？），西方世界有很多争论。显然，在设计界，更多的人容易走上迷信图形的道路，心中无数，惧怕数学。为了披上科学外衣，一个简单的加法也要用$A=\sum X_n$之类的式子表达；简单的比例也要用饼图、柱状图之类的来显示……又如有人说他用几十平方米的阳台储存了数十吨雨水，来供给这个阳台花园的灌溉，也不算算这个储水池需要多高，还剩下多高可供栽植和人的活动,爬上这个水池需要多长的阶梯,屋内外的视线如何,更不想想阳台的荷载能力。再如所谓生态格局(pattern),是从普适性生物繁衍机制中导出来的,也不能解决环境污染问题,而城市是以人为主的地域,又是污染的主要源头,单凭所谓格局来做各种规划是很有问题的。

上述文字如有不当,热望大家批评指教。

Phase 11, 2015

We are very grateful to Prof. Han Lili, Director of the Institute of Landscape Architecture Planning and Design of Beijing Landscape Science Research Institute, for organizing the theme topic "Green Roof" of this month.

Lately I attended the "2nd China International Digital Landscape Architecture Symposium" organized by Southeast University. Southeast University has been in the forefront of the world in the frontier field of digital landscape and gained international recognition, which is a promising start in the landscape architecture circle of China. I believe more institutions will bravely head for the leading edge of world landscape architecture and ascend the academic heights, based on the profound understanding of Chinese culture of the human-nature relationship, and good intelligence genes of the Chinese people.

The rapid increase of papers involving digitization and mathematical models in our journal and other occasions also reflects this trend of the academic circle of China, which is gratifying. On the other hand, since most institutions are just putting their first steps into this field, rigorous research procedures and inspection systems are still under construction, and various problems will inevitably appear. Here I would like to make a preliminary sort of frequently asked questions for your reference.

1)Not understanding the conditions. Not understanding the application conditions of foreign models, nor the real conditions of their own projects, just blindly copying. All models are worked out under some real or assumed conditions, not a panacea, and the more broadly applicable, the more we need to make a check according to site conditions prior to application, especially for those models deduced with linear regression. Unfortunately, the current common models are mostly linear regression ones, so we should only strengthen the check and verification before application. As the popularization of science in the past, our journal published some introductory articleson models, while in the future, we will no longer publish the models applied without check and verification, unless it has been proved that they can be applied in the cases of the articles themselves. Also a reminder, our journal will generally not recognize the papers published in those low-level magazines as the basis.

2)Not understanding the mechanism. Not understanding the principles of foreign existing models, what problems they are to solve, nor how they are deduced, just putting them into unlimited promotion and application. For example, the general law of runoff itself is a S-shaped curve, while some runoff models are the linear or quadratic exponential curves, which are apparently for very localized problems, and are misleading if using them to solve the rainfall and flood problems. Another example is that in some seemingly top-level papers, the units of many concepts are wrong in careful study, which shows that the authors actually do not understand the substances of the concepts or the meanings of the units. In short, the mechanism is very important. We must start from the understanding of the mechanismsif China's disciplines want to lead the world.

3)Intellectual property rights. Apart from the published models, other models spread from research institutions may have copyright issues. Therefore, the use of someone else's models (including the secondary development) should obtain the copyright, and those "Based on…" papers should pay attention to this. In fact, the landscape architecture models are just starting, with many loopholes (or still quite extensive), most of them not difficult, why are our studies always based on someone else's stuff? Do the Chinese people not have confidence to create a better model system? For example, why should we rely on GPS, GIS, and how about starting from the BeiDou Navigation System? Some satellite images or GIS information are misleading, such as the Sichuan Basin is a pale yellow area, the tree coverage rate of Jiangsu Province is under 2%, and some so-called master has planned the southeastern part of the whole land of China, and all these have been far from the truth.

4) Dreading mathematics, and having blind faith in graphs. As to the nature of the world (number? graph? substance? soul?), there are many arguments in the Western world. Obviously, in the design community, more people tend to have blind faith in graphs, no numbers in mind, and dreading mathematics. In order to be seemingly scientific, a simple addition is also expressed with the equation like $A=\sum X_n$; a simple ratio is also shown with pie charts, histograms and so on… Another example is that someone said he stored dozens of tons of rainwater with his dozens-of-square-meters balcony, to supply the irrigation of the balcony garden; the things neglected are the required height of the water tank, the room left for planting and human activity, the length of the ladder to climb up the tank, the sight inside and outside the house, and most important, the load capacity of the balcony. As another example, the so-called ecological pattern is deduced from the universal biological reproduction mechanism, not to solve the environmental pollution problem, while city is the people- prioritized region, not the main source of pollution, and it is very problematic to make all kinds of planning solely relying on the so-called pattern.

We earnestly welcome your comments if the above contents are inappropriate.

2015 年 12 期

2015年第11期《主编心语》的中心是谈科研工作的基本要求和规矩。老实说，对于处于科学与艺术之间的人居环境学科来说，要想完全彻底地遵照自然科学的一套规矩去对待我们的工作对象，基本不可能，如果这样了，艺术就从我们领域中被驱逐了。

我想从另一角度观察一下科学的另一条边界。

任何成功的文化，必然内含互补的因素。其实，人都有两种基本生活方式：与自然隔离和与自然接触。中式院落与西式别墅，是普通人梦想生活方式的两种代表，它们同时都拥有了两种生活方式的条件，只不过别墅是用房屋把人保护起来，同时用花园与自然环境接通；而中式院落用墙把人保护起来，同时在小院内部尽量创造人与自然相通的条件。

但是追溯源头，两类文化的源起却正好相反。西方文化中，人类初始被置于伊甸园（Eden）或乐园（Paradise），是犹太教、基督教、伊斯兰教的经典所供奉，人因违背神的意旨而被放逐荒野，故而他们对荒野（wild）没有好感。中国人的神话则说人从诞生起就与各种以怪力乱神形态存在的多样物种共生在一起，人是靠奋斗和能够认识天道才从万物中脱颖而出，从而获得与大自然友好共处的能力。

西方文化有两个源头：希腊，希伯来，所谓"两希"（这是研究西方文化的中国人发明的词儿）。这两希是互补的，缺一不成气候。例如只有"希腊—科学"这条线，没有了"希伯来—宗教"的纠偏，这个地球恐怕已经被西方式对发展和个人欲望的崇拜毁灭了。反之，如果只有宗教，就不可能有文艺复兴。于是，两条线的纠结形成了西方文化的主线，但它们都基于一元逻辑主义（一神论）。

中国文化也有两条线索：入世与出世，或儒家的齐家治国平天下，或释道的出家避世归隐。但二者皆有共同的哲理基础——二元（阴阳或有无）辩证法，辩来辩去，许多事就逐渐模糊起来。

发展到现在，如何看待城市与自然的关系上又出现了两类方案，一类主张用科技的发展将人类封闭在一个巨大的人造空间里，在地球或其他星球上造成城市，在宇宙空间中则成为飞船，壳外才是自然本身。这是科学主义者和技术至上主义者所信奉的目标，是科幻小说的热门题材；一类则主张只有地球才是人类的家园，人们应该学会与自然共处，将自然引进城市，让人们日常都拥有和自然接触的机会。后面这种人的态度比较谦虚，自信力和野心较小，因而常常感受到科技主义者的蔑视眼光。

两类方案各有优缺点，至少现在还不能断定何者更正确。我比较倾向于后者，主要理由是人如果长期脱离自然，就会不理解自然，就会滋生蔑视自然的心态，这非常危险，即使人类在征服宇宙空间中获得某些成功，必然是处处要与大自然作对才能生存，从而带来的只能是无穷烦恼。所以，科学的远景不一定是幸福，很可能是麻烦。

如果不那么急功近利，许多事就不一定非搞得清清楚楚，"难得糊涂"真的不是一句玩笑语。

感谢北京林业大学李雄教授为本期主题组稿。

Phase 12, 2015

In the chief editor's words of 11th issue of 2015, I mainly talked about the basic requirements and rules of scientific research. Honestly, for the human habitat discipline between science and art, it is basically impossible to deal with our target groups by fully and completely complying with a set of rules of natural science; if so, art is then banished in our discipline.

I want to observe another border of science from another angle.

Any successful culture inevitably contains complementary factors. In fact, people have two basic ways of life: isolation from the nature and contact with the nature. The Chinese- style courtyard and the Western-style villa are two representatives of ordinary people's dream lifestyles. Both of them have the conditions of the two ways of life, and the difference lies in that the villa protects people with house and keeps contact with the natural environment through garden, while the courtyard protects people with walls and creates conditions for human-nature communications within the yard.

However, if tracing the source, the origins from the two types of cultures are just the reverse. Mankind was initially placed in Eden (or paradise), which is worshiped in the Judaist, Christian, and Islamic classics; mankind was against the God's will and was exiled in the wilderness, and therefore they do not like the wild. But Chinese mythology says mankind coexisted with bizarre forms of diverse species since their birth, people stand out from all things because they struggled and recognized the natural laws, and then they obtain the ability of friendly coexistence with the nature.

The Western culture has two sources, the Greek and the Hebrew, and the two are complementary to each other and mutually indispensable. For example, if there is only the development of the "Greek science" line, without the rectification by the development of the "Hebrew religion" line, this earth would have been destroyed by the Western worships of development and personal desire. Conversely, if only religion, there would be no renaissance. Thus, the tangled two lines formed the main thread of the Western culture, but they are both based the monadic logic (monotheism).

The Chinese culture also has two clues: the mundane (the Confucian) and the seclusion (the Buddhism and Taoism). But both of them have the common philosophy basis-the dual dialectics (yin and yang, or presence or absence), and many things gradually blurred in debates.

To the present, two types of concepts on the relationship between city and nature emerge. One advocates to enclose the mankind in a huge artificial space along with the development of science and technology, building cities on the earth or other planets and spaceships in the space, and the nature is outside the shell. This is the goal of the science and technology supremacists, as well as the hot themes of popular science fictions. Another argues that only the earth is the human home, and people should learn to coexist with the nature, introduce the nature into the city and enable people to keep contact with the nature in everyday life. People with the latter concept are relatively modest, with smaller self-confidence and ambition, and they often feel contempt from science and technology advocates.

The two concepts have their advantages and disadvantages, and at least for now we cannot determine which is more correct. I prefer the latter, and the the main reason is that if people are separated from the nature, they will not understand the nature and will have contempt for the nature, which is very dangerous. Even though people get some success in the conquest of the space, they could only survive by fighting against the nature, and this can only bring endless troubles. So, the prospect of science might be trouble, not necessarily happiness.

If we do not seek quick success and instant benefits, many things would not necessarily be made definitely, and "being woolly-headed" is not so ridiculous as it seems.

I want to express my thanks to Professor LiXiong of Beijing Forestry University for collecting the papers of the topic of this issue.

2016年

2007年　2008年　2009年　2010年　2011年　2012年　2013年　2014年　2015年

2016年1期

深圳的渣土山滑坡，联系到此前一系列令人痛心的城市事件，都表现出很多城市规划设计和城市管理的混乱，更显现了不少人员自然科学知识的片面和贫乏，才会见到危机苗头却毫无反应，当然，还有可恶的贪腐在里面起着作用。我觉得，有必要再次呼吁中国改变用口号和理念指导城市建设的现状，而立即开始对城市化发展道路的顶层设计。

从网上发表的Google地图历史照片可以看到，这个光明新区红坳渣土场2014年11月还是个山谷内的采石场大坑，充满着一池水（意味着这些水渗漏的可能性极小）。2015年3月，就变成了一塘泥，12月，已是个高达数十米（一说近百米）的大土山。这个大山的基础，就是那塘泥，即使上面按照土壤安定角进行堆填，也防止不了滑坡。如果仅仅是干土崩塌，不可能冲出那么远，造成那么大危害，所以，事件肯定不是塌方。这就说明，我们对待土壤含水量，必须慎重。

除此之外，土壤含水量还有什么意义呢？

1）它决定着场地上生长的植物命运。改变了土壤含水量，那些原生态的植物就可能适应不了，多采用本地物种的方针就变成一句空话。

2）它决定着场地的功能和用途，凡需要一定承载力的土壤，含水量不可太高。特别是城市需要大量的防灾避险场地，一般都是绿地，如果被用来储水，灾难特别是地震来了市民到哪里去？

3）它决定着山区城市的地质安全。所谓地质灾害，除了地震，大多数是由于地表土层或地层夹层的含水量过多造成的。

4）它决定着大暴雨时土壤缓解城市防洪压力的能力，如果因为到处都过分积极地蓄水而很容易水饱和，城市土地在大暴雨来临时将对减小洪水不起任何作用。

5）在台风来临时，大树被连根拔起的重要原因，是土壤已经稀软。

6）在干旱缺水地区，土壤的吸水功能又具有很大的正面意义，可以雨季蓄水，旱季释放。此时，需要注意的是另一面问题，即不能形成到处截水，下游无水的局面。

起初，我还以为深圳的这次事件又是"三通一平"惹的祸：挖填不平衡，造成那么多弃土。后来得知，这个新区开始时并没有多少弃土，甚至还缺土，只是近年来城市建设风气大变，大量增加了地下工程，包括地铁、地下管廊、地下商场、地下通道、地下停车场等，产生了许多弃土（可以推算一下：如果城市都是高楼，因为是紧缩型城市，下面都设计有3层停车场，加上其他地下设施，共占据城市面积的1/3，则30km^2的城区就需要占地1km^2平均堆高100m的弃土堆。光明新区有将近60km^2的建设用地，幸好还不是紧缩型规划，但也可以想象需要多大的弃土场！）遗憾的是，当初并没有规划那么多的弃土容纳场地，填海和占林又受到环保法规和团体的抵抗，以致最终造成如此大灾。可见，没有顶层设计的城市建设，后患无穷。

我们学科是综合性学科，然而很多做规划设计的人热衷于玩弄图形和理念，对于自然规律的知识准备不足，不能做到上知天文，下知地理，中通生物群。深圳事件教育我们，这是很危险的。我不知道，我们是否需要一场科普补课。

本期主题"风景园林小气候"，是我们向自然科学进军的一次试验。

Phase 1, 2016

Shenzhen residue landslide, and a series of distressing urban incidents previously, have shown the chaos of urban planning and urban management of many cities, and have even more shown the insufficient natural and scientific knowledge of many people, resulting in only seeing the signs of crisis but without any reaction. Of course, there are abhorrent corruption acting inside. I think it is necessary to call on China to change the status quo of guiding the urban construction with slogans and ideas, but to immediately set about the top design of the urbanization road.

With the historical photographs of Google Maps, it can be seen that this Hong'ao residue field of Guangming New District was just a quarry pit in the valley in November 2014, filled with a pool of water (which means the possibility of water leakage is very small). In March 2015, it became a muddy puddle, and in December, it was already a big heaped-up hill up to tens of meters (someone said nearly one hundred meters). The base of this big hill is the muddy puddle, and the landslide could not be prevented even if the up-level part was filled within the soil stability angle. If it was only the dry soil collapse, it was impossible to run so far out and cause so much harm. So the incident was certainly not a collapse. This shows that we should treat the soil moisture content with prudence.

In addition, what is the significance of soil moisture content?

1)It determines the fate of plants growing on site. The change of soil moisture content may bring inadaptability of the original ecology plants, and then the approach of using more native species approach will become an empty word.
2)It determines the function and purpose of the site, and the moisture content of the soil bearing capacity is not too high. Especially the urban disaster prevention requires a lot of space, and they are generally green space. If it is used for the storage of water, when in disasters, especially earthquakes, where should people go?
3)It determines the geological safety of the mountainous city. The so-called geological disasters, apart from earthquakes, are mostly caused by the excessive moisture content in surface soil or stratum.
4)It determines the ability of soil to ease urban flood control pressure in heavy rain. If the ground and soil easily become water-saturated after too aggressive water retaining, the urban soil will have no effect in reducing flood when the heavy rain comes.
5)When the typhoon comes, an important reason for trees to be uprooted is that the soil has been too soft.
6)However, in arid regions, the water absorption function of soil is of great positive significance, storing water in the rainy season and releasing in the dry season. At this point, the issue of the other side should be noted that the situation of cutting off water in the upstream and leaving no water in the downstream should not be formed.

At first, I thought this incident in Shenzhen is caused by "three supplies and one leveling": the imbalance excavation and filling results in so much spoil. It was learned later that there was not much spoil in the new area at the beginning, and even short of soil. But in recent years the trend of urban construction is drastically changing, with a significant increase in the underground works, including subways, underground pipes, underground shopping malls, underground tunnels, and underground parking, generating a lot of spoil (it can be calculated: if cities are tall buildings, because it is the compact city, with three-story underground parking, plus other underground facilities, occupying 1/3 of the total area of the city, then the urban area of 30km^2 will need an area of 1km^2 to heap up the averagely 100m-high spoil. Guangming New District has nearly 60km^2 of land for construction, fortunately not the compact planning, and it can be imagined how big area is needed for spoil!) Unfortunately, not many sites were planned to accommodate the spoil, and land reclamation and occupying the forest are opposed by the environmental laws and regulations and groups, so it eventually caused such a big disaster. As you see, no top-level design of urban construction brings endless trouble.

Our discipline is a comprehensive one, but many people in planning and design are interested in playing with the graphics and ideas, lack of knowledge of the natural laws, knowing little of the heaven above, the earth underneath and the biology and ecology. The Shenzhen incident teaches us that it is very dangerous. I do not know whether we need a remedial popular science lesson.

The theme of this month, "Landscape Architecture Microclimate", is an experiment of our march towards the natural science.

2016年2期

2015年第12期的《主编心语》中我提到："中国文化也有两条线索——入世与出世，或儒家的齐家治国平天下，或释道的出家避世归隐。但二者皆有共同的哲理基础——二元（'阴阳'或'有无'）辩证法，辩来辩去，许多事就逐渐模糊起来。"结果有些读者就误以为我是二元论者。

其实，我的观点是：世界在原点处，非一不二。

中国人称这种状态为"太极"：你说他是一，但明明是由阴和阳组成；你说他是二，又的确阴阳是互相纠缠互相渗透无法彻底分开的一体。你无法拆分出单纯的彻底的阳，也无法找出单纯的彻底的阴。换句话说，此时此处，一是不纯粹的一，二是拆不开的二。"非一"，是宇宙多样性之源；"不二"，是宇宙统一性之根，"多样统一"本是一体。在最极限处，"一分为二"与"合二而一"没有区别，你无法说清楚它是一还是二。南宋印肃禅师曾长期追问"一归何处"？当他阅读《华严合论》至"达本忘情，知心体合"一句时，豁然大悟，作偈曰："捏不成团拨不开，何须南岳又天台。"

由此看来，在人天关系上，"人必胜天"和"天人合一"若追溯到原点，也是非一不二的。应该说在自然生物界一直保持着这种状态，但后来出现了人类，是人类能力和思维的发展将二者逐渐拆开，人的能力越强大，"人必胜天"就跳得越凶，这种拆分就越厉害。这不等于说我主张不发展主义，只是说发展还需要另外一个东西来平衡，或是善意，或是宗教，或是哲思。在中国，传统上基本是用"天人合一"文化来平衡的，至于实事求是地描述事物的真实状态，则《管子》的"人与天调"或许更为确切。

我基本上倾向于中国文化，不同之处在于我认为：人虽然来自自然，但他是自然的异化，人是自然异化的产物。人对自然的改造，是自然无数"正反合"过程发展到高级形态的一种必然，对其（即人对自然的改造）没必要无条件地反对，更没道理无原则地赞美。人类对自然改造的最终结果，还在于人类自己能否自我进行约束（而不是对个人主义和无限发展的无条件吹捧），以图尽量延长自己的寿命。至于是否能达到合，还很难说，一旦人类做得过分，自然只有毁灭人类，然后再启动另一个正反合过程。

我真诚地向读者特别是年轻读者推荐朱钧珍先生的《咀嚼一生——关于中国传统园林自然观的思考》一文。朱先生是清华大学老教授，中国首届园林专业毕业，现居香港，她以一生咀嚼园林，凝结成对"中国传统园林自然观"的思考，借本刊的版面发表出来，是本刊的荣耀。在朱先生的中国传统园林自然观里，我分明看到了中国园林实际上存在两个平行的世界：一个实，自然之实；一个虚，人文之虚。将一实一虚揉捏成一团，其所达到的高度，的确是以实论实的欧美园林理论很难达到的。正因如此，朱先生才在结束处充满感情地写道："研究传统中之可继承者，发挥新时代所需之创意，以求实现伟大的复兴之梦，并保持'中国（园林）为世界园林之母'的光环，应是指日可待的期望。我作如是观。"

Phase 2, 2016

In last year's December edition of our journal, I mentioned: "The Chinese culture also has two clues: the mundane (the Confucian) and the seclusion (the Buddhism and Taoism). But both of them have the common philosophy basis—the dual dialectics (yin and yang, or presence or absence), and many things gradually blurred in debates." And it results that some readers mistakenly thought I was dualist.

In fact, my view is: the world is at the origin, neither monistic nor dualistic.

Chinese call this state as "Tai Chi": you say it is monistic, but obviously it is composed of yin and yang; you say it is dualistic, but it is an inseparable whole that yin and yang mutually entangle and penetrate. You cannot split out the completely pure yang, nor the completely pure yin. In other words, here in this case, monism is not purely monistic, and dualism is the inseparably dualistic. "Neither monistic" is the source of the diversity of the universe, "nor dualistic" is the root of the unity of the universe, and the "unity of diversity" is actually the one. At the most extreme point, "one divides into two" and "two combined into one" have no difference, and you cannot say it is one or two. The Chan master Yinsu in the Southern Song Dynasty had long asked, "where is the place for unity?" When he read the sentence in *Huayan Theories* that "reaching to the original point and without emotions, we know that the heart and the body are the same", he suddenly understood and made a Buddhist verse that "we cannot make it as one nor separate it into two, so we don't have to distinguish it."

It can be seen that, in the human-universe relationship, "man can conquer nature" and "unity of heaven and man" are also neither monistic nor dualistic if tracing back to the origin. It should be said that the natural biosphere has maintained in this state for long, but subsequently human beings emerged. It is the development of human capabilities and thinking that gradually separate the two. The more powerful human capacities are, the more prevailing "man can conquer nature" is, and the more severe this split is. This does not mean that I do not advocate development, and I am just saying that development still needs another thing to balance, goodwill, religion, or philosophical thinking. In China, traditionally the culture of "unity of heaven and man" is used to make the balance; as to realistically describing the real state of things, the "harmony of man with nature" in *Guan Zi* might be more accurate.

Basically, I prefer Chinese culture, except that I think: though man is from nature, but man is an alienation of nature, the product of the alienation of nature. The transformation of nature by human is a necessity of the numerous "pros, cons and combinations" processes developing into the advanced form. It is not necessary to unconditionally oppose it (the transformation of nature by human), and it is also unreasonable to praise it without principle. The end result of the transformation of nature by human lies in their ability to self-constraining (rather than unconditionally flattering individualism and unlimited development) in an attempt to extend their lives. As to whether it can achieve combination, it is hard to say; once human overdo, the nature will destroy mankind, and then start another course of pros, cons and combination.

I sincerely recommend to readers, especially young readers, Professor Zhu Junzhen's *The Lifelong Savoring – Thinking about the Nature View of Chinese Traditional Garden*. She is a senior professor of Tsinghua University, was the graduate of the first class of landscape architecture in China, and now lives in Hong Kong. She has been savoring garden for a lifetime, and condenses her thinking about "the nature view of Chinese traditional garden". It is our honor to publish it in our journal. In Professor Zhu's nature view of Chinese traditional garden, I clearly see there are actually two parallel worlds in Chinese garden: a real one, the real nature; a virtual one, the virtual humanity. The height of integrating the real and the virtual into a whole one is really difficult to reach for the Western gardening theories which make concrete gardens with concrete means and theories. It is because of this that Professor Zhu ends her writing with full emotion: "Researching on the inheritable part of Chinese traditions and developing the creativity the era needs in order to achieve the great rejuvenation dream and maintain the aura of 'China (garden) as the mother of the world's gardens' is well in sight. I hold this belief."

2016年3期

我想谈一谈引导学科发展的基本思想问题——回到机理。

中国风景园林学会将学刊学术把关的任务交付于我，我们学刊实际上担负着很重的学科建设任务。何谓科学？就是找到那些（在一定的时空条件下）经得起检验的道理（规律、真理、机理），而不是靠拍脑袋发明一些新想法、新理念。创新当然是需要的，但新东西未经实践的千锤百炼的验证之前，只能是假说，而不是科学真理。

人居环境建设，是社会起飞阶段（即城市化）的最重要工作之一，也是百年大计。当初钱学森想搞大建筑学科，后来吴良镛、周干峙想搞人居环境学科门类，都是因为其重要，想将其放在科学的基础上，经得起历史检验。不巧的是，当中国开始城市化时，西方的城市化已经完成，人居环境学科在那里不再专注于追求真理，他们更关心追逐新奇，甚至抛开规律，抛弃真理，学科变成了知识生产，而知识是可以拍脑袋生产出来的，理念成为追逐目标。库哈斯、哈立德等就是代表人物（实际上只有这样才能让金融资本无中生有地赚取更多的利润，老实的科学家是做不到这点的）。我曾经不下二三十次问过一些新锐建筑师：在30m^2的阳台上做一个贮藏50t水的贮水池用来灌溉这个阳台花园，会是什么样子？竟然没有一个人显现出诧异的表情，有人还说这是个很好的事情啊，仅有几个人想了想荷载有没有问题，但也仅此而已。要知道，我们名牌大学的建筑系都是各地高考状元级的学生，可见在他们后来的学习中早已把中学的数理化忘掉了，这也反映出中小学教育不注重培养人，不注重培养学生联系实际和从变化规律来观察和思维事物的能力，而是以灌输知识点和解题方法为中心的体系是多么糟糕。

有的文章作者接到任务，在以盐碱地著名的城市作湿地规划。我提醒他们当心加重盐碱，开始他们还不理解：多蓄雨水就是增加淡水，不是淡水可以压盐吗？而且种植土层的下面已经铺设了石屑隔离层，阻止了盐分的上升。我告诉他们：减低盐碱的目标是指向农田绿地，不是湿地，由于湿地和附近土地的地下水具有连通器效益，所以湿地的水多了，其他土地的地下水位会升高，加剧盐碱化。土地种植层下面铺上石屑等，是为了隔断土壤的毛细管，如果石屑层充满了水，这个作用将消失，反而更有助于地下盐碱的上升。好消息是，文章作者欣然接受了我的意见。坏消息是，有些其他类似的文章不再理我，估计投向了别处。

所以，不从原理上进行思考，只是追随一些理念，抄袭一些做法，经常会南辕北辙，适得其反。可惜的是，当代的人居环境学科教育，已被五花八门的理念所控制，大多数人还误以为靠这个东西就可以解决人类的问题。而真实起作用的原理、规律、机理，已经很少被人关注。这难道是人类发展的正道？对此，我坚定地回答："NO！"

让我们的学科：Return to mechanism of action！

Phase 3, 2016

I would like to talk about the basic idea that guides the development of the discipline problems – return to mechanism.

CHSLA handing over the task of checking the journal to me actually bears the big task of disciplinary building. What is science? It is to find those truths (laws, truths, and mechanisms) that stand the test (under certain conditions of time and space), rather than inventing news ideas or concepts depending on experience. Innovation is certainly needed, but news things are only hypotheses, rather than scientific truth, before being tempered and verified with practice.

Human habitat construction is one of the most important work during the ascending stage (i.e. urbanization) of the society, and also the long-term goal. Originally Qian Xuesen wanted to set up an enlarged architecture discipline, and Wu Liangyong and Zhou Ganzhi wanted to have the discipline of human habitat, and it was because of its importance that they wanted to put it on the basis of science that it could with stand the test of history. Unfortunately, when China began to urbanize, the Western urbanization had been completed. The human habitat environment discipline there had no longer focused on the pursuit of truth, and they were more concerned about novelty, or even putting aside the laws and abandoning the truth, and the discipline became knowledge production, and knowledge can be produced depending on experience and ideas and concepts became the goal. Koolhaas, Khalid, etc., are the representative ones (in fact, this is the only way for financial capital to earn more profits out of thin air, while the honest truth-seeking scientist cannot do this). I have asked some cutting-edge architects over twenty or thirty times: how does it like to build a 50-ton capacity water storage tank on the 30m^2 balcony to irrigate the balcony garden? There is not even a person showing surprised expression, someone even said that it was a good thing, and only a few people thought just about the problem of load. You know, the architecture departments of our prestigious universities have many students with high performance in the college entrance exam, it can be seen that in the later study they have forgotten the scientific knowledge they learned in high school, and it also reflects the bad situation of the primary and secondary education that it is not to train people or pay attention to the training of students' ability of linking with the practice and observing and thinking about the laws of changes, but to instill knowledge and problem-solving approaches.

Someone received a task to make wetlands planning in the city known for saline land. I reminded them to watch out not to aggravate the saline situation, but they did not understand at the beginning: more rains would increase the storage of fresh water which could decrease salt, and the stone chips isolation layer was laid under the plating soil to prevent rising salinity. I told them: the goal of reducing salinity should be directed to farmland green space, not wetlands; the underground water below wetlands and neighboring land were connected, so the rising of wetlands water would result in the increasing of the underground water of other parts of land and aggravate salinization. The stone chips under the planting soil was to cut off the capillary of soil, but when the layer of stone chips was filled with water, this effect would disappear, more conducive to increasing underground salinity. The good news is that the author gladly accepted my advice. The bad news is that the authors of some other similar articles no longer respond me, and they might have turned to other journals.

So, not thinking from the principle, and just following some ideas and copying some practices of others, often puts poles apart and results in the opposite. Unfortunately, the contemporary human habitat environment education has been controlled by a wide variety of ideas, and most people mistakenly think that these things can be solve human problems, while the principles, laws and mechanisms that really function have rarely been concerned about. Is this the right path of human development? In this regard, my firm answer is: "NO!"

Let our disciplines return to mechanism of action!

2016年4期

我有个不好的感觉，学科开始乱动。

当前可以清楚地看到：在地上，林学打着生态的旗帜进入园林，要求城市绿地都要成为生态效益最高的顶级群落；在地下，水文打着治水的旗号进入园林，要求城市土壤首先服从于蓄水和净水；交通上，要求更多的城市地面用于行车和停车，于是小区和街区绿化可以不要，城区也回到密网道路的状态；经济上，要求从效益角度考察城市，结果发现园林能够算出来的钱很少，于是城市用地可以大大压缩……

这些都是打着科学的旗号。不要轻视科技的能力，即使电脑AlphaGo非常耗电，思维方法非常笨，一旦自我学习和量子计算机成功，智力便会很快超过人类，那时人类必会巨变，社会必然巨变。

但我始终有一个疑问：人类头脑和心理的综合能力，对美的感觉，特别是悟性，是现代科技能够解决的吗？比如，是西洋风景画的透视有理呢，还是中国山水画的散点观察有理？其实他们看的东西是一样的，但感受和判断却有巨大的差别。思来想去，我还是觉得中国人有理，因为即便是把风景看成固定的，可人是活的，不可能长期一动不动地死盯着一个东西。在没有西洋画和摄影术以前，中国人对世界的认知和现在肯定不一样：一切都是多维且变动的，只不过有快有慢而已。

世界是一系列固定画面的拼接，还是本身就在不断地流变，这真是两种思想方法。依照后者，把片面的、固定的画面无限放大是荒诞的，宇宙图景中所有的元素、个体、格言、口号都仅仅是局部的反映，事物的总体和趋势才更重要。而对于庞大的宇宙，人们只有通过领悟，才可理解、才能亲近。对于靠大数据来掌握世界的电脑来说，亲近、美感、悟性，有意义吗？人类思维的本质就是计算吗？但在人类认知的初级阶段（分辨阶段），在事物变化的基础层面，严格的逻辑和死硬的规律还是必需的，不能违反的，否则人类就无从开始思维，或者思维的结果直接就是一团糟。举例来说，当下最突出的思维乱象之一是"偏正颠倒"：把一个偏正词组颠倒过来，偏变正、正转偏，而且以为如果原来词组是个正确事物，颠倒过来也当然正确。比如"城市农业"，城里有条件的人家和场所，在其地盘上种些蔬菜瓜果，自古就有，只要产品没有被污染，当然是好事。但有人将其颠倒过来了，叫作"农业城市"。古今有这样的城市吗？即使一个城市的服务对象都是农民，这个城市的市民和设施也基本都是从商从工的。同样，我们从"城市森林"、"城市经营"、"城市海绵"的演变中，也可以发现这个现象：一些本来属于局部性质或功能的东西，现在要坐正了。例如，城市管理当然有经营问题，不算账怎么能行？但是把整个城市当作经营工具，城市就已经不是为了人类的福祉而存在了，而变成金融的走狗。

北京大学王志芳副教授为本期主题"风景园林的科学性"进行了辛勤的组稿，在此深表谢意！

Phase 4, 2016

I have a bad feeling that the discipline is messing up.

It can been clearly seen that: on the ground, the forestry study enters the landscape architecture field with the banner of ecology and requires urban green space to become the top community with the highest ecological benefits; under the ground, the hydrology study enters the landscape architecture field with the banner of water control and requires urban soil to subject to water storage and water purification first; in traffic, more urban grounds are required for driving and parking, so the community and neighborhood greening becomes unnecessary and the urban area returned to the state of dense road networks; economically, the city is reviewed through economic benefits, resulting in that little money can come out of landscape architecture and urban land can be greatly compressed…

These are all with the banner of science. Do not underestimate the ability of science. Even though the power consumption of the AlphaGo system is very high and its way of thinking is very simple, once the self-learning and quantum computer succeeds, its intelligence will soon surpass human, and at that time mankind will have a big change, and the society will also have a big change inevitably.

But I always have a doubt: can the comprehensive mind and mental ability of human, and sense of beauty, especially the savvy, be solved by modern science and technology? For example, the perspective of the Western landscape painting, and the scattered observation of Chinese landscape painting, which one is reasonable? In fact, they look at the same things, but the feelings and judgments actually have a huge difference. I still think that the Chinese one is reasonable, because even the landscape is regarded as being fixed, people are alive and cannot stare at one thing motionlessly for a long time. In the absence of Western painting and photography, the Chinese perception of the world was certainly not the same as now: everything was multi-dimensional and changing, just being fast or slow.

The world is the collage of a series of fixed screens, or It is constantly changing itself, are really two kinds of thinking. In accordance with the latter, the Infinite amplification of the one-sided, fixed screen is absurd, all the elements, individuals, proverbs, and slogans in the picture of the universe are merely a partial reflection, and the overall and general trend of things are more important. For the vast universe, it is only through comprehension that people can understand and be intimate. For computers that master the world with big data, are the intimacy, sense of beauty, and comprehension meaningful? Is the nature of human mind calculation?

But in the early stages of human cognition (the distinguishing phase), and in the basic level of changes of things, strict logic and inflexible laws are quite necessary, not to be violated, otherwise human beings can never start thinking, or the result of thinking is directly a mess. For example, one of the most prominent contemporary chaos of thinking is "the reverse of the attributive structure": putting an attributive structure phrase reversed, and thinking it as correct just because the original one is the right thing. Such as "city agriculture", urban residents who have the conditions plant some vegetables and fruits in their places, which has been so since ancient times, and as long as the products are not contaminated, it is of course a good thing. But someone put it reversed, and called it "agricultural city". Has there been such kind of city? Even if the city serves farmers, citizens and city facilities are basically for business and industry. Similarly, we can also find this phenomenon from the evolution of "city forest", "city operation", and "city sponge": the originally partial or functional things now tend to play a leading role. For example, city management of course has operational issues, or how can it do? But if the whole city is taken as a business tool, the city is no longer for human well-being, but the lackey of capital instead.

Our heartfelt gratitude goes to Associate Professor Wang Zhifang of Peking University for her hard work in organizing papers for the current theme "Scientificity of Landscape Architecture".

2016年5期

2016年4期讨论道"世界是一系列固定画面的拼接,还是本身就在不断地流变,这真是两种思想方法"。恰好,前些日子在电视上看到一个小制作电影,只记得片尾是一首河北儿歌:"娃娃呢?上山了。山呢?雪盖了。雪呢?化水了。水呢?和泥了。泥呢?糊墙了。墙呢?猪拱了。……高粱呢?喜鹊叼了。喜鹊呢?飞——了!"

听起来,这里像是堆砌了一组毫无逻辑的混乱关系,是一系列无厘头的链接。其实,朴素的儿歌唱出的是中国人,不,应该是一切原始生民的世界观:一切都在流淌,一切都在演变。人,只不过是这条长河中的一个成分,甚至只是某个阶段中的一个部分。

摆脱僵硬的规律和逻辑的束缚,儿歌中所叙述的一系列关联,哪个不曾在现实中发生过?只不过人类精英们的思维能力还无法将这些关联组成一个模型,于是就指斥落后民族思维幼稚,缺乏逻辑,不懂科学。至今,我们还经常听到有人批评说:中国文化中根本就没有逻辑,有点逻辑影子的"因明"也是从印度传过来的。诸如此类的言论,并不出现在一般的讲坛上(这里的听众们对逻辑、因明之类的词儿大多不感兴趣),而是出现在高级场合,所谓精英雅集的场合。

在儿歌所描述的一系列转变中,有的是可以被现代科学所描述的,例如雪化成水;但更多的是无法处理的,如水是流走了还是被人用桶装起来拿去和泥了呢?高粱是发了芽还是被鸟儿叼去吃了?在当下的教育和科研体制下,绝大多数人只懂得研究前者,所谓课题也集中于前者,日积月累,他们对后者甚至养成了视而不见的习惯,后者已经从他们的思维中被彻底排除。

规律(在人脑就是逻辑)和干扰(在人脑就是悖论),本来就并存于这个宇宙中,二者都是"道"的显现,都是"道"的必然。这就是那首儿歌所揭示的真理:复杂的流变条件下精准预测是不可能的。

有生命的地方一定是条件最复杂的地方,因此可以推断:地球属于存在着巨量无法精准预测的事物的地方。简单说,科学是依靠着可以重复验证(注意,不是靠逻辑!)确定了自己的权威。但是这个世界上大量存在着不可重复的事情,比如天文灾难,比如秦始皇,比如一个人的命运。那么,我们该怎么办?难道就死了这条探求未来的心吗?

生活在地球上的人类,肯定不能只依靠科学来满足自己的本性,实现自己的价值观必须寻找另一种途径,那就是智慧——设法如实地认知这个充满变化的世界,并依此来做出自己的决策。这,就是中国人老祖宗讲的:实事求是。

很不幸的事实是,所谓"科学思维"当下已经一统天下,而生活仍然按照自己的步履行进,于是许多"公式"、"模型"、"主义"一方面不断在事实面前碰壁,一方面还在继续误导众人。想要矫正西方思维模式的这个毛病,在中国恐怕至少还得100年,其根子在于如果教育体系不改,问题没可能改善,至于全世界,恐怕200年都不够。

Phase 5, 2016

In the issue of April of 2016, we discussed "The world is the collage of a series of fixed screens, or It is constantly changing itself, are really two kinds of thinking". Just a few days ago, I watched a film on TV ending with a Children's song of Hebei: "Where is the child? Up on mountain. Where is the mountain? Under snow. Where is the snow? Having melted into water. Where is the water? Into mud. Where is the mud? Molded to build wall. Where is the wall? Pushed down by pigs... Where is the sorghum? Eaten by magpies. Where are the magpies? Having flown away!"

It sounds like a pile of illogical confused relationships, and a series of nonsensical links. In fact, the simple Children's song shows the world view of the Chinese people–no, it should be all the primitive people: everything flows, and everything is evolving. Man is just one component of this long river, or even just a part of a certain stage.

Out of the shackles of the rigid laws and logic, have the links described in the Children's song never happened in reality? It is just because the thinking abilities of human elites cannot associate them to form a model that they denounce the less-developed nations' thinking as childish, illogic, or not scientific. So far, we have often heard the criticism that: There is not logic in the Chinese culture, and the Hetuvidya–seemingly sort of logic–is also passed over from India. This kind of remarks did not appear on the general forums where the audience are not interested in the words like logic and Hetuvidya, but in high-level occasions–the so-called elite-gathering occasions–instead.

In a series of transitions described in the Children's song, some can be described by modern science, such as snow melting into water; but more cannot be handled, such as water is running away or is bottled up to mix with mud? Do the sorghums sprout or are eaten by birds? Under the current system of education and research, most people only know how to study the former, and the so-called research projects are also concentrated on the former, and over time, they even develop a habit of turning a blind eye to the latter, which has been completely excluded from their thinking.

Law (logic in the human brain) and interference (paradox in the human brain) co-exist in this universe, and both of they are the appearance and inevitability of "way". This is the truth the Children's song revealed: It is impossible to predict accurately under the complex changing conditions.

Where there is life, it must be the place with the most complex conditions, so it can be inferred: The Earth is the place where a huge amount of things cannot be accurately predicted. Simply speaking, science relies on the repeated verification (note, not by logic!) to determine its own authority. But in this world there are a lot of things that cannot be repeated, such as astronomy disaster, Qin Shi Huang the First Emperor, and a person's fate. So, what can we do? Should we just end up the aspiration to explore the future?

Human on the Earth certainly not only rely on science to meet their own nature, and must find another way to realize their own values, which is the wisdom to try faithfully to cognized the ever changing world, and so to make their own decisions. This is what the Chinese people's ancestors said: seeking truth from facts.

Unfortunately, the fact is that the so-called "scientific thinking" has now been dominating the world, while the living is still moving forward in its own paces. So many of the "formulas", "models", and "isms" on the one hand continue to run into a wall in the face of the facts, and on the other hand still continue to mislead people. To correct this thinking patten of the western word, I am afraid it will take at least a hundred years, and if the roots in the education system does not change, the problem is not likely to improve. As to the whole world, I am afraid two hundred years is not enough.

2016年6期

不久前，在中国建筑业协会古建筑与园林施工分会、安徽省风景园林学会和清华大学建筑学院景观系的支持下，87岁高龄的朱钧珍教授在滁州主持了一个"亭文化与园理学研讨会"。朱先生提出应创立"园理学"作为中国风景园林学理论的核心，我觉得可行：园之理源于景之理，但又融合了大量社会生活与文化，比西方人单纯从生态和视觉角度出发的景观学，要深厚得多。而且，我觉得把亭文化与园理学一起讨论，有其内在联系。我讲了2件事，证明园林里的亭其实是祖国大地古老景观的映射，是协调人与自然的中华古文化的遗迹。第一件，近30年前我初到广东江门侨乡，惊讶地发现路边、桥旁、空场还保留着不少亭子，这些亭子全是由基层乡里自发建的，而这个中国千年以前的传统在内地已经基本失传；第二件，是云南腾冲的洗衣亭。正好，本期魏成等的《传统聚落乡土公共建筑营造中的生态智慧——以云南省腾冲市和顺洗衣亭为例》作了很好的分析，有兴趣者可以从中获益良多。本期主题"传统村镇保护与发展"是华南理工大学肖大威教授组稿的，特此感谢。

张杨的《基于因子分析的上海城市社区游憩机会谱（CROS）构建》作了一个很有意思的探讨。我觉得文章在背景层次上提出了一个重要问题：国家公园和社区公园可以采用同一个评价系统吗？国家公园的宗旨是保护自然优秀资源，所以其游憩机会频谱应是"以我（公园）为主"，是以我能提供什么为主，而不是以你游客需要什么为主。此时，游客满意度评价不能用于指导公园工作，仅有管理上可改进服务工作的参考作用。而社区是为市民服务的，所以其游憩机会频谱应以满足市民需求为主。这二者的区别，反映出风景园林两大领域的重要区别。

郭竹梅等的《困境与突破——对北京平原造林工程总体规划核心问题的思考》讲述了北京市政府用了很大气力在做一件对北京和华北地区很有意义的大好事——百万亩平原大造林，并如实记录了实际工作中遇到的技术难点和作出的技术突破，必须予以肯定。但是掩卷沉思，感到有一个更基本的问题：如此布局的机理是什么？文章提了几个理念，难道理念等于大气、水体、生物活动的规律吗？其次，百万亩，照北京2000万人估计，每人有30多m^2，当初要是融进北京城市规划用地的话，每人至少可拥有60~70m^2绿地，许多矛盾都可以解决，那中国真的可以给世界提供一个规划样板了。

象伟宁教授邀我加入"生态智慧学社"微信群时，作为见面礼我送上了2道思考题：1）有没有对一切生物都适用的生态条件？如果有，是什么？2）如果没有，那我们整天忙碌是为了什么？其实，这正是我看到黄瑞等的《基于阻力指数的屋顶斑块生态网络规划研究》一文后引发的思考。当下弥漫整个业界的，是掌握了几种所谓景观生态学的parttern（模式），脱离具体的场地条件，脱离生物与环境、生物与生物之间相互关系，脱离能量、水分、养分的运动与分配，仅凭脑袋的平面构思，就可以完成对地球表面的规划。其次，人与生物的关系，是有两面性的，相当复杂，并非只有一片美好。本文如果能准确把握这种关系，就是站在泛生态主义者之上了。

Phase 6, 2016

Recently, with the support of the Ancient Building Branch of China Construction Industry Association, Anhui Society of Landscape Architecture, and the Department of Landscape Architecture of Tsinghua University, the 87-year old Professor Zhu Junzhen hosted the "Pavilion Culture and Garden Science Seminar" in Chuzhou. Professor Zhu put forward that the "garden science" should be the core of Chinese landscape architecture theories, and I think it is feasible: the science of garden originated from the science of landscape, and integrated a large number of social life and culture, which is far profounder than the Western landscape architecture that originated from ecology and vision. And I think there are internal relations to discuss the pavilion culture and garden science together. I can tell two things to prove that the pavilion in Chinese garden is actually the reflection of ancient Chinese landscape and the remain of ancient Chinese culture coordinating human and nature. The first, when I went to the overseas Chinese hometown in Jianmen of Guangdong Province about thirty years ago, I was surprised to find that there were a lot of pavilions along the roads, bridges and squares, built spontaneously by towns and villages, and this tradition of thousands of years ago had been basically lost in the inland of China. The second one in about the Laundry Pavilion in Tengchong of Yunnan Province. Precisely, in this issue, Wei Cheng made a very good analysis in his paper *Ecological Wisdom in the Construction of Rural Public Buildings: Case Study of Laundry Pavilion in Heshun Historic Town, Yunnan Province*, and those interested can benefit a lot. And I would like to express our gratitude to Professor Xiao Dawei of South China University of Technology for organizing the current theme topic "Protection and Development of Traditional Township".

Zhang Yang's *The Construction of Community Recreation Opportunity Spectrum (CROS) in Shanghai Based on Factor Analysis* made a very interesting discussion. I think the article raises an important question on the background level: can the national park and community park use the same evaluation system? The purpose of national park is to protect the outstanding natural resources, so the recreation opportunity spectrum should be "dominated by me (park)", by what "I" can offer, not by what you tourists need. In this case, the tourist satisfaction evaluation cannot be used to guide the work of the park, just being reference for improvements in services. But the community is to serve the public, so the recreation opportunity spectrum should be mainly to meet the public demand. The difference between the two reflects the important distinction between the two fields of landscape architecture.

Guo Zhumei and her coauthors' *Plight and Breakthrough – A Thinking about the Core Problem of Afforestation Project Master Plan in Beijing Plain Area* told about a very good thing that is meaningful for Beijing and North China and that Beijing Municipal Government put a great deal of efforts into—Million-mu Plain Afforestation, and truthfully recorded the technical difficulties and technological breakthroughs, which should be affirmed. But there is a more fundamental question after meditation: what is the mechanism of this layout? Several ideas were mentioned in this article, but do these ideas equal the laws of atmosphere, water, or biological activities? Secondly, a hundred mu, with the estimated population of 20 million in Beijing, there are over $30m^2$ per person. If it had been integrated into the Beijing urban plan land at the beginning, there could be $60-70m^2$ green space per person, many conflicts could be solved, and then China can really provide the world with a planning model.

When Professor Xiang Weining invited me to join the "Ecological Wisdom Society" WeChat group, I presented two questions first: 1) is there an ecological condition applicable to all living things? If yes, what is it? 2) if no, then what are we busy all day long for? In fact, this is my thinking after reading Huang Rui and his coauthors' paper *Using Resistance Index for Urban Roof Patch Network Planning*. At this moment, the prevailing situation in this industry is that the planning of the Earth's surface could be done with just several so- called landscape ecology patterns and plain idea in mind, without specific site conditions, without the relationships between living things and environment and between living things themselves, nor the movements and distribution of energy, water, and nutrients. What's more, the relationship between people and living things has two sides, quite complex, not just a better place. The paper would stand out of the pan-ecologists if it could precisely handle this relationship.

2016年7期

从申请一级学科起，我们就在思考：学科的核心是什么？是外形与好看？是心情的抒发？是小园的舒适？是侍弄花草的本领？是打造周围景色的能力？是进行地表形态肌理的研究？是对人类控制自然能力的赞赏（或是忏悔）？是对生物生存环境的许诺？是室外人类文化成果的集成？……显然，它们都是片面的，都只是学科的一个部分，甚至是一个角落。这也说明，我们学科很容易被"一叶障目，不见泰山"的思维所误导，特别在这个"理念优先"的时代。

终于，我们找到"人与自然的关系"这个核心。但是只谈这点还是不准确的，因为范畴过大，炼钢、造船、种田、养牛、治水、发电等活动也是在处理人与自然的关系。我们学科是在为人类的一切生活与生产活动（包括上述这些活动）提供最佳的室外场所环境。这里的关键词是"环境"，我们是在环境这个层面上从事调节人与自然关系的工作，而且涉及室外环境的各个层面，从气候的环境，到地形的环境，到水文的环境，到植被的环境，到环保的环境，到环卫的环境，到建筑的环境，到艺术的环境等。在这个层面上，我们应该相当于乐队指挥。或许，"绿色人居环境"可以作为比"绿色基础设施"更好的候选学科名称。

人与自然的关系是生物与自然关系的一种，但很特殊，因为生物是自然演进的产物，而且完全顺应自然而存在，人类却背离了"完全顺应自然"，走上了"异化"道路。这种异化为人类带来了巨大的发展，也带来了无穷的烦恼。烦恼的核心，是人类违背了地球生命的根本原则——"绿色"原则。

现今地表大气和水分的形成，离不开绿色生物的功劳。至今，叶绿素是自然中能够利用阳光让CO_2和水分变成有机物和O_2的唯一中介（催化剂）。所以，绿色原则是维系地球生命的最重要原则，背弃绿色原则就是在威胁生命，威胁人类。（即使人类未来有能力仿造叶绿素的功能，基于对人性与人类能力的怀疑，慎重的学者都不认为完全靠人力可以创造一个长期独立于自然、脱离自然监督的人类世界。）

如果采用"碳足迹"方法来考量，人类的绝大部分活动都是在增加碳排（或减少碳汇），包括人工制氧、整个农业和部分林业，也就是都在违背绿色原则。如果将城市绿化的整个营造和维护过程的资源和能源消耗都计算在内，特别是在工业化和自动化时代，城市绿化的碳汇贡献也可能是一个负值，这就对整个学科提出了严重的挑战，也将学科与人类整体思维的命运紧紧捆绑在一起。

人类的整体思维能力到底如何？这是个非常复杂和艰难的问题，它取决于：（1）我们的测量是否准确？（2）我们的视野是否全面？（3）我们的思维结构是否合理？（4）我们的价值判断是否可信？搞不好，就是"失之毫厘，差之千里"的判断结果。无论如何，整体上毁坏绿色，违背绿色原则的事情，是错误的。

2016年6月5日是第45个世界环境日，《中国青年网》发表了《习近平绿色发展理念引领中国环境治理新实践》，我认为"绿色发展"这个概念非常好，平衡了绿色与发展，应该成为人类发展的根本原则。

Phase 7, 2016

We have been thinking since applying for the first-level discipline: What is the core of the discipline? Is it the shape and good look? The expression of the mood? The comfort of a small garden? The gardening skills? The ability to build the surrounding scenery? The study of the morphology and mechanism of the surface? The appreciation (or confession) of human ability to control the nature? The promise to the biological living environment? The integrated outdoor cultural achievements of mankind?... Obviously, they are one-sided, just a part of the discipline, or even a corner. This also shows that our discipline is apt to be misled by the thinking pattern of "not seeing the wood for trees", especially in this "concept-first" era.

Finally, we find the core of "the relationship between man and nature". But it is not accurate to talk about this point only, because the scope is too large, and the activities of steel-making, shipbuilding, farming, raising cattle, flood control, power generation, etc., are also in the relationship between man and nature. Our discipline is to provide the best possible outdoor environment for all human life and production activities (including above-mentioned activities). The key word here is "environment", and we are engaged in regulating the relationship between man and nature at this level of environment, involving all levels of outdoor environment – the climatic environment, the terrain environment, the hydrological environment, the vegetation environment, the environmental-protection environment, the sanitation environment, the architectural environment, the artistic environment, etc. At this level, we should be the equivalent of an orchestra conductor. Perhaps "green habitat environment" is the better candidate name for the discipline than "green infrastructure".

The relationship between man and nature is one kind of relationships between living things and nature, but a very special one, because the living things are the product of the natural evolution, and also exist fully in harmony with nature, but mankind has deviated from the complete harmony with nature and embarked on the "alienation" road. This alienation brought great development for mankind, and brought endless troubles, too. The core of troubles is that mankind goes contrary to the fundamental principle of human life on the Earth – the "green" principle.

Today the formation of surface atmosphere and water cannot do without green living things. Up to now, the chlorophyll is the only media (catalyst) in the nature to transform CO_2 and water into organic matters and oxygen through sunlight. So, the green principle is the most important principle to maintain the life of the Earth, and abandoning the green principle threatens lie and mankind. (Even if in the future mankind has the ability to mimic the function of chlorophyll, based on the doubts about humanity and human capacity, cautious scholars do not think we can completely rely on manpower to create a human world independent upon the nature and out of the oversight of nature for a long time.)

If judging with the "carbon footprint" approach, the vast majority of human activities are increasing carbon emissions (or reducing carbon sinks), including artificial oxygen, the entire agriculture and forestry, that are contrary to the green principle. If accounting the resource and energy consumption in urban landscaping and maintenance, especially in the era of industrialization and automation, carbon sequestration contribution of urban greening may also be a negative value, which puts a serious challenge on the whole discipline, and also closely tie the discipline with the fate of the whole human thought.

How on earth is the overall thinking ability of humankind? This is a very complex and difficult issue, and it depends on: 1) Are our measurements accurate? 2) Is our vision comprehensive? 3) Is the structure of our thinking is reasonable? 4) Are our value judgments credible? The judgement might result in "a miss is as good as a mile". In any case, it is wrong to destroy the green or go contrary to the green principle.

On June 5, 2016, the 45th World Environment Day, *The New Practice of China Environmental Governance Led by Xi Jinping's Green Development Concept* was published on *China Youth Network* . I think the "green development" concept is very good, which balances the green and development and should be a fundamental principle of human development.

2016年8期

这次超级厄尔尼诺现象，造成了2016年几乎横扫中国整个东半部的洪涝灾害。

在全国到处"城市观海"的风潮中，大部分人明白了，城市"小海绵"只能调节城市中小型雨洪问题，碰到数十年乃至上百年一遇的大洪水，还是得另想办法。

长江中下游的大洪水，在我的记忆中大概每过几十年就闹一次：1935、1954、1998、2016年。汉口一带，海拔原本只有18m多一点，1998年江面最高达到29m多，也就是说，此时江面高于汉口10m以上。如果此时大堤毁了，洪水将淹掉多大的地方呢？1935年是真的垮堤了，淹掉640万亩的地方（4200km²，相当于今天武汉市行政辖区面积的一半），死亡和失踪8万多人。这还只是相当于50~100年一遇的大水，据说，1870年闹过一次千年一遇的大水，情况到底如何，我就不清楚了。

如果把这个水淹面积（再加上其他水灾面积）平摊到武汉以下的长江两岸，大约两侧江岸需要各后退两三千米。但这都是些什么地方啊！这是中国人口最密集，经济最发达的一条地带，大致估计了一下，每年怎么也有以万亿元计的GDP，加上千百年发展积累的财富，至少得有几十万亿的价值。所以这个方案不行。

这样一算，我曾设想过的疏浚长江的方案也有了眉目。疏浚武汉到长江口的1 000km长的河段，假设宽1 000m，挖深4m（相当于1935年受淹地区的水量），总计约40亿m³泥沙。就算按1 000元/m³的成本估计，总造价也就4万亿元，与"小海绵"规划的总价是一个数量级。如果同时造出一条永久性的排沙管道（须知中国造涵管打隧道的本领也是世界领先），供常年维护，再加上有意识地规划出一些分洪保护区（此时基本上数十上百年才用一次），则可保千年无虞。何况，这些泥沙足够再造一个崇明岛。

如果再增加20%的经费，把这项工程上延到洞庭湖口，就可以开展另一项伟大工程——洞庭湖恢复工程（须知，我年轻时中国最大的淡水湖是洞庭湖），这样湖南和荆江地区的水害，也就有了解决的基础。即使将来三峡水库没有了，长江也不会无望。

我还曾设想过另一个方案，在城市较少的地方为长江另开一条河道，那里的居民可以迁入现在的河道。新河道是一条自然主义的任由长江奔流的生态河道，很理想、很浪漫，其中包括一个"引长入淮"的设想，因为向南引是不行的，到处是山，而且大量泥沙会毁掉杭州湾以南珍贵的港口资源。不过这个方案问题更大：其一，大规模改变了中国的水资源分布；其二，毁掉了现有的长江黄金水道，也毁灭了一个经济带；其三，动迁上亿的民众，不但成本巨大，而且会造成巨大的社会震荡。

所以，生态智慧绝不是诗人般的随意想象。真正的生态智慧，必须以谦虚、善意和韧性为内核，以联系、流变和辩证为方法，以确凿的科学为基础，建立在感悟之上。进入21世纪后，有一些专家好像把基本数学都忘光了，到处宣扬新理念，制造新理论，却从不进行估量和计算，更不从正反两方面对前因后果进行考量，甚至不会设计、组织和检验科学实验，仅凭着idea就作出判断或建议。他们让我很怀念过去一些虽然理论水平有限但满脑子是经验和规范的老专家，因为他们的意见总是闪耀着实践的光芒，非常实事求是，尽管有时他们的意见未见得最好，但我还没见过他们出馊主意。

Phase 8, 2016

The super El Nino phenomenon resulted in the floods almost sweeping the entire east part of China this year. In the trending "city view of the sea" everywhere in China, most people understand that the city "little sponge" can only adjust the small and medium urban storm water problems, and others ways should be thought of when encountering once-in-several-tens-of-years or once-in-a-century floods.

The floods in the middle and lower reaches of the Yangtze River, I remember about once every tens of years: 1935, 1954, 1998, 2016. Hankou area is about 18m above the sea level. In 1998, the river was up to 29m and more, that is, the river is over 10m higher than Hankou. If at that time the levee was destroyed, how large area would the flood drown? In 1935 the embankment collapsed, and 6400000 acres (4200km^2, equivalent to half of the administrative area of today's Wuhan City) were drowned, more than 80000 people died and were missing. This was just the equivalent of once-in 50-to-100-years flood. It is said the flood in 1870 was a once-in-1000-years one, and I do not know what was the situation then.

If allocating this water logging area (together with other flooded area) to the both sides of the Yangtze River beyond Wuhan, the river banks should go backward about two to three thousand kilometers both sides. But what is this place? This is China's most populous, most economically developed area, with roughly estimated trillions of dollars of GDP a year, coupled with the accumulation of wealth during the development of thousands of years, and it should worth at least tens of trillions. So this program does not work.

After such a calculation, the Yangtze River dredging program I once envisaged is also taking shape. Dredging the 1000km-long section of the Yangtze River from Wuhan to the estuary, assuming the width as 1000m, and the depth as 4m (equivalent to the amount of water in the flooded region in 1935), there would a total volume of about 4 billion cubic meters of sediment. Even if the estimated cost per cubic meters is 1000 yuan, the total cost would be 4 trillion yuan, around the same order of magnitude of the total cost of the "little sponges" planning. If creating a permanent desilting pipeline (please note that China is a world leader in pipe-shaped culvert and tunnel technology) for annual maintenance, plus consciously planning of flood protection area (in this case substantially for once in tens or hundreds of years), the safety of thousands of years could be guaranteed. Moreover, the volume of sediment is enough to rebuild a Chongming Island.

With an additional 20% of funding to extend the project up to the outlet of Dongting Lake, another great project – the Dongting Lake Restoration Project – could be carried out. We should know that Dongting Lake was China's largest freshwater lake when I was young. This way would also provide the basis for a solution to the flooding damage in Hunan and Jingjiang River area. Even if there would be no Three Gorges Reservoir in the future, the Yangtze River would not be hopeless.

I had also imagined another program: opening another river course for the Yangtze River in the less urbanized area, and moving the residents there into the current river course. The new river course would be an ecological one enabling the Yangtze River to rush naturally, very ideal and romantic. The idea of "leading the Yangtze River water to Huaihe River" is included, since the southward leading is infeasible, there are mountains, and the large volume of sediment would destroy the precious port resource beyond the south of Hangzhou Bay. But this program is more problematic: first, the large-scale change in the distribution of water resources in China; second, destroying the existing golden waterway along the Yangtze River, as well as the destruction of an economic belt; third, the relocation of hundreds of millions of people would cause huge cost and also give rise to huge social turmoil.

Therefore, the ecological wisdom is never the random poetic imagination. The real ecological wisdom must take modesty, good will and toughness as the core, relationship, rheology and dialectics as the approach, and solid science as the basis, and be founded upon apperception. In the 21st century, some experts seem to have forgotten the basic math, promote new ideas, create new theories, but never carry out estimates and calculations, never consider both positive and negative aspects of the causes and consequences, even incapable of designing, organizing and testing scientific experiments, and just make judgments or recommendations simply based on idea. They made me miss the old experts with limited theoretical level but full of experience and norms, as their opinions are always shining the light of practice, very realistic; although sometimes their opinions might not be the best, I have never seen they misadvise.

2016年9期

为什么我国在自然科学上的投入增长很快，但自主突破不多？除了投资效益后滞效应，还有一个不可忽视的重要原因是目前的论文和职称的评审制度造就了一个很坏的学风——热衷于抓小题目、专门化的题目和可实操的题目。于是，研究者和学生都越搞越专，思维越来越狭窄。既具有扎实基础，又能够横跨多学科，思维发散，联想活跃的人已越来越少。

大部分人从来没有想过，在别人给定的大圈子里，本身就可能有不少谬误，搞不好就越钻越错。比如，对于"让自然做功"这个理念，我只敢十分谨慎地采用。自打1975年从事环保事业以来，我就对许多环保措施的基本思维——低于一定浓度后排放到自然——产生怀疑。真的通过自然过程就可以免除污染危害了吗？的确有些物质会进入常态下的自然循环过程，从而解决污染问题，但很多人造物质不是这样。根据"物理化学"的基本原理（任何化学反应都是在一定条件下发生的，其最后结果是原料和产物达到这种条件下的平衡），就地球表面的自然条件来讲，很多污染过程的逆反应条件是不具备的。而物理变化更可怕，几乎没有这个化学平衡限制，例如PM2.5，很难过滤，也不沉降，不下雨其浓度可以拼命地增加，只有下雨可以洗一洗，靠风吹走只是转移了污染。而纳米，就更可怕，比PM2.5还小得多，到处钻，谁知道进了细胞会怎样！

过去，人造的东西很少，大自然的宽容度给人一种错觉，觉得自己任意糟蹋自然没关系，个别地方有害物体多了，就交给大自然，这就是所谓"让自然做功"，没有想到自然对人的宽容度也有限。最典型的是CO_2。在一定条件下大气的CO_2是定值。是人把贮藏在地下的碳大量变成CO_2破坏了原有平衡，而且速度远远超过了大自然逆反应（通过自然增加叶绿素来缓慢实现）的速度，这是人类第一次遇到的大自然宽容度问题。资本的力量让西方不是走限制碳开发（油气公司主要掌握在西方资本手中）的路子，而是选择了处罚使用碳（这是发展中国家的需要）的路子，由此，所谓的碳足迹方法得到飞快发展。

西方目前在提倡大规模地把污染水灌入土壤，以获得净水，参见本期《绿色基础设施经济收益评估的综合成本收益分析法研究：以美国费城为例》（顺便说一下，在美国费城CSOs系统中，种树是微微增加PM2.5的，似乎精确到连这个都要算进去），但长期后果会如何？直至目前都没有系统地大规模研究，为何？资本不感兴趣。年复一年，土壤和地下水污染总会积累到一定程度，一旦成为资本社会注意的问题，我估计离人类整体危机也就不远了。因为通过农业补贴压低农产品价格，打压不发达国家，维持全球剥削体系，是这个资本世界基本结构的核心部分。只有抬高农业价值，才能获得资本对农业的重视，而这又和资本的根本利益是相违背的。

所以，我觉得，与其提倡"让自然做功"，不如提倡"人（对自然）欠的债，人自己还"，这是自然之主与自然之子在伦理上的根本区分。

Phase 9, 2016

Why does China's input on natural sciences grow rapidly, but few independent breakthroughs come out? In addition to the lag effect of investment returns, an important reason is that the current paper and professional title review system brings a very bad style of study—keen to catch small, specialized, or practical topics. Thus, researchers and students have become increasingly specialized, and the thinking is more and more narrowing down. There are fewer ones with solid foundation, multi-disciplinary spans, divergent thinking, and active association.

Most people have never thought in the large circle given by others, there may have a lot of fallacies, the more one continues, the more wrong it will be. For example, for the idea of "let the nature act", I only dare to use very carefully. Ever since 1975 I engaged in the cause of environmental protection, I have been suspicious upon the basic thinking of many environmental protection measures—discharge into nature below a certain degree of concentration. Can the pollution hazard be really exempted through the natural processes? Indeed some substances will enter into the natural cycle in the normal state, so as to solve the pollution problem, but a lot of artificial substances are not the case. According to the basic principle of the "physical chemistry" (any chemical reactions are occurring under certain conditions, and the final result is the raw material and the product reach equilibrium under these conditions), as to the natural conditions of the earth's surface, the conditions of the reverse reactions of a lot of pollution processes are not available. And the physical change is more terrible, almost no limit for this chemical equilibrium. Such as PM2.5, difficult to filter, no settlement, its concentration increasing desperately if no rain, only the rain can wash, and blowing away by wind is just transferring pollution. And nano materials are even more frightening, hundreds of times smaller than PM2.5, drilling everywhere, who knows what will happen if enters into the cell!

In the past, there were not many artificial things, the tolerance of nature gave people an illusion that it did not matter if they spoiled the nature at will, in some places with more harmful objects, handing over to the nature, which was called to let the nature act, and they did not expect the nature's tolerance of people is limited. The most typical one is CO_2. Under certain conditions, the atmospheric CO_2 is at a fixed value. People changed large amounts of carbon stored underground into CO_2, destroyed the original equilibrium, and the speed was much faster than the speed of the reverse reaction of the nature (achieving slowly through natural increase of chlorophyll), which is the first time people encounter the tolerance issues of the nature. The power of capital makes the West not go the way of limiting carbon development (oil and gas companies are largely in the hands of Western capital), but the way of penalizing the use of carbon (the need of developing countries), and whereby the so-called carbon footprint method develops fast.

The West is currently promoting the large-scale polluted water poured into the soil, in order to obtain clean water, seen in *Comprehensive Cost-Benefit Analysis on Economic Benefits of Green Infrastructure: A Case Study of Philadelphia, USA* (By the way, in Philadelphia CSOs system, planting trees bring a slight increase of PM2.5, so precisely that even this seems to be included for calculation), but what will the long-term consequences be? Up to now no systematic large-scale study, why? The capital is not interested. Year after year, soil and groundwater contamination will always accumulate to a certain extent. Once it becomes a problem drawing the attention of the capital, I guess the crisis of the whole human race will be not far away. Lowering agricultural prices through agricultural subsidies to suppress the developing countries and to maintain the global system of exploitation, it is a core part of the basic structure of the capital world. Only raising agricultural value will obtain the capital's attention to agriculture, which is contrary to the fundamental interests of capital.

So, I think, it is better to promote "people should pay their debts on the nature by themselves" rather than "let nature act", which the fundamental ethical distinction between the child of the nature and the owner of the nature.

2016年10期

本期主题"中国·杭州2016年G20峰会景观设计"由杭州园林设计院股份有限公司风景园林规划设计一院院长李勇主持，非常感谢。G20的历史意义，在于它是人类开始改变极不合理的世界经济秩序的转折点，这种改变可能要经历数十年甚至上百年之后才会越看越清楚。

近日见到新浪微博有个《山西竟有个300万岁的南极，不畏火山，震惊世界的不解之谜！》的帖子，说：山西宁武县有个万年冰洞，洞内始终保持在-3~-4℃，上下5层，越往深处走，冰层越厚！这也是冰洞最神奇的地方。有人提出了地热负异常说，因为这与"越向地心走，温度越高"的地热正常说刚好相反！而且这里的"冰雕"一年四季都能看到，简直就是藏身于华北的一个"南极"！这里根本就没有一年四季都结冰的气候条件，为什么会形成这万年不化的冰洞呢？这至今还是个谜团。距此几百米外还有一座煤层自燃形成的"火山"，火烧水，火旺冰融，这冰火两重天是如何共存的？简直就是好莱坞科幻片！博客还附了一个旅游推广片，大体内容如上所述。

依我看，这东西没有什么神秘，奥秘就在隔山的这个自燃的煤层。这个煤层下面必定有孔洞与冰洞连通，由于燃烧的热气上升，下面的空气就得上来补充，带动山那边的空气不断进入冰洞，而这个冰洞的洞口又很小，比洞内的口径小很多，根据物理化学原理，空气进洞后就体积膨胀，自然就降温了，再加上数十米的地下已经基本恒温，海拔又高达2 000多m，可以常年结冰。然后，就是冷空气下沉，所以越向下，气温越低。这与越近地心越热是两回事。不信你就由此向下钻，钻个3 000m，看看热不热。可笑的是，还有人由此发明了所谓的"地热负异常说"，这类专门发明新概念的人怪不得被群众讥笑为"砖家"。

气体压缩则升温、膨胀则降温，这是大自然的基本规律之一。曾经看过一个外国神话片，国王王座后面是个装了扇叶的大转轮，有仆人摇动摇把，扇风取凉。其实这很笨。岭南"四大园林"的东莞可园里有个"双清室"，地板上有2个孔洞通到隔壁的风车，盛夏客人来时，仆人摇动风车，双清室室内气温可以下降4~5℃，就是利用了这个原理。这比空调简单多了，热效率也应很高，搞绿色建筑的人可以研究一下，利用起来。

风景园林人最好扎实一下自己的自然科学功底，多读点书，多懂点基本原理，并学会综合平衡地对待复杂事物，这有助于坚持自己的基本信念（其中主要是许多代前人经过考验的积累，或曰生态智慧），而不是热衷于追逐时髦。生态这个东西，是大自然造就的，需要的主要不是创新，而是了解和遵从。懂得不多就去创新，后果堪虞。近些年学科的很多事情，说到底，或是忘记了科学规律，或是理念掩盖了科学。比如景观热的时候，什么都以景观为中心（背后可能是为经营城市卖地服务），于是，在某些地方领导的支持下，大量调蓄湖变浅，改造成湿地公园；本来行洪的河道，为了景观种上了树林，添造了栈道和亭台等等（湖南某市在已经硬化的河床上直接填土，搞滨水生态景观，今年（2016年）不仅影响了防洪，而且这些泥土连同景观都被洪水一锅端掉了）。等到雨洪管理的风气一来，又忙着把绿地改造成蓄水场地，把土壤换成沙石，把原生树种挖掉改成"既耐水又耐旱"的树种，或者干脆只种些草。下次再来个什么风，不知还能生出什么新花样来。

切盼这类事以后尽量少些重现。

Phase 10, 2016

I would like express our gratitude to have Li Yong, Dean of the 1st Landscape Architecture Planning and Design Department of Hangzhou Landscape Design Institute to host the theme topic of "Landscape Design for G20 Summit 2016, Hangzhou, China". The historical significance of the G20 is that it is a turning point to change the world's increasingly unreasonable economic order, and it may take decades or even a hundred years for it to become clearer.

On the Sina microblog web, a recent post about *A 300-year-old Antarctic in Shanxi Province, Defying Volcanos, A World-shocking Mystery!* said, there is a million-year ice cave in Ningwu County of Shanxi Province, maintaining -3/-4 degrees Celsius, five layers vertically, from top to bottom, the deeper the go, the thicker the ice! This is the most amazing point of the ice cave. It was suggested as the geothermal negative anomaly, because it is opposite to the geothermal theory that closer to the center of the earth means higher in temperature. And here the ice can be seen throughout the year, simply an "Antarctic" hiding in North China. There are no climate conditions for all year round ice, then how does this ice cave form? It is still a mystery up to now. A few hundred meters away from this there is a "volcano" formed through the spontaneous combustion of coal, how comes the coexistence of ice and fire? Really like a Hollywood science fiction film! The post also attached a tourism promotion film, and the general content was the same.

In my opinion there is no mystery of this thing, and the secret lies in the seam by the spontaneous combustion of coal. There must be holes under the seam to connect the ice cave. Because of the rising of burning hot air, the following air will have to come upwards to supplement, driving the air from the other side of the mountain to enter the ice cave continuously. Since the entrance of the ice cave is very small, much smaller than the diameter of the inner space, according to the physical and chemical principles, the volume of air will expand after entering the cave, and it will naturally cool down. And tens-of-meters underground has been in basically constant temperature, and the elevation is up to 2,000 meters high, so the ice can exist all year round. Then, there is the cold air sinking, so the more downwards, the lower the temperature. Comparing with the geothermal phenomenon, they are two different things. If you do not believe, you can drill down to 3,000 meters to see if it is hot. Ironically, there are people who invented the so-called "geothermal negative anomaly", and it is these new concept inventors are often ridiculed.

The gas temperature goes up under compression and down under expansion, and it is one of the basic laws of nature. I once saw a foreign myth, in which a fan wheel is installed behind the throne of the king, and the servants swing the fan handle to make cool wind. In fact, this is stupid. There is a "Shuangqing Room" in Ke Garden in Dongguan, one of the "four famous Lingnan gardens". There are two holes in the floor to the windmill in the next room. In the midsummer when guests come, the servants swing the windmill, the indoor temperature of Shuangqing Room can drop four to five degrees, which used this principle. This is much simpler than air conditioning, and the thermal efficiency should also be high, and the people engaging in green building can study and use it.

People in landscape and gardening should better enrich their scientific knowledge, read more books, understand more basic principles, and learn to treat complex things in balance and comprehensively, which helps to adhere to their basic beliefs (mainly the accumulation of experience of generations, or called ecological wisdom), rather than keen to chase fashion. Ecology, created by the nature, needs not innovation, but understanding and compliance. Knowing not much and just going to innovate, the consequences are serious. Many bad things in the discipline in recent years, in the final analysis, is because the laws of science are forgotten, or ideas cover up science. For example, in the landscape architecture fever, the landscape was always put in the center (the underlying purpose might be for selling land), so, with the support of some local leaders, a large number of regulation lakes became shallow and were transformed into wetland parks; trees were planted and corridors and pavilions were built in the river way originally for flood draining (A Hunan city directly filled soil on the hardened riverbed to create waterfront landscape, which affected the flood draining this year, and the soil and landscape were wiped totally by the flood.); and so on. When the atmosphere of rainwater management and flood control came, they were busy transforming the green space into water storage site, replacing soil with gravel, digging the original tree species for "both water-enduring and drought-enduring" tree species, or simply planting some grass. I do not know what new tricks will come out when new tendencies come.

I really hope this kind of things should better not happen again.

2016年11期

文人，从来不是一个整体，总会分裂成若干集团，这背后，有利益之争、学派之争，更有深层的思想方法和哲学体系上的差别。撇开春秋战国和魏晋时代的百家争鸣不谈，从西汉独尊儒术开始，东汉的新旧经学之争，唐代的牛李党争，宋代的朋党之争，明代的阉党与东林……文人内部一直争斗不断。对于这种社会现象，有两种态度：一是用政权甚至军事手段高压来统一化，比如元代和清代，我们知道其结果只能导致社会的僵化；二是视之为一种社会常态，但这需要一种极高的智慧和心境。南宋印肃禅师曾长期追问："一归何处？"当他阅读《华严合论》至"达本忘情，知心体合"一句时，豁然大悟，作偈曰："捏不成团拨不开，何须南岳又天台。"这里的本，用现代语言可理解为真理、本体；这里的情，可理解为人性、心性。印肃终于理解到，所谓"合一"，本质是"拨不开"，并非是捏成没有区分的一团。

正因如此，所谓文人园也不会只有一种模式。过去我们讨论文人园，基本上被《园冶》派统一了，乃至有过将《园冶》的书名翻译为《造园圣经》的建议。依我看，计成属于将园林奉为理想境界的一派，而社会上更多的人，其实是将园林作为生活的一种工具。本期《闲情偶寄中的造园思想研究》一文，相当全面地阐述了李渔的园林思想，回归到了园林的生活本性。

这么说，是否天下就"无是无非"了呢？这要从两方面看。从自然本身来说，一切都是自然过程，本来就无是无非。但从人类立场观察，凡事都有个是非问题：所谓是，指对人类有利；所谓非，指对人类不利。需要提醒的是，这里强调人类，不是个人。群体与个体，有时利益一致，有时利益相反。我认为，习近平主席提倡"以人民为本"，较好地处理了这个问题。

任何思想的恰当与否，都是与当时的社会和自然条件相关联的。李渔的人本主义，表面是一种个人主义，实质上是对清初的民族压迫和思想禁锢（虽然还不及雍正乾隆）的一种无声反抗，也离不开明末文艺的思想解放（这也许是作为理学的对立面出现的）的余波。而如今，是不是我们可以对现代做一个同理但相对的理解：社会已有些过分放纵？过分提倡自私？当下不少看不惯世俗的人提倡王阳明，是否也与此相关？到此，有人会说，怎么正向反向都是你的理？这就离不开对当时现实的调查研究，我们还得回归到《实践论》的基本原理。

与此相关的另一篇文章是《"静寄"的意与象——静寄山庄释名》。"性静，情动"的命题，很有启发。具有稳定性，才能算是性，故曰性静。既然是本性，无所谓善恶，必须接受。性的时空越广大，道行越深。时空最广大的性，就是大道。情，是心态，常人心态多变，故曰情动。然情可控，可修（修的成果是回归大道，摆脱情动）。恶向胆边生，是不控，不修，因之固执地任性必将导致恶果。——此为中国体系伦理之基础。显然，它与鼓吹绝对自由才是善的理念，结论的差别很大。

中国是文人园，故有"造园如作文"之说。写文章与源于形体思维的设计，应该有较大的区别。我们现在基本上是后者主导，然而这一套真的是绝对真理吗？

Phase 11, 2016

Scholars, never an integral whole, always split into several groups, and behind this there are disputes of interests and schools of thoughts, and even deeper differences in methods and philosophical systems. Leaving aside the likely "hundreds of schools of thoughts contending" in the Spring and Autumn and Warring States periods and the Wei and Jin Dynasties, the scholars have never stopped fighting, such as the domination of Confucianism in the Western Han Dynasty, the dispute of the new and old studies of Confucian classics, and the fighting between the castrated gang and the Donglin school. For this social phenomenon, there are two kinds of attitudes: the one, political power and even military means were used to reach unification, such as in Yuan Dynasty and Qing Dynasty, and we know that only led to social ossification; the other, it was regarded as a social normality, but this required high-degree wisdom and state of mind. The Buddhist master Yinsu in the Southern Dynasty had been questioning, "Where will it return?" He suddenly understood when he read the sentence "reaching the basis beyond affection and the integration of mind and body" in the book *Comments on Huayanjing*, and wrote the Buddhist verse, "It could not be combined or separated, so there is no need to make the difference." Here the "basis" in modern language can be understood as truth, reality; the "affection" can be understood as human nature, temperament. The so-call "integration" means "inseparable" in the essence, and instead of forming into a group without distinction.

For this reason, the so-called scholar gardens would never be only one mode. In the past we discussed the scholar gardens and were basically unified by the school of *Yuan Ye*, and there was even a proposal to interpret *Yuan Ye* as the *Bible of Gardening*. In my opinion, its author Ji Cheng garden belongs to the school that regards garden as the ideal realm, while in fact more people in the society in fact regard garden as a tool for life. The paper of this month *Research on Gardening Thought of Li Yu in Xian Qing Ou Ji* makes a comprehensive exposition of Li Yu's gardening ideas, namely returning to life – the nature of garden.

So, is there "no difference of right or wrong"? We should consider from two aspects. From the nature itself, everything is a natural process, and there is of course no difference of right or wrong. However, from the human standpoint, everything is right or wrong: the so-called right refers to the benefit for mankind, and the so-called wrong refers to disadvantage for humankind. It need to be reminded that the emphasis here is mankind, not the individual. Group and individual sometimes have the same interests, but sometimes not. In my opinion, the "people-oriented" idea advocated by President Xi Jinping better deals with this problem.

The appropriateness of any thought is associated with the social and natural conditions of the time. Li Yu's humanism was a kind of individualism in the surface, but in essence it was the silent resistance to the national oppression and ideological imprisonment in the early Qing Dynasty (although less than in Yongzheng and Qianlong reigns), and also inseparable from the aftermath of the literature and ideological liberation (which might appear as the opposite of Neo-Confucianism) in the late Ming Dynasty. Now, is it possible for us to make a similar but relative understanding of the modern: has the society been somewhat over-indulgent? Too much selfishness? Is it also associated with this that nowadays many people who distain the mundane advocate Wang Yangming? At this point, some people will say, how are the opposite aspects both your reason? This is inseparable from the investigation of reality at that time, and we have to return to the basic principles of the *Theory of Practice*.

Another related article is *The Implication and Landscape of Jingji Mountain Resort* . The proposition of "peaceful life and emotional action" is very enlightening. With stability, it can be regarded as a property or nature, so it is called peaceful life. Since it is nature, there is no good and evil in it, and we must accept, such as wolves eat sheep. The wider the time and space of the property is, the profounder the meaning is. The biggest property of time and space is the mainstream Tao. The emotion is a state of mind, and ordinary people often change their minds, so it is called emotional action. However, the emotion is controllable, and cultivatable (the result of cultivation is to return to the mainstream Tao, and get rid of emotional action). The surge of the evil is not to control or cultivate, so the stubborn indulgence will certainly lead to evil consequences. This is the basis of Chinese system ethics. Obviously, it is quite different from the idea advocating absolute freedom as good.

The Chinese scholar gardens bring the wording that "gardening is like composition". There should be a big difference between composition and design originated from physical thinking. At present the latter is basically dominant, but is this set really the absolute truth?

2016年12期

孟兆祯先生的《把建设中国特色城市落实到山水城市》一文，对于解答很多地方出现的"如何把握当前城市建设路向"的疑惑，很有价值。怎样理解中央"尊重自然、顺从自然、保护自然。宁要绿水青山、不要金山银山；绿水青山就是金山银山。要让市民望得见山，看得见水，勿忘乡愁"的政策，关键是抓住一切为了人民、顺应自然规律的整体思想，而不是只看个别字眼，更不是在城市中无视人民，歪曲自然，甚至随心所欲。否则，如管子所言："人之所为，与天顺者天助之，与天逆者天违之，天之所助，虽小必大；天之所违，虽成必败。"

生态修复是当下被业界看好的领域。刘秀晨先生的《在"京津冀一体化"和"建设北京城市副中心"两大战略背景下加快京津冀生态功能区规划的联动实施》，结合京津冀地区的实际，重点提出生态功能区的规划和实施战略要实行区域联动，对风景园林专业进军生态修复领域具有战略与现实意义。

思维的混乱，好像是现代人的特征。本来城市河岸的设计，在人多地少、人与江河争地的条件下，应该是水利专家主持的事，景观是必须服从水利规范的。但在一些所谓专家的怂恿和一些地方官员的瞎指挥下，这里变成了景观为主的事物，这些人既不懂水文和流体力学，也不懂工程和动物，瞎指挥一通，行洪道上种植树木灌丛来让洪流变慢，大堤上栽大树、种浆果引来老鼠打洞，已经硬质化的河床里填土搞景观造成洪水一来全部冲光……只能造成更大的隐患。

20世纪国务院本来打算分立《城市园林条例》和《城市绿化条例》，后来受到某些阻力，只通过了《城市绿化条例》。实际上我国城市公园还缺乏国家立法，以致各地公园可以被随便什么专业的人或机构利用和指挥。我想，如果在法治社会，一拍脑袋就想把主要供市民公众使用的城市公园绿地大面积地改造成沙池、湿地或蓄洪区，是做不到的。《关于城市公园可持续发展的法制思考》比较深入地探讨了我国城市公园立法的问题，值得向各级城市园林绿化管理有关部门推荐。

中国的北京、上海、广州、深圳、天津、重庆、杭州、武汉、成都等城市中心城区的密度都超过了人口密度1.5万人/km^{2}或建筑容积率2.0，属于高密度城区。这里也是我国单位面积GDP产值最高的地域，笼统地研究一般城市常常不能解决这一问题。我与蔡云楠博士相识已久，他一直注重规划的科学依据，《高密度城市绿色开敞空间的建设误区和优化策略》一文很有深度，关注大城市的人可以关注"开敞空间规划设计"这一组文章。

中国人提出天人和融的风景概念已将近2000年，西方人关注文化景观，不过一二十年的事，对于他们的多数人，需要做一个脑筋的大转弯，否则大格局还是人是人、天是天。让我惊讶的是，我发现有些学生把中国的"风景"列在景观发展史的尾巴上，而且还附骥在日本人创造的"景观"（用来翻译landscape）的后面，不由得脊背发凉：没文化，真可怕。感谢同济大学韩锋教授，多年从事这方面的研究，在世界遗产组织对扭转西方人的脑筋、维护中国的利益方面做了大量工作，并为本刊组织"文化景观"专题，此专题文章今后陆续发表，以让中国读者更全面、准确地理解文化景观和掌握它的发展。

Phase 12, 2016

Mr. Meng Zhaozhen's paper *Implementing Building Chinese Characteristics Cities with Shanshui City is* of great value for many places with doubts on "how to grasp the current direction of urban construction". As to the understanding of the Central Government's guiding principles of "respect for nature, to comply with nature, and to protect nature... prefer the good environment rather than economic gains, and the good environment is truly a kind of economic gains... let the public see the mountains and water and remember their hometowns...", the key is to grasp the general thinking of all for the people and complying with the natural law, instead of just eyeing on individual words, nor ignoring the people, distorting the nature, or doing whatever one wants in cities. Otherwise, as Guanzi said, "God helps those who go along with the nature, and abandons those who go against the nature. Helped by God, it is meaningful although small; abandoned by God, it is a failure although succeeded."

Ecological restoration is regarded as an increasing field in the landscape architecture industry. Mr. Liu Xiuchen's paper *The Coordinated Implementation of Ecological Function Zones under the Strategy of Integration of Beijing-Tianjin-Hebei and the Construction of Beijing Sub-Center* combines the reality of Beijing-Tianjin-Hebei area, puts forward the regional linkage in the ecological functional areas planning and implementation strategies, which is of strategic and practical significance for the landscape architecture discipline to enter the ecological restoration field.

The chaos of thinking seems to be the characteristics of modern people. Originally, the design of urban riverside, under the conditions of more people and less land, should be mastered by hydraulic experts, and landscape must be subject to hydraulic norms. Under the encouragement of some so-called experts and the blindness of some local officials, it becomes something dominated by landscape architecture, and those knowing nothing of hydrology or fluid mechanics, nor engineering and animals, made blind command: flood channel were planted with trees and shrubs to slow the torrents, rats were attracted by the trees and berried planted along the levee and burrowed in the levee, the landscape made with soil on the hardened river beds was washed away by floods... which can only cause greater risks.

In the last century, the State Council had intended to separate the *Urban Landscape Architecture Regulation and the Urban Greening Regulation*, but finally only the latter was passed through due to some resistance. In fact, China's urban parks lack of national legislation, as the result that parks can be utilized and manipulated by any so-called professional persons or institutions. where around any professional person or organization to use and command. I think, in the society ruled by law, it would impossible to simply transform the large-scale green space of urban parks that mainly serve the public into the sand pond, wetlands or flood storage areas. The paper *Thinking on the Sustainable Development of the City Park from the View of System of Law* made an in-depth discussion of China's urban park legislation, and is worth to be recommended to all levels of urban landscape management departments.

The density of urban centers in China's city areas (such as Beijing, Shanghai, Guangzhou, Shenzhen, Tianjin, Chongqing, Hangzhou, Wuhan, Chengdu, etc.) is more than the population density of 1.5 million people per km^2 or the building volume rate of 2.0, belonging to high-density urban areas. They are also the areas with the highest GDP per unit area in China, and the general study of general cities often cannot solve this problem. I have known Dr. Cai Yunan for a long time, and he has been focusing on the scientific basis for planning. His paper *Misunderstanding and Optimization Strategies of High-Density Urban Green Open Space Construction is* very profound, and people concerned about big cities can pay attention to the papers collected under the theme of "open space planning and design".

It has been nearly two thousand years since the Chinese people proposed the concept of the integration of human and nature, while it is but one or two decades since the Westerners concerned about the cultural landscape. For most of them, there should be a big change in their mind, otherwise the human and nature are still separated. To my surprise, I found that some students recently listed Chinese landscapes on the tail of the history of landscape architecture, and even attached to the back of the Japanese-created "landscape", which sends shivers down my spine: it is terrible without knowledge. Thanks to Professor Han Feng of Tongji University, who has done research in this field for many years, and has done a lot of work in reversing the Westerners' minds and safeguarding the interests of China in the World Heritage Organization. He helped organize the "Cultural Landscape" topic and related papers in this and following months, for Chinese readers to have a more comprehensive and accurate understanding of cultural landscape and its development.

后 记

《王绍增文集》在王绍增先生逝世一周年后结集出版，是值得庆慰的一件要事。书稿得以出版，首先感谢风景园林学界、业界的同仁们在先生病重期间慷慨捐助的爱心善款，提供了本书的出版经费。是你们深厚的爱心使得这部文集得以很快付梓与大家见面。在此谨代表家属向你们致以诚挚的敬意和浓浓的谢意！

王绍增先生在1998~2017年先后任《中国园林》杂志编委、副主编、常务编委、主编，发表了风景园林学科内涵、规划原理和理论、设计方法、中外园林历史的交流、比较与发展、教育教学及人才培养、生态理论等诸多方面文章，现将它们结集出版，如果没有《中国园林》杂志社的鼎力支持，书稿是不可能在短时间内整理完成的，在此深表感谢！

感谢暨南大学原副校长纪宗安教授、《中国园林》杂志社及金荷仙社长为本文集提供了宝贵的资料，感谢中国建筑工业出版社，感谢北方工业大学建筑与艺术学院傅凡教授，华南农业大学校、院领导给予的大力支持，以及林学与风景园林学院高伟副教授、古德泉老师、邱巧玲老师等为本文集的出版花费了大量的时间和精力。所有的感激之情难以言表，将化作传承先生的思想、继承先生的遗愿，推进风景园林学科与事业更快更好地发展！